Lecture Notes in Statistics **123**

Edited by P. Bickel, P. Diggle, S. Fienberg, K. Krickeberg,
I. Olkin, N. Wermuth, S. Zeger

Springer
New York
Berlin
Heidelberg
Barcelona
Budapest
Hong Kong
London
Milan
Paris
Santa Clara
Singapore
Tokyo

D. Y. Lin
T. R. Fleming
(Editors)

Proceedings of the First Seattle Symposium in Biostatistics: Survival Analysis

Springer

D. Y. Lin
Department of Biostatistics, Box 357232
School of Public Health and Community Medicine
University of Washington
Seattle, WA 91895

T. R. Fleming
Department of Biostatistics, Box 357232
School of Public Health and Community Medicine
University of Washington
Seattle, WA 91895

CIP data available.
Printed on acid-free paper.

Softcover reprint of the hardcover 1st edition 1997

Camera ready copy provided by the editors.

9 8 7 6 5 4 3 2 1

ISBN-13: 978-0-387-94992-5 e-ISBN-13: 978-1-4684-6316-3 SPIN 10557562
DOI: 10.1007/978-1-4684-6316-3

Preface

The First Seattle Symposium in Biostatistics: Survival Analysis was held on November 20 and 21, 1995 in honor of the twenty-fifth anniversary of the University of Washington (UW) School of Public Health and Community Medicine. This event was sponsored by Amgen and co-sponsored by the UW School of Public Health and Community Medicine and the Division of Public Health Sciences, the Fred Hutchinson Cancer Research Center (FHCRC). The symposium featured keynote lectures by David Cox, Richard Gill and Ross Prentice, as well as invited talks by Norman Breslow, David Clayton, John Crowley, Susan Ellenberg, Mitchell Gail, Nicholas Jewell, Peter Lachenbruch, Jerald Lawless, Kung-Yee Liang, David Oakes, Margaret Pepe, Steven Self, Anastasios Tsiatis, Lee-Jen Wei, Jon Wellner and Zhiliang Ying. It was attended by 437 statisticians from 16 countries. In addition, 163 people attended a two-day short course taught by Thomas Fleming, David Harrington and Terry Therneau on *Survival Analysis Methods and Software* on the weekend preceding the symposium.

When the UW School of Public Health and Community Medicine was formed in 1970, biostatistics as a discipline was only a few years old. In the subsequent twenty-five years, both the field and the UW Department of Biostatistics have evolved in many exciting ways. The Department had only seven faculty when it moved from the School of Medicine to the new School of Public Health and Community Medicine in 1970. The faculty roster currently lists 39 regular and research faculty and 20 adjunct and affiliate faculty. Ed Perrin was the Department Chair in 1970, succeeded by Donovan Thompson, Norman Breslow and presently Thomas Fleming. The faculty have been actively involved in methodological and collaborative research in addition to graduate teaching. The choice of *Survival Analysis* as the theme for the *First Seattle Symposium in Biostatistics* was a tribute to the significant contributions made by the UW and FHCRC faculty to this important area of statistical science over the past twenty-five years.

The Symposium Organizing Committee consisted of Lloyd Fisher, Thomas Fleming, Alan Forsythe, Danyu Lin (Chair), Art Peterson and Ken Resser. The staff of the Biostatistics Department, including Diane Ames, Velta Benson, Vickie Graybeal, Ernest Leland, Cynthia Marks and Elaine Peterson, coordinated the symposium. The UW President Richard McCormick and the School Dean Gil Ommen delivered the opening remarks. The scientific sessions were chaired by Norman Breslow, Danyu Lin, Paul Meier and Jon Wellner. We are grateful to the aforementioned people as well as all the speakers and participants for making the symposium a great success.

This volume contains most of the papers presented at the symposium plus one paper from the short course. These papers encompass recent methodological advances on several important topics, such as multivariate failure time data and interval censored data, as well as innovative applications of the existing theory and methods. This collection serves as a reference for those working in the area of survival analysis.

Each of the 15 papers in this volume was refereed by two peer reviewers, and their comments were incorporated by the authors into the final versions of the papers. The referees are listed at the end of this book. We are indebted to them for their time and efforts. We also appreciate the guidance and assistance by John Kimmel of Springer-Verlag as well as the secretarial support by Alex MacKenzie during the preparation of this volume.

Finally, we would like to acknowledge the scientific and financial contributions by Amgen. Without their generous support, it would have been impossible to hold this symposium.

D. Y. Lin
T. R. Fleming
Seattle, December 1996

Contents

Prcfacc v

Some remarks on the analysis of survival data 1
D. R. COX

Multivariate failure time data: Representation and analysis 11
R. L. PRENTICE and L. HSU

Analysis of multivariate survival times with non-proportional hazards models 23
L. CHEN and L. J. WEI

Analysis of mean and rate functions for recurrent events 37
J. F. LAWLESS, C. NADEAU and R. J. COOK

Extending the Cox model 51
T. M. THERNEAU

Model-based and/or marginal analysis for multiple event-time data 85
D. OAKES

Artificial insemination by donor: discrete time survival data with crossed and nested random effects 99
D. CLAYTON and R. ECOCHARD

Interval censored survival data: A review of recent progress 123
J. HUANG and J. A. WELLNER

Singly and doubly censored current status data with extensions to multi-state counting processes 171
N. P. JEWELL and M. VAN DER LANN

Additive hazards regression models for survival data 185
D. Y. LIN and Z. YING

Some exploratory tools for survival analysis 199
J. CROWLEY, M. LeBLANC, J. JACOBSON and S. E. SALMON

Survival analysis in the regulatory setting 231
S. S. ELLENBERG and J. P. SIEGEL

Proposed strategies for designing and analyzing sepsis trials 243
P. A. LACHENBRUCH and J. TIWARI

Coarsening at random: characterizations, conjectures, counter-examples 255
R. D. GILL, M. J. VAN DER LAAN and J. M. ROBINS

Sequential models for coarsening and missingness 295
R. D. GILL and J. M. ROBINS

Addresses for Contact Authors 307

List of Referees 310

Some Remarks on the Analysis of Survival Data

D. R. Cox

PREAMBLE

This paper is based on a talk given in Seattle, Washington, November 1995 at a Conference celebrating the 25th anniversary of Department of Biostatistics, University of Washington. It thus provided an opportunity of congratulating current and previous members of the Department on their achievment in establishing such a fine reputation for the high international standards of their work and of expressing confidence in the future of the Department as a focus for work of excellence.

ABSTRACT The implications for survival analysis are explored of various general criteria for statistical models. Extensions to more complex kinds of data are briefly discussed.

1 Introduction

The analysis of survival data has a long history going back at least to the systematic study of life table methods by German actuaries at the turn of the century. One important aspect of the field is the wide range of types of application; thus work in Seattle in the 1960's at Boeing introduced ideas from industrial reliability and more recently there have been interesting econometric (Lancaster, 1990) and sociological (Blossfeld and Rohwer, 1995) applications under the general name *event-history analysis*. The recent explosion of interest in the subject is a welcome sign of vigour and ever-widening application, although it is important that the topic does not become isolated from the main stream of statistical thinking.

The object of this paper is to make a few miscellaneous comments staying, as far as possible, away from technical detail but considering the implications for survival analysis of general desiderata that can be put forward for statistical models.

2 Model interpretation

There are advantages to models that have some link with an underlying generating process. An example is the use of the Weibull distribution with its link with extreme value theory. In terms of models for dependence on explanatory variables such considerations suggest a preference for an accelerated life model over a corresponding proportional hazards model; see Cox (1972a, p.200) for some discussion. The preference stems from the representation of the accelerated life model directly for random variables in the form

$$T_z = a^{-1}(z)T_0, a(0) = 1,$$

where T_z denotes the survival time that would be observed under explanatory variables z, with $z = 0$ denoting a baseline level. The corresponding properties for survivor function and hazard are

$$S_z(t) = S_0(ta(z)), h_z(t) = a(z)h_0(ta(z)).$$

While the expression of the proportional hazards model is, of course, simpler in terms of the hazard there is no simple corresponding representation for the random variables.

Disadvantages of the accelerated life model, however, include the greater difficulty of making a semi-parametric analysis and that the extension of the accelerated life model to deal with time-dependent explanatory variables (Cox and Oakes, 1984, p.66) is somewhat contrived.

When is the proportional hazards model indicated by a generating process? If there are a number of independent types of failure then the baseline hazard function is

$$h_0(t) = \Sigma h_{0j}(t), \tag{1}$$

where $h_{0j}(t)$ refers to the jth failure type. Suppose for simplicity that there is a single binary explanatory variable, z, with $z = 0$, baseline, corresponding to control and $z = 1$ to treatment. Suppose that the baseline hazard $h_0(t)$ corresponds to the full sum in (1). Now if the change to $z = 1$ suppresses a fixed number of terms in (1) that are effectively constant, corresponding to an implicit exponential distribution, the result is an additive hazard model. On the other hand, if there are a large number of effective terms in (1) and the shift to $z = 1$ effectively suppresses a random sample of them, then a proportional hazards model results.

The only systematic empirical study I know of, concerning radiation hazards, tends to show that a proportional hazards model fits better than an additive model (Darby and Muirhead, 1987).

In the context of laboratory accelerated life testing it is often argued that the testing procedure is reasonable and in effect the standard accelerated life model holds if and only if there is a single dominant failure mode. It is therefore interesting to see whether a proportional hazards model could reasonably emerge from an accelerated life model with several failure modes.

Suppose then that the j th failure mode has baseline hazard

$$\rho_{0j} = t^{\gamma}\rho_j(1 + \epsilon\delta_j t),$$

where $\Sigma\delta_j = 0$ and ϵ will be supposed small. That is, we suppose that there are departures from an average Weibull shape with a common γ. Suppose that the jth failure mode has an acceleration factor $a_j(z)$. Then the observed failure time under explanatory variable z has the form

$$T_z = \min\{a_1^{-1}(z)T_1, \ldots, a_p^{-1}(z)T_p\},$$

say, and provided the component times are independent, we have that to the first order in ϵ

$$\frac{h_z(t)}{h_0(t)} = \frac{\Sigma a_j^{\gamma+1}(z)\rho_j}{\Sigma\rho_j}\left\{1 + \epsilon t\left[\frac{\Sigma a_j^{\gamma+2}(z)\rho_j\delta_j}{\Sigma a_j^{\gamma+1}(z)\rho_j} - \frac{\Sigma\delta_j\rho_j}{\Sigma\rho_j}\right]\right\}.$$

The leading term is independent of t, a reflection of the property of the Weibull distribution of inducing a model that is both of accelerated life and proportional hazards form. The coefficient of t vanishes only exceptionally. It will do so approximately, however, if there is rough orthogonality between the variation with j in the $a_j(z)$, the ρ_j and the δ_j. This is an interesting but presumably very special situation.

3 Parameter interpretation

A second requirement for a statistical model is that the parameters of interest should individually have clear interpretations, for example in terms of properties of an assumed generating process or the presumed effect of some intervention in the system. That is, as far as feasible, parameters of interest should be one-dimensional. Of course in any study different questions will focus on different parameters of interest. In a sense this in the spirit of R.A.Fisher's emphasis in analysis of variance on partitioning into single degrees of freedom.

The interpretability of individual parameters is important, so that it is helpful if parameters in some sense keep their meaning under perturbations of the secondary features of the model. This leads to consideration of the question: what happens if a proportional hazards model is fitted when an accelerated life model is appropriate, or vice versa? In particular can regression coefficients estimated in one setting be interpreted in the other?

This was discussed by Solomon (1984) and by Struthers and Kalbfleisch (1986). Their results are related to a much more general result about models involving linear combinations of explanatory variables. Suppose that we fit a model assigning the jth response variable Y_j the density $f(y; z_j^T\beta)$ whereas the true density of Y_j is $g(y; z_j^T\gamma)$. The maximum likelihood estimate $\hat{\beta}$ converges under the true model to β_γ, depending on the true value of γ and defined by the equation (Cox, 1961)

$$E(\nabla\Sigma \log f(y; z_j^T\beta_\gamma); \gamma) = 0,$$

where ∇ denotes the vector of partial derivatives with respect to β. While this equation rarely has an explicit solution, a solution near a null model, in which all parameters except those representing a general mean are small, can be found by Taylor expansion. The result is that, excluding the general mean, locally

$$\beta_\gamma = k\gamma,$$

component by component, where k is a constant. That is, considering for simplicity two components (z_1, z_2), the ratio of the regression coefficients from the model with parameters β estimates approximately the corresponding ratio of the components γ. That is, ratios of regression coefficients are more stable under this kind of model perturbation than are individual regression coefficients. The qualitative explanation of this is that the ratio estimates the change in z_2 required to produce the same change in response as a unit change in z_1 and this specification is relatively insensitive to how the response is measured. Brillinger (1983) gave a strong exact version of this result under the usually unrealistic assumption that the explanatory variable z has a multivariate normal distribution. Although these conclusions are less satisfactory than total invariance of the parameter under model perturbation, they do imply that conclusions about the rel ative importance of different explanatory variables tend to be preserved.

4 Error structure

A further requirement of a statistical model is that it allows a reasonable assessment of the uncertainty in the primary conclusions. Moreover this should be done without introducing unnecessary elaboration and complication. The arguments against complication are nowadays not so much to reduce the burden of computation but rather to make the path between the data and the conclusions more direct and transparent so that sensitivity to assumptions and data deficiencies is easier to assess.

Put rather qualitatively, the requirement is that all appreciable sources of variance that are not balanced out by design should be represented in the model and their contribution to uncertainty assessed; of course in some contexts the measures of variability, measured in normal theory problems by components of variance, may be of intrinsic interest. Assessment of uncertainty is to some extent an issue of choice of appropriate distributional form, but often more importantly of avoiding unjustified assumptions of independence.

While the introduction of a common frailty term into a series of positively correlated failure times is a simple device for studying simple forms of intraclass correlation, especially in a simple parametric setting, most analyses reported in the literature make quite strong assumptions of the independence of different individuals. In some contexts it would be helpful to have methods involving more complex types of correlation, for instance corresponding to the usual time series structures.

One simple method of checking on the appropriateness of error measures with extensive data is to divide the data into a fairly large number of meaningful subgroups such that interaction of properties of interest with the subgroups is considered uninteresting. Then we calculate estimates of parameters of interest within each subgroup, together with an internal estimate of error. The appropriateness of the internal error estimates can now be tested and if necessary an additional component of variance inserted.

More generally the analogue for survival data of the study of relatively complex component of variance problems is an underdeveloped topic.

5 Complex data collection procedures

A further requirement of statistical models is that they should represent special features of the data collection procedure that might, if not properly represented, induce misleading conclusions. The most obvious such feature is uninformative right censoring and the closely related left truncation. Other special schemes, especially as they arise in studying the AIDS epidemic are reviewed by Brookmeyer and Gail (1994).

In some ways the most difficult possibility is that of informative censoring or drop-out, associated in particular with the challenging issue of how to address noncompliance in clinical trials.

When there is a population of individuals in statistical equilibrium studied by cross-sectional sampling a number of special problems arise according, for example, as the property measured is the life-time of a sampled individual or the time from sampling back to birth or forward to death (Cox, 1955, 1969).

Another important possibility is that the system is observed only at discrete time points with incomplete information about what happens in between observation points.

Allowance for errors of measurement in explanatory variables, and occasionally in response variables, is sometimes needed (Prentice, 1982).

6 More complex forms of data

The discussion above refers to the simple situation where one failure-time, possibly censored, is observed for each individual. There are many generalizations including the following.

1. There may be single failure-times of two or more different types for each individual calling for the specification of a multivariate distribution.

2. Single failure-times on different individuals may be correlated, arising, for example, from genetic effects.

3. There may be several failure-times of the same type occurring in sequence on an individual.

4. There may be a sequence of events of the same or different types occurring on an individual.

5. Closely associated with the previous possibility is that we may have individuals moving between a discrete set of states, points of transition defining point events of type determined by the destination state.

6. At the end of the life of an individual a vector of further responses may be measured, defining for example quality of life in some sense, the joint distribution of failure-time and the new response being of interest.

7. An extension of the previous possibility is that a response vector is measured, in principle continuously in time, throughout the life of the individual. There is a distinction between such intermediate responses being of intrinsic interest and their being merely indicators of remaining life-time.

In all these situations explanatory variables, fixed or time-dependent, are likely to be present as well as such complications as censoring. There is some general relation both with the study of multivariate point processes (Cox and Lewis, 1972), although there the emphasis is usually on the analysis of a single long stationary series, and with formal methods for inference in discrete-state Markov and semi-Markov processes. Aspects of the process that are regarded as purely explanatory are typically regarded as ancillary and not modelled. The connection with work on graphical models for multivariate systems remains largely unexplored (Cox and Wermuth, 1996).

In general the difficulty with such extended situations is not so much the formulation of models as a basis for analysis as that of choosing between a rich variety of types of dependence that can scarcely be distinguished on empirical grounds.

A few of these points will now be illustrated in a little more detail on the second of the above schemes, where there is a sequence of events of the same type on each individual. The simplest possibility is that the points form a Poisson process and the next simplest that they form a renewal process. In the present context a renewal process has a hazard at time t depending only on the backwards-recurrence time u_t, i.e. the time measured from t back to the immediately preceding event or to the time origin. If an explanatory variable z is available this suggests using the hazard

$$\exp(\beta^T z) h_0(u_t),$$

forming a modulated renewal process (Cox, 1972b). In a semi-parametric setting a partial likelihood for β can now be formed, although the rigorous justification via martingale theory (Andersen et al, 1991) is not available and indeed a careful justification has been given only recently (Oakes and Cui, 1994).

If there are a small number of individuals with at least a modest number of events per individual, individual-specific parameters can be introduced. If there are a large number of individuals and especially if there are only a small number of events per individual, then individual-specific random terms, so-called frailties, are one route to the introduction of dependencies. The analysis is probably best done in a fully parametric setting.

If the underlying renewal process is replaced by some more complicated point process model, the argument extends directly provided the baseline hazard,or what in point process terminology is called the complete intensity function, can be expressed in terms of a one-dimensional function of the history. Otherwise dependencies must be introduced as time-dependent regression terms. For example, the length of the interval preceding the one under study might be introduced as an additional explanatory variable.

Some of the difficulties are illustrated by potential non-stationarity. This could be examined by introducing as a further explanatory variable either time since the process began or the number of events experienced by the individual since the start of the process. It is known that in some special cases it is hard or even impossible to discriminate between these; see, for example, Cane (1977).

There are many challenging problems here, probably best studied initially via intensive study of specific applications.

7 Objectives of analysis

Underlying much of the above discussion is the implicit notion that the objective of analysis is best regarded as the formulation of a model specifying the effect of explanatory variables in a way that preferably has a potential interpretation in terms of a process that might have generated the data. Such a model is hopefully more likely also to be a stable basis for comparing the results of the study under analysis with other similar studies. This underpins the gentle preference expressed in Section 2 for accelerated life models. If, however, the objective is more modestly summarization for largely descriptive purposes a formulation directly in terms of hazard functions has attractions, especially if different risk groups have nonintersecting hazard functions. Such models have also, as noted above, the considerable advantage of allowing the incorporation of time-dependent explanatory variables.

In some contexts the objective may better be formulated as prediction for individual patients. Here a fully parametric hazard-based analysis has considerable attractions, especially if from time to time updating in the light of new information is likely to be required.

References

Andersen, P. K., Borgan, O., Gill, R. D. and Keiding, N. (1991). *Statistical Models Based on Counting Processes.* New York: Springer.

Blossfeld, H.-P. and Rohwer, G. (1995). *Techniques of Event History Modeling.* Hillsdale, N. J.: Lawrence Erlbaum.

Brillinger, D. R. (1983). A generalized linear model with Gaussian regressor variables. In *Festschrift for Erich L. Lehmann*, pp 97-114. Belmont, CA: Wadsworth.

Brookmeyer, R. and Gail, M. H. (1994). *AIDS Epidemiology: A Quantitative Approach.* New York: Oxford University Press.

Cane, V. R. (1977). A class of non-identifiable stochastic models. *J.Appl. Prob.* **14**, 475-482.

Cox, D. R. (1955). Some statistical methods connected with series of events. (with discussion). *J. R. Statist. Soc.* B **17**, 129-164.

Cox, D. R. (1961). Tests of separate families of hypotheses. *Proc. 4th Berkeley Symp.* **1**, 105-123.

Cox, D. R. (1969). Some sampling problems in technology. In *Recent Developments in Sampling*, Eds. N. L. Johnson and H. Smith, pp. 506-527. New York: Wiley.

Cox, D. R. (1972a). Regression models and life-tables (with discussion). *J. R. Statist. Soc.* B **34**, 187-220.

Cox, D. R. (1972b). The statistical analysis of dependencies in point processes. In *Stochastic Point Processes*, Ed. P. A. W. Lewis, pp. 55-66. New York: Wiley.

Cox, D. R. and Lewis, P. A. W. (1972). Multivariate point processes. *Proc. 6th Berkeley Symp.* **1**, 401-448.

Cox, D. R. and Oakes, D. (1984). *Analysis of Survival Data.* London: Chapman and Hall.

Cox, D. R. and Wermuth, N. (1996). *Multivariate Dependencies.* London: Chapman and Hall.

Darby, S. C. and Muirhead, C. R. (1987). Modelling the relative and absolute risks of radiation-induced cancer (with discussion). *J. R. Statist. Soc.* A **150**, 83-118.

Lancaster, T. (1990). *The econometric Analysis of Transition Data.* Cambridge University Press.

Oakes, D. and Cui, L. (1994). On semiparametric inference for modulated renewal processes. *Biometrika* **81**, 83-90.

Prentice, R. L. (1982). Covariate measurement errors and parameter estimation in a failure time regression model. *Biometrika* **69**, 331-342. Amendment (1986), **73**, 245.

Solomon, P. J. (1984). Effect of misspecification of regression models in the analysis of survival data. *Biometrika* **71**, 291-298.

Struthers, C. and Kalbfleisch, J. D. (1986). Misspecified proportional hazards models. *Biometrika* **73**, 363-369.

Multivariate Failure Time Data: Representation and Analysis

Ross L. Prentice
Li Hsu

ABSTRACT While univariate failure time methods, including Kaplan-Meier survivor function estimators, censored data rank tests, and Cox regression procedures are well developed, corresponding flexible, standardized tools are not available for multivariate failure time analysis. This paper considers methods for the modeling and analysis of clustered failure times, with a focus on the estimation of marginal hazard functions and pairwise cross-ratio functions. First some representations of bivariate failure times are reviewed, along with corresponding nonparametric estimators and summary measures of pairwise dependence. Then procedures are outlined for simultaneous estimation of marginal hazard ratio, and pairwise cross-ratio parameters, for use in more general multivariate failure time regression problems. These estimation procedures are somewhat restrictive concerning the form of pairwise dependencies between failure times. Some approaches to relaxing these restrictions are briefly mentioned.

1 Introduction

A range of marginal, conditional, or partly conditional associations may be of interest in the analysis of multivariate failure time data. Such data may arise in various forms, including multiple failure times on individual study subjects, and single failures on individuals that are related or grouped in some fashion. In the former data type, the failure time data on each subject constitutes a point process and it is natural to model the process intensity in order to examine, for example, the pattern of dependency among successive failure times, or the manner in which the intensity depends on fixed or evolving covariates. With the latter data type interest may focus on marginal intensities, and their dependence on covariates, while dependencies among the failure times in a cluster may be of secondary importance. In some contexts, however, such as in studies of disease patterns within cohorts of family members in genetic epidemiology, the nature of dependencies among the failure times in a cluster may be of primary interest. In such circumstances simple summary measures of pairwise dependence are likely to be most useful.

This paper focuses on multivariate failure time methods for the analysis of marginal hazard rates and selected pairwise measures of dependence, particularly those based on the pairwise cross-ratio function. In addition to genetic epidemiologic studies, potential application areas include studies of event times in paired organs, littermate animal experiments, and group randomized clinical trials.

Let $T_1 \geq 0, T_2 \geq 0$ be failure time variates, and denote their joint survivor function by $F(t_1, t_2) = \text{pr}(T_1 \geq t_1, T_2 \geq t_2)$. The marginal distributions for $T_i, i = 1, 2$ can be represented in various ways. The hazard differential for T_i at t_i can be written $\Lambda_i(dt_i) = E\{N_i(dt_i)|T_i \geq t_i\}$, where N_i is the failure time counting process corresponding to T_i, defined by $N_i(t_i) = 1$ if $T_i \leq t_i$ and $N_i(t_i) = 0$ otherwise. The marginal distribution of T_i is determined by the cumulative hazard function $\Lambda_i(t_i) = \int_0^{t_i} \Lambda_i(ds_i)$, or by the survivor function

$$F_i(t_i) = \text{pr}(T_i > t_i) = \prod_{s_i \leq t_i} \{1 - \Lambda_i(ds_i)\}.$$

Corresponding to the counting process N_i one can also define the marginal martingale M_i by $M_i(t_i) = N_i(t_i) - I(T_i \geq t_i)\Lambda_i(t_i), i = 1, 2$, where I denotes an indicator function. See Fleming and Harrington (1991) and Anderson et al (1993) for comprehensive accounts of the use of counting process convergence results to develop distribution theory for univariate failure time tests and estimators.

2 Bivariate Failure Time Data Representation and Estimation

Various 'dependency' functions can specify the joint survivor function for (T_1, T_2), in conjunction with the marginal survivor functions F_1, F_2. For example, Dabrowska (1988) develops a useful product integral representation

$$F(t_1, t_2) = F_1(t_1)\ F_2(t_2) \prod_{s_1 \leq t_1} \prod_{s_2 \leq t_2} G(ds_1, ds_2),$$

where the differential $G(ds_1, ds_2)$ essentially involves the covariance between the counting process differentials $N_1(dt_1)$ and $N_2(dt_2)$ among pairs having $T_1 \geq t_1$ and $T_2 \geq t_2$, and can be written

$$G(ds_1, ds_2) = 1 + \text{cov}\{N_1(ds_1), N_2(ds_2)|T_1 \geq t_1, T_2 \geq t_2\}$$

$$\prod_{i=1}^{2}[1 - E\{N_i(\Delta s_i)|T_1 \geq s_1, T_2 \geq s_2\}]^{-1}.$$

Similarly, Prentice and Cai (1992) provide a representation

$$F(t_1,t_2) = F_1(t_1)F_2(t_2)$$

$$\left[1+\sum_{n=1}^{\infty}\int\cdots\int_{\substack{0\le s_{11}<\cdots<s_{1n}\le t_1\\ 0\le s_{21}<\cdots<s_{2n}\le t_2}}\prod_{j=1}^{n}\frac{A(ds_{1j},ds_{2j})}{\{1-\Lambda_1(\Delta s_{1j})\}\{1-\Lambda_2(\Delta s_{2j})\}}\right]$$

where the differential function A is given by

$$A(ds_1,ds_2) = E\{M_1(ds_1)M_2(s_2)|T_1 \ge s_1, T_2 \ge s_2\}.$$

These representations yield corresponding nonparametric bivariate survivor function estimators $\hat{F}$ by plugging in estimators $\hat{F}_1, \hat{F}_2, \hat{G}$ (or $\hat{A}$) that are specified by replacing expectations by simple averages over study subjects satisfying the conditioning event. These estimators have been shown to be consistent, and to behave well in moderate sized samples, in simulation studies. These results assume that censorship is independent. Specifically, one observes $X_1 = T_1 \wedge C_1, X_2 = T_2 \wedge C_2$, along with censoring indicators $I(X_i < T_i), i = 1,2$, where the potential censoring variates (C_1, C_2) are assumed to be independent of (T_1, T_2). Recently Gill et al (1995) have shown both of the bivariate survivor function estimators just described to have asymptotic Gaussian distributions, and to be nonparametrically efficient under the independence of T_1 and T_2. The lack of full efficiency of these estimators more generally presumably arises because the estimators of the marginal survivor functions (F_1 and F_2) do not acknowledge the simultaneous estimation of the dependency functions (G or A) and visa versa. In fact, one could consider an argument along the lines of the original development of the Kaplan-Meier univariate survivor function estimator to produce an overall nonparametric maximum likelihood estimator of F. Specifically, one could seek $\hat{F}$ that maximizes

$$L = \prod_{i=1}^{I}\prod_{j=1}^{J}\Big[\{F(t_{1i}^-,t_{2j}^-)-F(t_{1i}^-,t_{2j})-F(t_{1i},t_{2j}^-)+F(t_{1i},t_{2j})\}^{n_{ij}^{11}}\, F(t_{1i},t_{2j})^{n_{ij}^{00}}$$

$$\{F(t_{1i}^-,t_{2j}) - F(t_{1i},t_{2j})\}^{n_{ij}^{10}}\, \{F(t_{1i},t_{2j}^-) - F(t_{1i},t_{2j})\}^{n_{ij}^{01}}\Big]$$

where $t_{1i}, i = 1,\cdots,I; t_{2j}, j = 1,\cdots,J$ are the observed uncensored failures on T_1 and T_2, and $n_{ij}^{11}, n_{ij}^{10}, n_{ij}^{01}$ and n_{ij}^{00} are, respectively, the frequencies of doubly uncensored, uncensored on T_1, uncensored on T_2, and doubly censored observations at (t_{1i}, t_{2j}). For example, one could consider maximizing L by setting

$$F(t_{1i},t_{2j}) = \prod_{\ell=1}^{i}(1-\lambda_{1\ell})\prod_{m=1}^{j}(1-\lambda_{2m})\prod_{\ell=1}^{i}\prod_{m=1}^{j}\gamma_{\ell m}$$

and maximizing over all identifiable parameters of $\lambda_{1i}, \lambda_{2j}$ and γ_{ij}. This approach would generally require an iterative calculation, typically involving a very large number of parameters. It is unclear whether the potential efficiency gain compared to the previously mentioned plug-in estimators is sufficient to merit such an effort, and numerical aspects of such a nonparametric maximum likelihood procedure have yet to be developed.

The cross-ratio function provides a particularly useful description of pairwise dependency. This function (e.g., Oakes, 1989) can be written

$$c(t_1, t_2) = \frac{F(dt_1, dt_2)F(t_1, t_2)}{F(dt_1, t_2)F(t_1, dt_2)},$$

explaining the cross-ratio terminology. It can also be written

$$\begin{aligned} c(t_1, t_2) &= \lambda_1(t_1|T_2 = t_2)/\lambda_1(t_1|T_2 \geq t_2) \\ &= \lambda_2(t_2|T_1 = t_1)/\lambda_2(t_2|T_1 \geq t_1) \end{aligned}$$

providing a very natural interpretation, especially in the type of genetic epidemiologic setting mentioned above. Furthermore, one can write c in terms of the differential processes mentioned above as

$$c(t_1, t_2) = \frac{E\{N_1(dt_1)N_2(dt_2)|T_1 \geq t_1, T_2 \geq t_2\}}{E\{N_1(dt_1)|T_1 \geq t_1, T_2 \geq t_2\}\; E\{N_2(dt_2)|T_1 \geq t_1, T_2 \geq t_2\}}$$

leading to the simple empirical 'estimator'

$$\hat{c}(t_1, t_2) = r(t_1, t_2) \sum_{k \varepsilon R(t_1,t_2)} N_{1k}(dt_1)N_{2k}(dt_2)$$

$$\left\{ \sum_{k \varepsilon R(t_1,t_2)} N_{1k}(dt_1) \sum_{k \varepsilon R(t_1,t_2)} N_{2k}(dt_2) \right\}^{-1},$$

on the grid formed by the uncensored failure times, where $R(t_1, t_2)$ denotes the risk set at (t_1, t_2) and $n(t_1, t_2) = \#R(t_1, t_2)$. Note that T_1 and T_2 can be said to be positively related, negatively related, or unrelated at (t_1, t_2) according to whether $c(t_1, t_2)$ exceeds, is less than, or equals, unity.

A representation for F in terms of F_1, F_2 and c has not been presented to date, except under frailty model special cases (Oakes, 1989). However, if T_1 and T_2 are discrete one can establish an equivalence between c and G, given F_1 and F_2, via

$$G(\Delta t_1, \Delta t_2) = 1 +$$

$$\frac{[1 - \{1 - \lambda_1(\Delta t_1)\} \prod_{s_2 < t_2} G(\Delta t_1, \Delta s_2)][1 - \{1 - \lambda_2(\Delta t_2)\} \prod_{s_1 < t_1} G(\Delta s_1, \Delta t_2)]}{\{1 - \lambda_1(\Delta t_1)\}\{1 - \lambda_2(\Delta t_2)\} \prod_{s_1 < t_1} G(\Delta s_1, \Delta t_2) \prod_{s_2 < t_2} G(\Delta t_1, \Delta s_2)}$$

$$\{c(t_1, t_2) - 1\},$$

making it plausible that F_1, F_2 and c determine F for absolutely continuous and mixed variates as well. Note that recursively plugging in $\hat{c}$, $\hat{\lambda}_1$, $\hat{\lambda}_2$ with the above expression provides an alternate estimator of G, and hence of F, that may be somewhat adapted to the marginal distribution estimators $\hat{F}_1$, $\hat{F}_2$. However, the properties of this estimator have not yet been studied.

In most applications in addition to a dependency function, such as G, A or c one would like to have a simple summary measure of the strength of dependence between T_1 and T_2. Hsu and Prentice (1996) propose the correlation between the standardized, unit exponentially distributed, cumulative hazard variates as a summary measure that does not depend on the marginal distribution shapes; that is, $\rho = \text{corr}\{\Lambda_1(T_1), \Lambda_2(T_2)\}$. While this correlation is readily estimated nonparametrically in the absence of censorship it will usually be necessary to impose additional constraints to estimate ρ under general independent censorship, as clearly is the case if the support of (X_1, X_2) is restricted. Hsu and Prentice consider absolutely continuous (T_1, T_2) and suppose that

$$F(t_1, t_2) = F_0\{F_1(t_1), F_2(t_2); \theta\}$$

for a fixed function F_0 and scalar parameter θ that is in 1–1 correspondence with ρ via

$$\rho = \rho(\theta) = \int_0^\infty \int_0^\infty F(t_1, t_2)dt_1, dt_2 - 1.$$

One can estimate ρ by first estimating θ as solution to

$$K^{-1} \int_0^{\tau_1} \int_0^{\tau_2} r(t_1, t_2)\{\hat{A}(dt_1, dt_2) - \hat{A}(dt_1, dt_2; \theta)\} = 0,$$

where (τ_1, τ_2) is in the support of (X_1, X_2), $\hat{A}(dt_1, dt_2)$ is the plug-in nonparametric estimator mentioned above, and $\hat{A}(dt_1, dt_2; \theta)$ is a corresponding model restricted estimator. For example, for the important Clayton (1978) model

$$F(t_1, t_2) = [\{F_1(t_1)^{-\theta} + F_2(t_2)^{-\theta} - 1\} \vee 0]^{-\theta^{-1}},$$

which is absolutely continuous for $\theta > -0.5$, one has

$$\hat{A}(dt_1, dt_2; \theta) = A_0\{\hat{F}_1(t_1), \hat{F}_2(t_2); \theta\}\hat{\Lambda}_1(dt_1)\hat{\Lambda}_2(dt_2),$$

where $A_0(y_1, y_2; \theta) = (\theta + 1)\ y_1^{-\theta}\ y_2^{-\theta}\ (y_1^{-\theta} + y_2^{-\theta} - 1)^{-2} - (y_1^{-\theta} + y_2^{-\theta} - 1)^{-1}$. Note that the Clayton model is characterized by constancy of the cross-ratio; specifically, $c(t_1, t_2) = 1 + \theta$. The solutions $\hat{\theta}$ and $\hat{\rho}$ to this estimation procedure have been shown to be consistent and asymptotically normally distributed with readily estimated variance, under the assumed semiparametric model $F_0(t_1, t_2; \theta)$. Furthermore, in simulation studies the efficiency of $\hat{\theta}$ appeared to be similar to that of the more computationally intensive generalized maximum likelihood estimator considered by Nielsen et al (1992).

These estimators of θ and ρ are limited by their dependence on the particular semiparametric model F_0. Hsu and Prentice also suggested weighted cross-ratio summary measures of dependence over the region $[0, \tau_1] \times [0, \tau_2]$ of the form

$$C(\tau_1, \tau_2) = \int_0^{\tau_1} \int_0^{\tau_2} c(t_1, t_2)\ w(dt_1, dt_2).$$

For this measure to have a clear interpretation it seems necessary that the weight function w depend only on the failure time, and not the censoring, distribution. For example, one could specify

$$w(dt_1, dt_2) = F(dt_1, dt_2) \left\{ \int_0^{\tau_1} \int_0^{\tau_2} F(ds_1, ds_2) \right\}^{-1}$$

in which case $C(\tau_1, \tau_2)$ is simply the average cross-ratio over $[0, \tau_1] \times [0, \tau_2]$. However, the plug-in estimator of $C(\tau_1, \tau_2)$ does not appear to be consistent, so that instead one can consider the reciprocal of an average reciprocal cross-ratio

$$\left[\int_0^{\tau_1} \int_0^{\tau_2} c^{-1}(t_1, t_2) F(dt_1, dt_2) \left\{ \int_0^{\tau_1} \int_0^{\tau_2} F(ds_1, ds_2) \right\}^{-1} \right]^{-1}$$

which has corresponding plug-in estimator

$$\left[\int_0^{\tau_1} \int_0^{\tau_2} \hat{F}(t_1, t_2) r^{-2}(t_1, t_2) \sum_{k \varepsilon R(t_1, t_2)} N_{1k}(dt_1) \sum_{k \varepsilon R(t_1, t_2)} N_{2k}(dt_2) \Big/ \right.$$
$$\left. \{1 - \hat{F}_1(\tau_1) - \hat{F}_2(\tau_2) + \hat{F}(\tau_1, \tau_2)\} \right]^{-1}.$$

The asymptotic and moderate sample properties of this estimator is under investigation.

3 Multivariate Failure Time Regression Analysis

Now suppose that there are n absolutely continuous failure time variates $(T_1, \cdots, T_n)$ and that each T_i is accompanied by a regression p-vector $Z_i, i = 1, \cdots, n$. Much of the work on the regression analysis of multivariate failure time data assumes the T_i to be independent conditional on covariates and on the value of a hypothetical frailty variate, that is usually assumed to affect the hazard function in a multiplicative manner (e.g., Andersen et al, 1993, Chapters 9 and 10; Bickel et al, 1993, Chapters 4, 6 and 7). The joint survivor function for $(T_1, \cdots, T_n)$ is obtained by integrating out the frailty variable; and dependency among the failure times are characterized by the parameters of the frailty distribution. Cox (1972) models are typically specified for the hazard functions, conditional on frailty and covariates. Though frailty models undoubtedly have a place in multivariate failure time analysis they are limited in the flexibility with which dependencies can be modeled, since the parameters of a single frailty variate characterize the entire dependence structure for $T_1, \cdots, T_n$. More complicated frailty constructions with certain frailty variates shared by some, but not all, failure times in a correlated set may be able to partially overcome this limitation. A second problem concerns the form of the marginal distributions. Specifically, if Cox model forms are assumed for intensities given the frailties, the marginal intensities will generally no longer be of the Cox model form. Hence, in using a frailty model approach for multivariate data one may find oneself fitting marginal hazard models that differ from those that would be used if only the data on a specific margin were available. For these three reasons a modeling approach that focuses on marginal survivor functions and pairwise dependence functions, such as G, A or c defined previously, may be preferred.

Consider observed, possibly censored, failure times $(X_{k1}, \cdots, X_{kn}), k = 1, \cdots, K$ and assume Cox model marginal hazard functions

$$\lambda_{ki}(t) = \lambda_{0i}(t)\exp(Z_{ki}^T\beta), \ \ i = 1, \cdots, n,$$

where Z_{ki} corresponds to X_{ki}. Wei et al (1989) propose that the hazard ratio parameter β be estimated as solution to the independence model partial likelihood score equation

$$\sum_{k=1}^{K} \sum_{i=1}^{n} Z_{ki}^T \ \hat{M}_{ki}(X_{ki}) = \sum_{k=1}^{K} Z_k^T \ \hat{M}_k(X_k) = 0$$

where $Z_k = (Z_{k1}, \cdots, Z_{kn})$ and $M_k(X_k) = \{M_k(X_{k1}), \cdots, M_k(X_{kn})\}^T$ and the $\hat{M}_{ki}$ denotes that the standard baseline hazard function estimator (Breslow, 1974; Andersen and Gill, 1982) given by

$$\hat{\Lambda}_{0i}(dt_i) = \sum_{\ell \varepsilon R_i(t_i)} N_{\ell i}(dt_i) \Big/ \sum_{\ell \varepsilon R_i(t_i)} e^{Z_{\ell i}^T \beta}$$

has been inserted, where $R_i(t_i)$ is the T_i risk set at time t_i. Wei and colleagues show the solution $\hat{\beta}$ to this estimating equation to be consistent and asymptotically normal, and they provide a 'sandwich-type' variance estimator that adapts to dependencies among the failure times $(T_1, \cdots, T_n)$. Cai and Prentice (1995) examined whether the efficiency of hazard ratio parameter estimates could be improved by introducing a weight function into the estimating equation. Specifically, they considered $\hat{\beta}$ solving

$$\sum_{k=1}^{K} Z_k^T \ \hat{W}_k^{-1} \ \hat{M}_k(X_k) = 0$$

where $\hat{W}_k$ is an estimator of the correlation matrix $W_k = \text{corr}\{M_k(X_k)|Z_k\}$. Solutions to this more general estimating equation were also shown to be consistent and asymptotically normal and a sandwich variance formula was provided. As with other types of response data, random variation in the (consistent) estimator $\hat{W}_k$ does not affect the asymptotic distribution of $\hat{\beta}$. A nonparametric estimator of $\hat{W}_k$ can be calculated if the covariate matrix Z has finite support. However, theoretical calculations and simulations under the Clayton and Cuzick (1985) bivariate regression model, indicate the efficiency gains in using this weighted estimating equation to be fairly modest unless the failure time dependencies are quite strong. For example, the efficiency gain in using the weighted versus the unweighted estimator is only 26% at $\beta = 0$ and no censorship if the Clayton model parameter is as large as $\theta = 0.67$, corresponding to a correlation of 0.513 between cumulative hazard variates. Moreover, such efficiency gain tends to be reduced by censorship. Hence, the simpler unweighted estimating equation of Wei et al can be expected to be efficient enough for most applications. See Lee et al (1992) for corresponding unweighted methods under the more restrictive model in which baseline hazard functions are common (i.e., $\lambda_{0i} \equiv \lambda_0$).

Now consider regression on marginal hazard ratios and pairwise cross-ratios. Prentice and Hsu (1997) consider estimating equations along the lines of those used for mean and covariance parameter estimation with other types of response data. Denote $\Lambda_{ki}(t) = \int_0^t \lambda_{ki}(s)ds$, so that $\Lambda_{ki}(T_{ki}), i = 1, \cdots, n$ are standardized variates having unit exponential marginal distributions. Define the pairwise correlation vector

$$\rho_k = \rho_k(\beta, \alpha) = (\rho_{k12}, \rho_{k13}, \cdots, \rho_{k23}, \cdots)$$

where $\rho_{kij} = \text{corr}\{\Lambda_{ki}(T_{ki}), \Lambda_{kj}(T_{kj}); Z_k\}$. For example, one could model $\rho_k(\beta, \alpha) =$
$(\alpha_{12}, \alpha_{13}, \cdots, \alpha_{23}, \cdots)$ in order to allow each pair of cumulative hazard variates to have a distinct correlation. Each correlation ρ_{kij} could be relaxed further by allowing a dependence on the corresponding regression vector Z_{ki} and Z_{kj}.

Prentice and Hsu define cumulative baseline hazard variates

$$V_{ki} = \int_0^{T_{ki}} \lambda_{0i}(s)ds$$

and note that $E(V_{ki}) = e^{-Z_{ki}^T\beta}$ and $\text{cov}(V_{ki}, V_{kj}) = \rho_{kij}\, e^{-Z_{ki}^T\beta}e^{-Z_{kj}^T\beta}$. The 'quadratic exponential' estimating equations of Prentice and Zhao (1991) for mean and covariance parameters were then applied, and each V_{ki} was replaced by $\hat{V}_{ki} = \hat{\Lambda}_{0i}(X_{ki})$. The resulting estimating equations for β and α can be written

$$\sum_{k=1}^{K} Z_k^T\, \hat{\Omega}_k^{-1}\, \hat{M}_k(X_k) = 0; \sum_{k=1}^{K} E_k^T\, \hat{W}_k^{-1}\, \hat{L}_k(X_k) = 0,$$

where $E_k^T = \partial\rho_k/\partial\alpha^T$, $L_k(X_k) = \{L_{k12}(X_{k1}, X_{k2}), L_{k13}(X_{k1}, X_{k3}), \cdots, L_{k23}(X_{k2}, X_{k3}), \cdots\}$ and $L_{kij}(X_{ki}, X_{kj}) = M_{ki}(X_{ki})M_{kj}(X_{kj}) - \rho_{kij}$. Also in this uncensored situation $\hat{\Omega}_k$ is simply the correlation matrix for $M_k(X_k)$; that is, the matrix with (i, j) element $\rho_{kij}, i \neq j$, and $\hat{W}_k$ is a working specification of the correlation matrix for $L_k(X_k)$. As discussed above, it will usually be adequate to replace each $\hat{\Omega}_k$ by an identity matrix in the first of these equations, in which case $\hat{\beta}$ is again the estimator of Wei et al, while the use of identity weight matrices $\{W_k\}$ may well be adequate also in the second equation.

These joint estimating equations also apply under independent right censorship though additional assumptions are required on the pairwise survivor functions to estimate the correlations ρ_{kij}, just as was the case in the non-regression situation. Specifically, it is necessary to generalize the definition of $L_k(X_k)$ by setting

$$L_{kij}(X_{ki}, X_{kj}) = M_{ki}(X_{ki})M_{kj}(X_{kj}) - A_{kij}(X_{ki}, X_{kj})$$

where

$$A_{kij}(dt_i, dt_j) = E\{M_{ki}(dt_i)M_{kj}(dt_j)|T_{ki} \geq t_i, T_{kj} \geq t_j; Z_k\}.$$

Semiparametric models for the pairwise survivor functions

$$F_{kij}(t_i, t_j) = F_{0ij}\{F_{ki}(t_i), F_{kj}(t_j); \theta_{kij}\},$$

where the scalars θ_{kij} are in 1–1 correspondence with ρ_{kij}, then allow the A_{kij}'s to be specified and $(\hat{\beta}, \hat{\alpha})$ to be specified as solutions to the above pair of estimating equations. Prentice and Hsu show these estimators to be generally consistent and asymptotically normal, with a sandwich-type variance estimator. An important special case arises by assuming Clayton and Cuzick (1985) models for pairwise survivor functions, so that

$$A_{kij}(dt_i, dt_j) = A_0\{F_{ki}(t_i), F_{kj}(t_j); \theta_{kij}\},$$

all (k, i, j), with A_0 as given in Section 2. Under this specification β parameterizes the marginal hazard ratios, while α parameterizes the pairwise cross-ratios. The principal limitation of this regression estimation procedure arises from the fact that the pairwise cross-ratios $c_{kij}(t_i, t_j; Z_k) = 1 + \theta_{kij}$ are assumed to be independent of (t_i, t_j), for all (k, i, j). The estimated cross-ratios $\hat{\theta}_{kij} = \theta_{kij}(\hat{\beta}, \hat{\alpha})$ presumably have a weighted cross-ratio interpretation more generally, but with weights that depend on the censoring as well as the failure time distribution. Hence development of a more flexible approach to pairwise dependency modeling in this regression context would be desirable. See Prentice and Hsu (1997) for simulation results and illustrations to amplify these points.

4 Discussion

Dependencies among pairs of failure time variates can be characterized in various ways. The cross-ratio function provides a characterization having a clear and useful interpretation, particularly in epidemiologic contexts. Reciprocal cross-ratios, averaged over the bivariate failure time density, provide a simple summary measure of dependence over a failure time region. Corresponding variance estimation requires a variance estimator for the bivariate survivor function estimator which exists only in a very complex form for the plug-in joint survivor function estimators described in Section 2. Hence, at present, a bootstrap procedure may be required for average cross-ratio variance estimation.

Estimating equations and corresponding distribution theory are available for the simultaneous estimation of Cox model marginal hazard ratio parameters and for pairwise cross-ratio parameters, assuming such cross-ratio functions to be constant. A more flexible approach would allow a general cross-ratio function for each pair of standardized cumulative hazard variates $\{\Lambda_{ki}(T_{ki}), \Lambda_{kj}(T_{kj})\}$. Such cross-ratio functions can be estimated by applying the estimator $\hat{c}$, given in Section 2, to the estimated possibly censored variates $\hat{\Lambda}_{ki}(X_{ki}), \hat{\Lambda}_{kj}(X_{kj})$, leading to average reciprocal cross-ratio estimators for these standardized variates. These estimators are also currently being investigated.

Case-control sampling is very often a feature of the genetic epidemiologic studies alluded to above (e.g., Whittemore, 1995). The adaptation of these cohort study failure time estimation procedures to case-control sampling is an important outstanding research topic.

Acknowledgments

This work was supported by grant CA53996 from the U.S. National Cancer Institute. The authors would like to thank Jianwen Cai and Juan Juan Fan for helpful related discussions.

References

Andersen, P.K,, Borgan, O., Gill, R.D. and Keiding, N. (1993). *Statistical Models Based on Counting Processes.* New York: Springer-Verlag.

Anderson, P.K. and Gill, R.D. (1982). Cox's regression model for counting processes: A large sample study. *Ann. Statist.*, **10**, 1100–1120.

Bickel, P.J., Klassen, C.A., Ritov, Y. and Wellner, J. (1993). *Efficient and adaptive estimation for semiparametric models.* Baltimore, Maryland:

Johns Hopkins University Press.

Breslow, N.E. (1974). Covariance analysis of censored survival data. *Biometrics*, **30**, 89–99.

Cai, J. and Prentice, R.L. (1995). Estimating equations for hazard ratio parameters based on correlated failure time data. *Biometrika*, **82**, 151–164.

Clayton, D.G. (1978). A model for association in bivariate life tables and its application in epidemiological studies of familial tendency in chronic disease incidence. *Biometrika*, **65**, 144–151.

Clayton, D.G. and Cuzick, J. (1985). Multivariate generalizations of the proportional hazards model (with discussion). *J.R. Statist. Soc.* **A**, **148**, 82–117.

Cox, D.R. (1972). Regression models and life tables (with discussion). *J.R. Statist. Soc.* **B**, 187–220.

Dabrowska, D. (1988). Kaplan-Meier estimate on the plane. *Ann. Statist.*, **16**, 1475–1489.

Fleming, T.R. and Harrington, D.P. (1991). *Counting Processes and Survival Analysis.* New York: Wiley.

Gill, R.D., Van der Laan, M.J. and Wellner, J.A. (1995). Inefficient estimators of the bivariate survival function for three models. *Ann. Inst. Henri Poincare*, **31**, 545–597.

Hsu, L. and Prentice, R.L. (1996). On assessing the strength of dependency between failure time variates. *Biometrika*, **83**, 491–506.

Lee, E., Wei, L.J., and Amato, D.A. (1992). Cox-type regression analysis for large numbers of small groups of correlated failure-time observations. In *Survival Analysis: State of the Art*, Eds. J.P. Klein and P.K. Goel, pp. 237–247. Klewer Academic Publishers.

Nielsen, G.G., Gill, R.D., Andersen, P.K. and Sorensen, T.I.A. (1992). A counting process approach to maximum likelihood estimation in frailty models. *Scand. J. Statist.*, **19**, 25–43.

Oakes, D. (1989). Bivariate survival models induced by frailties. *J. Amer. Statist. Assoc.*, **84** 487–493.

Prentice, R.L. and Cai, J. (1992). Covariance and survivor function estimation using censored multivariate failure time data. *Biometrika*, **79**, 495–512.

Prentice, R.L. and Hsu, L. (1997). Regression on hazard ratios and cross-ratios in multivariate failure time analysis. *Biometrika*, in press.

Prentice, R.L. and Zhao, L.P. (1991). Estimating equations for parameters in means and covariances of multivariate discrete and continuous responses. *Biometrics*, **47**, 825–839.

Wei, L.J., Lin, D.Y. and Weissfeld, L. (1989). Regression analysis of multivariate incomplete failure time data by modelling marginal distributions. *J. Amer. Statist. Assoc.*, **84**, 1065–1073.

Whittemore, A.S. (1995). Logistic regression of family data from case-control studies. *Biometrika*, **82**, 57–67.

Analysis of Multivariate Survival Times with Non-Proportional Hazards Models

L. Chen
L. J. Wei

ABSTRACT In a clinical trial to evaluate treatments for a chronic disease, a commonly used regression method for analyzing multiple event times is based on a multivariate Cox model (Wei, Lin and Weissfeld, 1989). However, the Cox model may not fit the data well. For univariate survival analysis, a class of linear transformation models (Cheng, Wei and Ying, 1995a) provides many useful semi-parametric alternatives to the Cox model. In this paper, we take a similar approach as Wei et al. (1989) did for the multivariate case by modeling each marginal failure time with a linear transformation model and derive joint inference procedures for the regression parameters. In addition, we show how to check the adequacy of the fitted model graphically. We apply the proposed methods to data from an AIDS clinical trial and a cancer clinical trial for illustration.

1 Introduction

The Cox proportional hazards model (Cox, 1972) has been extensively used to analyze univariate failure time data owing to its flexible nature. The large sample properties of its inference procedures have been elegantly justified using martingale theory (Andersen and Gill, 1982). The proportional hazards model, however, may not fit the data well. Recently, Cheng, Wei and Ying (1995a) studied a class of linear transformation models, under which an unknown transformation of the survival time is linearly related to the covariates with completely specified error distributions. This class of regression models includes the proportional hazards and proportional odds models. The transformation models, coupled with the simple inference procedures proposed by Cheng et al. (1995a), provide many useful alternatives to the Cox model in univariate survival analysis.

For a long term clinical trial for treating a complex chronic disease, often there are a number of important endpoints involved in the study. The subjects may experience several *failures* which correspond to repeated occurrences of the same type of event or to the occurrences of events of entirely different natures. For example, in a recent AIDS trial to compare the safety and efficacy of continuing zidovudine (AZT) therapy with that of zalcitabine (ddC) alone or zalcitabine and zidovudine used together, 1001 patients with symptomatic human immunodeficiency virus (HIV) disease and 300 or fewer CD4 cells/mm^3 or asymptomatic HIV disease and 200 or fewer CD4 cells/mm^3 were randomly assigned to receive 600 mg of zidovudine per day; 2.25 mg of zalcitabine per day; or 600 mg of zidovudine and 2.25 mg of zalcitabine per day (Fischl, Stanley et al., 1995). These patients had previously received and tolerated at least 6 months of zidovudine. The primary endpoint for the study was time to disease progression or death. The most common disease progression events were *P. carinii* pneumonia (PCP), cytomegalovirus disease (CMV) and disseminated *Mycobacterium avium* complex (MAC). Quite a few subjects had more than one type of disease progression during the course of the study. It is important to know how to analyze such multiple event times efficiently to evaluate treatment differences. A commonly used regression method to handle this problem is a multivariate version of Cox's procedure proposed by Wei, Lin and Weissfeld (WLW) (1989). The idea of their approach is quite simple. First, the kth type of event times is modeled with the usual proportional hazards model, whose nuisance hazard function and regression parameters β_{0k} may vary with k, $k = 1, \ldots, K$. Next, the maximum partial likelihood estimate $\hat{\beta}_k$ for β_{0k} is obtained, and joint inferences about these regression coefficients with respect to all the distinct types of failures are made based on the large sample theory of $\{\hat{\beta}_k, k = 1, \ldots, K\}$. Computer software that implements the WLW method is available, such as MULCOX2, a FORTRAN code by D.Y. Lin, the SAS PHREG procedure, and the S macros by Terry Therneau. Although the WLW method is popular for the analysis of multiple failure times, the marginal Cox models may not fit the data well. The univariate accelerated failure time model (Wei, 1992), an alternative to the Cox model, has been generalized to the multivariate case (Lin and Wei, 1992). However, it is highly desirable to have a rich class of semi-parametric and non-proportional hazards models available for the analysis of such multiple event time data.

In this paper, we take the marginal approach as WLW did, but using the linear transformation models studied by Cheng et al. (1995a) for each marginal failure time variable. Inference procedures for regression parameters are derived. Furthermore, we show how to evaluate the adequacy of the fitted model graphically. All the proposals are illustrated with the above AIDS trial and a well-known bladder tumor study (Byar, 1980).

2 Modeling Failure Times Marginally with Linear Transformation Models

For the kth type of failure, $k = 1, \ldots, K$, let T_{ki} be the failure time of the ith patient, $i = 1, \ldots, n$. However, not all the T's can be completely obtained. Instead, one observes a bivariate vector (X_{ki}, Δ_{ki}), where $X_{ki} = \min(T_{ki}, C_{ki})$, C_{ki} is the censoring time, and $\Delta_{ki} = 1$ if $X_{ki} = T_{ki}$ and 0, otherwise. Let $z_{ki} = (z_{1ki}, \ldots, z_{pki})'$ be a $p \times 1$ vector of covariates for the ith patient with respect to the kth failure. We assume that the censoring variable C_{ki} does not depend on z_{ki}. This assumption can be easily relaxed for the case when the covariate vector z has a finite number of possible values. Furthermore, the failure time vector $T_i = (T_{1i}, \ldots, T_{Ki})'$ and the censoring vector $C_i = (C_{1i}, \ldots, C_{Ki})'$, $i = 1, \ldots, n$, are assumed to be independent. It is important to note that for the aforementioned AIDS example, the time to the kth type of disease progression event, for example, the MAC, might be censored by death. To avoid this kind of dependent censoring, one may define the kth event to be either the kth type of disease progression or death.

For the kth type of failure of the ith patient, let $S_k(\cdot|z)$ be the survival function of T_{ki} given z. The Cox model can be written as

$$\log[-\log\{S_k(t|z)\}] = h_k(t) + \beta_{0k}' z, \tag{2.1}$$

where $h_k(t)$ is a completely unspecified strictly increasing function and $\beta_{0k} = (\beta_{01k}, \ldots, \beta_{0pk})'$ is a $p \times 1$ vector of unknown regression coefficients. A natural generalization of (2.1) is

$$g_k\{S_k(t|z)\} = h_k(t) + \beta_{0k}' z, \tag{2.2}$$

where $g_k(\cdot)$ is a known decreasing function. If we let $g_k(x)$ be $-\text{logit}(x) = -\log(x/(1-x))$, (2.2) is the so-called proportional odds model (Bennett, 1983; Pettitt, 1982). It is easy to see that (2.2) is equivalent to the linear transformation model:

$$h_k(T) = -\beta_{0k}' z + \varepsilon_k, \tag{2.3}$$

where ε_k is a random error with distribution function $F_k = 1 - g_k^{-1}$. If F_k is the extreme value distribution, where $F_k(s) = 1 - \exp\{-\exp(s)\}$, (2.3) is the proportional hazards model; while if F_k is the standard logistic distribution, (2.3) is the proportional odds model. For univariate survival data, inference procedures for (2.3) have been proposed, for example, by Cuzick (1988) and Dabrowska and Doksum (1988a,b). Recently, Cheng et al. (1995a) have studied a set of simple estimating procedures for β_{0k} in (2.3) with possibly censored observations. In this article, we generalize the methods by Cheng et al. (1995a) to the multivariate case.

Let the estimating function (2.3) in Cheng et al. (1995a) for the kth type of failure time be denoted by

$$U_k(\beta) = \sum_{i=1}^{n}\sum_{j=1}^{n} w_k(\beta' z_{kij}) z_{kij} \left\{ \frac{\Delta_{kj} I(X_{ki} \geq X_{kj})}{\hat{G}_k^2(X_{kj})} - \xi_k(\beta' z_{kij}) \right\}, \tag{2.4}$$

where $w_k(\cdot)$ is a weight function, $z_{kij} = z_{ki} - z_{kj}$, $\hat{G}_k$ is the Kaplan-Meier estimate for the *survival* function of the censoring variable C_{ki}, and $\xi_k(s) = \int_{-\infty}^{\infty}\{1 - F_k(t+s)\}dF_k(t)$. Note that if there is no censoring, $\text{E}\{I(X_{ki} \geq X_{kj})\} = \xi_k(\beta_{0k}' z_{kij})$ and (2.4) is an unbiased estimating function. In the presence of censoring, one can show that $\text{E}\{\frac{\Delta_{kj} I(X_{ki} \geq X_{kj})}{G_k^2(X_{kj})}\} = \xi_k(\beta_{0k}' z_{kij})$, it follows that the estimating function (2.4) is asymptotically unbiased. Cheng et al. (1995a) showed that when $w_k(\cdot)$ is positive, the equation $U_k(\beta) = 0$ has, asymptotically, a unique solution $\hat{\beta}_k$. When $w_k = 1$, the above equation has a unique solution for any finite sample. They also showed through examples that the estimation procedure with $w_k = 1$ works well for the proportional odds model and the model with the standard normal error. If the error distribution F_k is the standard extreme value distribution, the weight function $w_k(\cdot)$ becomes 1. For this case, Cheng et al. (1995a) demonstrated through simulation that their approach is practically as good as the one based on the partial likelihood when censoring is moderate or heavy.

The estimators $\hat{\beta}_k$'s are correlated. For large n, $(\hat{\beta}_1', \ldots, \hat{\beta}_K')'$ is approximately normal with mean $(\beta_{01}', \ldots, \beta_{0K}')'$ and covariance matrix Γ (see Appendix A). With this large sample property, various kinds of inferences about the regression parameters can be made. For instance, suppose that we are interested in $\eta_{0k} = \beta_{01k}$, $k = 1, \ldots, K$, from a particular covariate. Let the estimator of the covariance matrix of $(\hat{\eta}_1, \ldots, \hat{\eta}_K)'$, where $\hat{\eta}_k = \hat{\beta}_{1k}$, be denoted by Λ, which can be obtained from Γ directly. Then the quadratic form $W = (\hat{\eta}_1, \ldots, \hat{\eta}_K)\Lambda^{-1}(\hat{\eta}_1, \ldots, \hat{\eta}_K)'$ may be used to test the joint null hypothesis $\{H_k : \eta_{0k} = 0, k = 1, \ldots, K\}$. Suppose that $\eta_{01} = \ldots = \eta_{0K} = \eta_0$, a common parameter, then one can construct an optimal linear combination $\hat{\eta}$ of the $\hat{\eta}_k$'s to estimate η_0, where $\hat{\eta} = \sum_{k=1}^K d_k \hat{\eta}_k$ with $\sum_{k=1}^K d_k = 1$, $d = (d_1, \ldots, d_K)' = (e'\Lambda^{-1}e)^{-1}\Lambda^{-1}e$ and $e = (1, \ldots, 1)'$.

The above procedures are valid when the distribution of the censoring variable C_{ki} is independent of the covariate z. Suppose that one can discretize the covariate z into M possible values. An analogue of the estimating function (2.4) that corporates dependence between C_{ki} and z is

$$U_k^*(\beta) = \sum_{i=1}^n \sum_{j=1}^n w_k(\beta' z_{kij}) z_{kij} \left\{ \frac{\Delta_{kj} I(X_{ki} \geq X_{kj})}{\hat{G}_k(X_{kj}|z_{ki})\hat{G}_k(X_{kj}|z_{kj})} - \xi_k(\beta' z_{kij}) \right\},$$

where $\hat{G}_k(\cdot|z)$ is the Kaplan-Meier estimator for the survival function of the censoring variable C_{ki} based on those pairs $\{X_{kl}, \Delta_{kl}\}$ whose $z_{kl} = z, l = 1, \ldots, n$. In Appendix B, we show that the distribution of the roots $\{\hat{\beta}_k^*, k = 1, \ldots, K\}$ to the simultaneous equations $\{U_k^*(\beta) = 0, k = 1, \ldots, K\}$ is approximately normal with mean $\{\beta_{0k}, k = 1, \ldots, K\}$ and covariance matrix Γ^*. Inference procedures similar to W and $\hat{\eta}$ for $\{\beta_{0k}, k = 1, \ldots, K\}$ based on the above large sample properties of $\{\hat{\beta}_k^*, k = 1, \ldots, K\}$ can be obtained accordingly.

3 Examples

In this section, we use two examples to illustrate our proposals based on the estimating function $U_k(\beta)$ in (2.4). The first one is the AIDS example cited in the Introduction. To make our illustration interesting, we consider a data set from two treatment groups of patients whose baseline CD4 cell counts were greater than or equal to 125 cells/mm^3. These two treatments are AZT and the combination of AZT and ddC. It is important to note that one should be

Table 1: Number of disease progression events and deaths for patients whose baseline CD4 $\geq$ 125 cells/mm^3 in the AZT+ddC and AZT groups of the AIDS example

	PCP	CMV	MAC	Other disease progression	Total number of death	(Death without disease progression)
AZT+ddC	9	9	5	17	10	(4)
AZT	13	12	5	19	12	(6)

careful to draw conclusions from such subgroup analyses in practice. For this subgroup of patients, there are 146 patients treated by AZT mono-therapy and 194 treated by the combination drug. Seventy five patients either had at least one type of disease progression or died during the study. To avoid dependent censoring, we define four types of failures: PCP or death, CMV or death, MAC or death, and other disease progression events or death. The number of events for the four types of disease progression and the number of deaths in the two treatment groups are given in Table 1. The Kaplan-Meier estimates for the AZT and combination groups with respect to each of the above four types of failures (either disease progression or death) are given in Figure 1. We also compute the two sample *marginal* logrank test statistics and report their two-sided p-values in Figure 1. Since patient's baseline CD4 cell count is an important predictor for disease progression, we use regression models (2.2) or (2.3) to make inferences about the treatment differences with this covariate adjustment. To this end, let $z_{1ki} = 0$, if this patient was assigned to the combination drug, 1, otherwise; and let z_{2ki} be the baseline CD4 cell count. Our main interest is in the treatment differences: $\eta_{0k} = \beta_{01k}, k = 1, 2, 3, 4$. In Table 2, we give the estimates $\hat{\eta}_k$ under the proportional hazards and odds models using the estimating function (2.4) with $w_k = 1$. For comparison, we also provide the estimates for η_{0k} based on the WLW method. For each model, the omnibus test statistic W and the estimate $\hat{\eta}$ for the common treatment difference η_0 are computed and the results are reported in Table 2. Note that under the proportional hazards model, the estimates for η_{0k} based on the WLW method seem rather different from those based on the new method. This indicates that the assumed proportional hazards model may not fit the data well. However, for testing the hypothesis that the common parameter $\eta_0 = 0$, we find that the p-values based on $\hat{\eta}$ under various models are quite similar.

Next, we show with an example how to use the new proposal to analyze recurrence times of a specific event. Consider the well-known bladder tumor study conducted by the Veterans Administration Cooperative Urological Research Group (VACURG) (Byar, 1980). In this study, all the patients had superficial bladder tumors when they entered the trial. These tumors were removed transurethrally and patients were randomly assigned to one of the three treatments: placebo, thiotepa or pyridoxine. Most patients had multiple recurrences of tumors during the study. The new tumors were removed at each visit. The

Figure 1: Kaplan-Meier estimates for different types of disease progression/death for the AZT+ddC (solid lines) and AZT (dotted lines) groups of the AIDS example

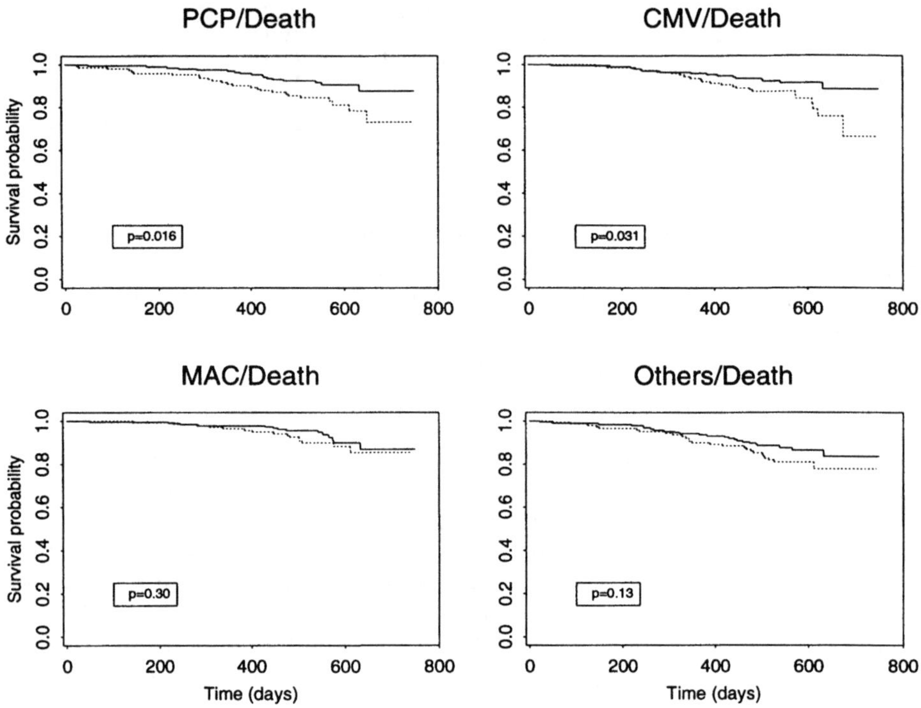

recurrence times of tumors for patients in placebo and thiotepa groups were reported by Wei et al. (1989). Each recurrence time was measured from the beginning of the patient's treatment. We assume that the patient's follow-up time is independent of his/her event times. Furthermore, for each patient, some potentially important covariates were also recorded, for example, the number of tumors at the patient's entry. To evaluate the thiotepa effect, one may conduct the usual survival analysis based on the first recurrence times. However, this univariate procedure may not be efficient and cannot be used to test for temporal treatment effect from thiotepa. Since there are very few patients who had fourth recurrences, we only use the first three recurrence times in our analysis with the linear transformation models. Here, $n = 86$ and $K = 3$. Let $z_{ki} = (z_{1ki}, z_{2ki})'$, where $z_{1ki} = 0$ if the ith patient was in the thiotepa group, and 1, otherwise; and z_{2ki} is the number of initial tumors. Again, our main interest is in the treatment effects: $\eta_{0k} = \beta_{01k}, k = 1, 2, 3$. The estimates for η_{0k} using (2.4) with weight $w_k = 1$ are given in Table 3 with proportional hazards and odds models. Furthermore, we report the results based on the corresponding omnibus test W and $\hat{\eta}$ for a common treatment difference η_0. For comparison, the results based

Table 2: Inferences about treatment differences $\eta_{0k}^{\dagger}$ (AZT+ddC vs. AZT) for patients with baseline CD4 $\geq$ 125 cells/mm^3 of the AIDS example

	WLW	New Method	
		Proportional hazards model	Proportional odds model
Type of failure		Estimate for η_{0k} (s.e.)	
PCP/Death	0.709(0.315)	0.286(0.174)	0.429(0.260)
CMV/Death	0.628(0.326)	0.432(0.238)	0.646(0.356)
MAC/Death	0.337(0.373)	0.033(0.157)	0.049(0.236)
Others/Death	0.403(0.283)	0.117(0.146)	0.176(0.219)
P-value for the omnibus test W	0.096	0.179	0.178
Estimate for a common η_0 (s.e.)	0.581(0.245)	0.170(0.092)	0.256 (0.137)

† Adjusted for baseline CD4 cell count; positive value of $\hat{\eta}_k$ indicates AZT+ddC is better than AZT.

Table 3: Inference about treatment differences $\eta_{0k}^{\dagger}$ (thiotepa vs. placebo) for the bladder tumor data

	WLW	New Method	
		Proportional hazards model	Proportional odds model
Recurrence		Estimate for η_{0k} (s.e.)	
1	0.504(0.313)	0.381(0.219)	0.567(0.326)
2	0.612(0.394)	0.449(0.251)	0.671(0.374)
3	0.711(0.463)	0.448(0.313)	0.668(0.465)
P-value for the omnibus test W	0.278	0.105	0.103
Estimate for a common η_0(s.e.)	0.540(0.287)	0.416(0.169)	0.621(0.251)

† Adjusted for number of initial tumors; positive value of $\hat{\eta}_k$ indicates thiotepa is better than placebo.

on the method by WLW are also summarized in Table 3. From a perspective of testing for a difference between the thiotepa and placebo groups, the new method with the proportional hazards and odds models gives quite similar results. On the other hand, quantitative assessments, such as interval estimates, for the treatment differences can be quite different under these two models.

4 Model Checking and Selection

The validity of the inference procedures proposed in Section 2 only depends on the adequacy of the marginal models (2.2) or (2.3). There is no need to check the correlation structure among distinct types of failure times. Here, we present a simple graphical method to check the appropriateness of the fitted model. Suppose that the link function g_k in (2.2) and the deterministic portion, $\beta_{0k}'z$, of the model are correctly specified. Then, from (2.3), $\{h_k(T_{ki}) + \beta_{0k}'z_{ki}, i = 1, ..., n\}$ should be a random sample from a population whose distribution function $F_k = 1 - g_k^{-1}$ is completely specified. However, T's may be censored and h_k is unknown. If h_k can be estimated reasonably well, say, by $\hat{h}_k$, one would expect that the Kaplan-Meier estimate $\hat{F}_k$ based on $\{\hat{h}_k(X_{ki}) + \hat{\beta}_k'z_{ki}, \Delta_{ki}, i = 1, ..., n\}$ is close to the distribution function F_k. Therefore, a P-P or Q-Q plot, constructed from $\hat{F}_k$ and F_k, can be used to examine visually the adequacy of the fitted model. Through an extensive empirical study, Cheng, Wei and Ying (1995b) show that the graphical method based on the P-P plots is a reasonable tool to examine the adequacy of the fitted model when the regression coefficients in (2.3) are not "too close to zero".

To obtain a consistent estimate for $h_k(t)$, one may use the following estimating function:

$$V(h_k(t)) = \sum_{i=1}^{n} \left\{ \frac{I(X_{ki} \geq t)}{\hat{G}_k(t)} - g_k^{-1}(h_k(t) + \hat{\beta}_k'z_{ki}) \right\}.$$

Cheng, Wei and Ying (1997) have shown that a solution, $\hat{h}_k(t)$, to the equation $V(h_k(t)) = 0$, is non-decreasing in t and is consistent on $(0, \gamma_k)$, where γ_k is a time point such that $P(X_{ki} \geq \gamma_k) > 0$.

We now use the bladder cancer data to illustrate this graphical method. Since we consider three distinct tumor recurrence times, we construct three sets of P-P plots under each fitted model. In Figure 2, we present the plots with proportional hazards and odds models. If the assumed model fits the data well, one would expect that the P-P plot is approximately a straight line through the origin. With respect to the first two recurrences, the patterns of the P-P plots between the proportional hazards and odds models are similar. For the third recurrence time, however, the P-P plot under the proportional odds model appears to be closer to the straight line than that under the proportional hazards model, indicating that the proportional odds model is a better choice for modeling the tumor recurrence data.

5 Remarks

Under the usual comparative randomized clinical trial setting, if we are interested in quantitative assessments for the relative merit between the two treatments or in predicting the survival probabilities of future patients, fitting the data with a "wrong" model may have detrimental effects. Although the graphical method discussed in Section 4 is useful for examining the adequacy of the fitted model, more research is needed on model checking and selection even in univariate survival analysis.

Figure 2: P-P plots for the bladder tumor data with proportional hazards and proportional odds models

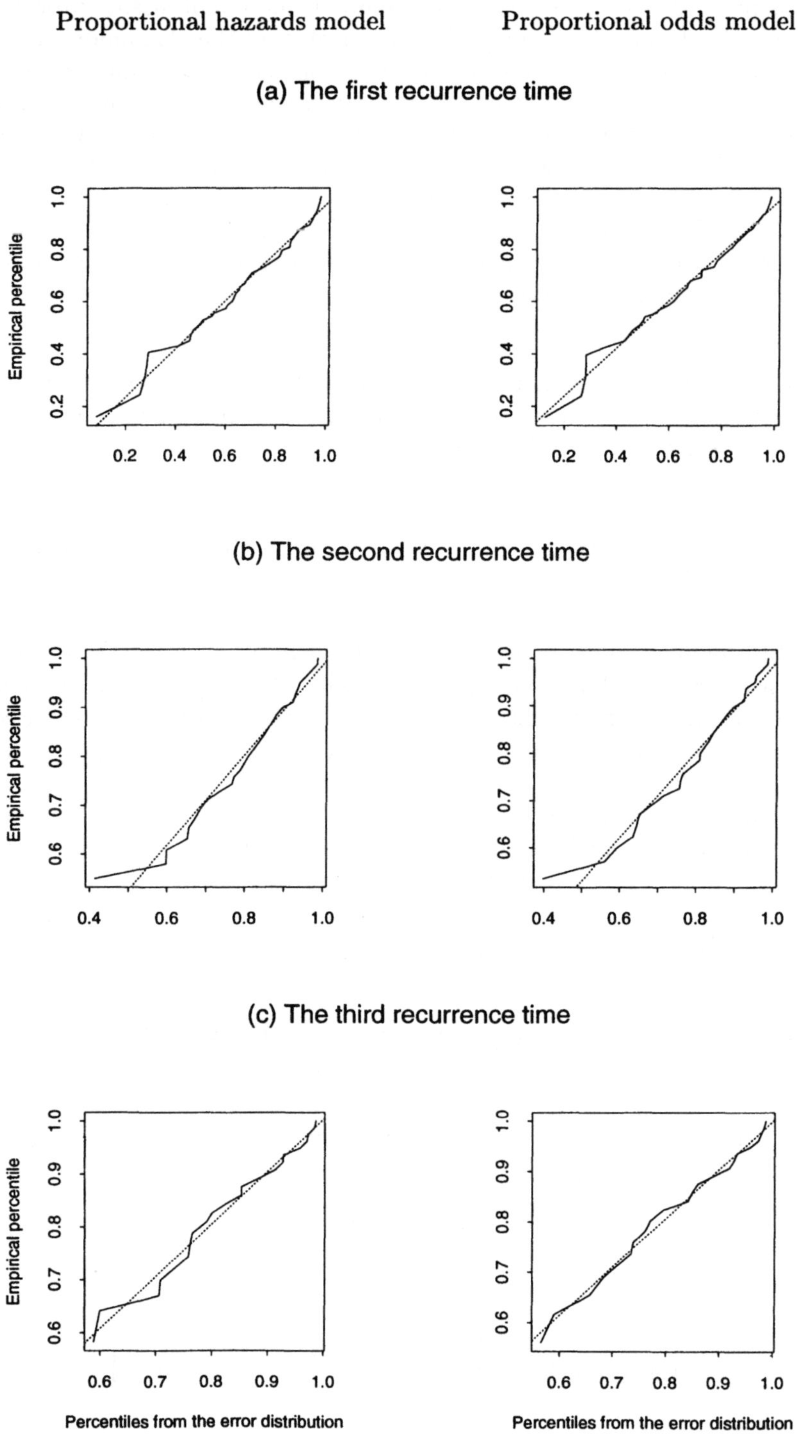

It is important to note that for the AIDS example, death may censor the time to disease progression. In this paper, we combine death and disease progression as the endpoint to avoid the competing risk problem. The Cox model can handle this problem differently by using the cause-specific hazard function for the disease progression time, but treating the death time as a censoring variable. However, it is not obvious how to generalize this idea to the linear transformation models. A possible way to deal with this competing risk problem is to replace $S_k(t|z)$ in (2.2) with, for example, the cumulative incidence function (Gray, 1988). Unfortunately, the resulting model (2.2) cannot be written as the form of (2.3), from which the estimating function (2.4) is derived without the presence of the competing risk.

Acknowledgments: The authors are grateful to Drs. Steve Lagakos, Su-Chun Cheng and Kenneth Stanley for their helpful comments on this paper. This research was supported by NIH Grant CA56844.

Appendix A: Asymptotic Joint Distribution of $\hat{\beta}_k$'s Based on U_k's

Using the argument given in Appendix 1 of Cheng et al. (1995a) for the univariate case, one can show that $\hat{\beta}_k$ is consistent. Next, we show that for large n the distribution of the random vector $(\hat{\beta}_1^{'}, \ldots, \hat{\beta}_K^{'})^{'}$ can be approximated by a pK-dimensional normal distribution with mean vector $(\beta_{01}^{'}, \ldots, \beta_{0K}^{'})^{'}$ and a covariance matrix which can be easily estimated.

First, we show that the joint distribution of $\{U_k(\beta_{0k}), k = 1, \ldots, K\}$ can be approximated by a normal distribution with mean 0. For the kth type of failure, the quantity $n^{-3/2}U_k(\beta_{0k})$ is asymptotically equivalent to

$$n^{-3/2} \sum_{i=1}^{n} \sum_{j=1}^{n} w_k(\beta_{0k}^{'} z_{kij}) z_{kij} \left\{ \frac{\Delta_{kj} I(X_{ki} \geq X_{kj})}{G_k^2(X_{kj})} - \xi_k(\beta_{0k}^{'} z_{kij}) \right\}$$

$$+2n^{-3/2} \sum_{i=1}^{n} \sum_{j=1}^{n} w_k(\beta_{0k}^{'} z_{kij}) z_{kij} \frac{\Delta_{kj} I(X_{ki} \geq X_{kj})}{G_k^2(X_{kj})} \frac{\{G_k(X_{kj}) - \hat{G}_k(X_{kj})\}}{G_k(X_{kj})}. \tag{A.1}$$

It follows from a martingale integral representation for $(\hat{G}_k - G_k)/G_k$ (Gill, 1980) that (A.1) is asymptotically equivalent to

$$n^{-3/2} \sum_{i=1}^{n} \sum_{j=1}^{n} \phi_{kij} + 2n^{-1/2} \sum_{i=1}^{n} \int_0^\infty \frac{q_k(t)}{\pi_k(t)} dM_{ki}(t), \tag{A.2}$$

where

$$\phi_{kij} = w_k(\beta_{0k}^{'} z_{kij}) z_{kij} \left\{ \frac{\Delta_{kj} I(X_{ki} \geq X_{kj})}{G_k^2(X_{kj})} - \xi_k(\beta_{0k}^{'} z_{kij}) \right\},$$

$$\begin{aligned} q_k(t) &= \lim_{n\to\infty} n^{-2} \sum_{i=1}^{n}\sum_{j=1}^{n} w_k(\beta_{0k}' z_{kij}) z_{kij} \frac{\Delta_{kj} I(X_{ki} \geq X_{kj})}{G_k^2(X_{kj})} I(X_{kj} \geq t), \\ \pi_k(t) &= \lim_{n\to\infty} n^{-1} \sum_{i=1}^{n} I(X_{ki} \geq t), \\ M_{ki}(t) &= I(X_{ki} \leq t, \Delta_{ki} = 0) - \int_0^t I(X_{ki} \geq u) d\Lambda_{G_k}(u), \end{aligned}$$

and $\Lambda_{G_k}(\cdot)$ is the common cumulative hazard function of C_{ki}'s.

Let $\zeta_k(t) = q_k(t)/\pi_k(t)$. Then the second term in (A.2) is

$$2n^{-3/2} \sum_{j=1}^{n}\sum_{i=1}^{n} \int_0^\infty \zeta_k(t) d\left\{ I(X_{ki} \leq t, \Delta_{ki} = 0) - \int_0^t I(X_{ki} \geq u) d\Lambda_{G_k}(u) \right\}.$$

Let the above quantity be denoted by $2n^{-3/2} \sum_{i=1}^{n} \sum_{j=1}^{n} \nu_{ki}$. It follows that $n^{-3/2} U_k(\beta_{0k})$ is asymptotically equivalent to $n^{-3/2} \sum_{i=1}^{n} \sum_{j=1}^{n} (\phi_{kij} + 2\nu_{ki})$, which is a U-statistic.

Using the standard asymptotic theory of the multivariate U-statistics (Wei and Johnson, 1985), one can show that the limiting distribution of $n^{-3/2}(U_1'(\beta_{01}), \ldots, U_K'(\beta_{0K}))'$ is asymptotically normal with mean 0. To estimate the above limiting covariance, let

$$\hat{\tau}_{kij} = \hat{\phi}_{kij} + 2\left\{ (1-\Delta_{ki})\hat{\zeta}_k(X_{ki}) - \sum_{r=1}^{n} (1-\Delta_{kr})\hat{\zeta}_k(X_{kr}) \frac{I(X_{ki} \geq X_{kr})}{\sum_{q=1}^{n} I(X_{kq} \geq X_{kr})} \right\},$$

where

$$\begin{aligned} \hat{\phi}_{kij} &= w_k(\hat{\beta}_k' z_{kij}) z_{kij} \left\{ \frac{\Delta_{kj} I(X_{ki} \geq X_{kj})}{\hat{G}_k^2(X_{kj})} - \xi_k(\hat{\beta}_k' z_{kij}) \right\}, \\ \hat{\zeta}_k(t) &= \frac{\hat{q}_k(t)}{\hat{\pi}_k(t)}, \\ \hat{q}_k(t) &= n^{-2} \sum_{i=1}^{n}\sum_{j=1}^{n} w_k(\hat{\beta}_k' z_{kij}) z_{kij} \frac{\Delta_{kj} I(X_{ki} \geq X_{kj})}{\hat{G}_k^2(X_{kj})} I(X_{kj} \geq t), \end{aligned}$$

and

$$\hat{\pi}_k(t) = n^{-1} \sum_{i=1}^{n} I(X_{ki} \geq t).$$

Then, a consistent estimator for the above limiting covariance matrix is $\mathbf{B} = \{B_{kl}\}$, where $B_{kl} = n^{-3}(\sum_i \sum_j \sum_{j'} \hat{\tau}_{kij} \hat{\tau}_{lij'}' + \sum_j \sum_i \sum_{i'} \hat{\tau}_{kij} \hat{\tau}_{li'j}'), k, l = 1, \ldots, K$.

To obtain the large sample distribution of $\{\hat{\beta}_k, k = 1, \ldots, K\}$, we simply take the Taylor expansion of $U_k(\hat{\beta}_k)$ around β_{0k}. The covariance matrix of

$\{\hat{\beta}_k, k = 1, \ldots, K\}$ can be estimated by Γ, which is

$$\begin{pmatrix} J_1 & \ldots & 0 \\ \vdots & \ddots & \vdots \\ 0 & \ldots & J_K \end{pmatrix} \mathbf{B} \begin{pmatrix} J_1' & \ldots & 0 \\ \vdots & \ddots & \vdots \\ 0 & \ldots & J_K' \end{pmatrix},$$

where

$$J_k^{-1} = n^{-3/2} \sum_{i=1}^{n} \sum_{j=1}^{n} w_k(\hat{\beta}_k' z_{kij}) \frac{\partial \xi_k(\beta_k' z_{kij})}{\partial \beta_k} \Big|_{\hat{\beta}_k} z_{kij}^{\otimes 2},$$

and $v^{\otimes 2} = vv^T$ for a column vector v.

Appendix B: Asymptotic Joint Distribution of $\hat{\beta}_k^*$'s with Discrete Covariates Based on U_k^*'s

Since the covariate vector z has a finite number of possible values, for the kth type of failure, the quantity $n^{-3/2} U_k^*(\beta_{0k})$ is asymptotically equivalent to

$$n^{-3/2} \sum_{i=1}^{n} \sum_{j=1}^{n} w_k(\beta_{0k}' z_{kij}) z_{kij} \left\{ \frac{\Delta_{kj} I(X_{ki} \geq X_{kj})}{G_k(X_{kj}|z_{ki}) G_k(X_{kj}|z_{kj})} - \xi_k(\beta_{0k}' z_{kij}) \right\}$$
$$+ n^{-3/2} \sum_{i=1}^{n} \sum_{j=1}^{n} w_k(\beta_{0k}' z_{kij}) z_{kij} \frac{\Delta_{kj} I(X_{ki} \geq X_{kj})}{G_k(X_{kj}|z_{ki}) G_k(X_{kj}|z_{kj})}$$
$$\times \left\{ \frac{G_k(X_{kj}|z_{ki}) - \hat{G}_k(X_{kj}|z_{ki})}{G_k(X_{kj}|z_{ki})} + \frac{G_k(X_{kj}|z_{kj}) - \hat{G}_k(X_{kj}|z_{kj})}{G_k(X_{kj}|z_{kj})} \right\}.$$

Using the martingale integral representation for $\hat{G}_k(\cdot|z)$, one can show that the above quantity is asymptotically equivalent to

$$n^{-3/2} \sum_{i=1}^{n} \sum_{j=1}^{n} \phi_{kij}^* + 2n^{-1/2} \sum_{i=1}^{n} \int_0^\infty \frac{q_k^*(t)}{\pi_k^*(t|z_{ki})} dM_{ki}(t|z_{ki}), \tag{B.1}$$

where

$$\phi_{kij}^* = w_k(\beta_{0k}' z_{kij}) z_{kij} \left\{ \frac{\Delta_{kj} I(X_{ki} \geq X_{kj})}{G_k(X_{kj}|z_{ki}) G_k(X_{kj}|z_{kj})} - \xi_k(\beta_{0k}' z_{kij}) \right\},$$

$$q_k^*(t) = \lim_{n \to \infty} n^{-2} \sum_{i=1}^{n} \sum_{j=1}^{n} w_k(\beta_{0k}' z_{kij}) z_{kij} \frac{\Delta_{kj} I(X_{ki} \geq X_{kj})}{G_k(X_{kj}|z_{ki}) G_k(X_{kj}|z_{kj})} I(X_{kj} \geq t),$$

$$\pi_k^*(t|z) = \lim_{n \to \infty} n^{-1} \sum_{i=1}^{n} I(X_{ki} \geq t, z_{ki} = z),$$

$$M_{ki}(t|z) = I(X_{ki} \leq t, \Delta_{ki} = 0) - \int_0^t I(X_{ki} \geq u) d\Lambda_{G_k(\cdot|z)}(u), \tag{B.2}$$

and $\Lambda_{G_k(\cdot|z)}(\cdot)$ is the common cumulative hazard function of C_{ki}'s whose covariate vector is z.

Let $\zeta_k^*(t|z) = q_k^*(t)/\pi_k^*(t|z)$. Then the second term in (B.1) is

$$2n^{-3/2}\sum_{j=1}^{n}\sum_{i=1}^{n}\int_0^\infty \zeta_k^*(t|z_{ki})d\left\{I(X_{ki} \le t, \Delta_{ki} = 0) - \int_0^t I(X_{ki} \ge u)d\Lambda_{G_k(\cdot|z_{ki})}(u)\right\}.$$

Let the above quantity be denoted by $2n^{-3/2}\sum_{i=1}^n\sum_{j=1}^n \nu_{ki}^*$. It follows that $n^{-3/2}U_k^*(\beta_{0k})$ is asymptotically equivalent to $n^{-3/2}\sum_{i=1}^n\sum_{j=1}^n(\phi_{kij}^* + 2\nu_{ki}^*)$, which is a U-statistic. This implies that the limiting distribution of $n^{-3/2}(U_1^{*'}(\beta_{01}),\ldots,U_K^{*'}(\beta_{0K}))'$ is asymptotically normal with mean zero. The limiting covariance matrix of $\{\hat\beta_k^*, k=1,\ldots,K\}$ can be estimated by Γ^*, which is

$$\begin{pmatrix} J_1^* & \cdots & 0 \\ \vdots & \ddots & \vdots \\ 0 & \cdots & J_K^* \end{pmatrix} \mathbf{B}^* \begin{pmatrix} J_1^{*'} & \cdots & 0 \\ \vdots & \ddots & \vdots \\ 0 & \cdots & J_K^{*'} \end{pmatrix}$$

where

$$J_k^{*-1} = n^{-3/2}\sum_{i=1}^n\sum_{j=1}^n w_k(\hat\beta_k^{*'}z_{kij})\frac{\partial\xi_k(\beta_k^{*'}z_{kij})}{\partial\beta_k^*}\Big|_{\hat\beta_k^*} z_{kij}^{\otimes 2},$$

$$\mathbf{B}^* = \{B_{kl}^*\} = \{n^{-3}(\sum_i\sum_j\sum_{j'}\hat\tau_{kij}^*\hat\tau_{lij'}^{*'} + \sum_j\sum_i\sum_{i'}\hat\tau_{kij}^*\hat\tau_{li'j}^{*'})\}, k,l = 1,\ldots,K,$$

and

$$\begin{aligned}\hat\tau_{kij}^* &= \hat\phi_{kij}^* + 2\Big\{(1-\Delta_{ki})\hat\zeta_k^*(X_{ki}|z_{ki}) \\ &\quad - \sum_{r=1}^n(1-\Delta_{kr})\hat\zeta_k^*(X_{kr}|z_{ki})\frac{I(X_{ki}\ge X_{kr}, z_{ki}=z_{kr})}{\sum_{q=1}^n I(X_{kq}\ge X_{kr}, z_{kq}=z_{ki}=z_{kr})}\Big\}.\end{aligned}$$

Here, $\hat\phi_{kij}^*$ and $\hat\zeta_k^*(\cdot|z)$ are obtained by replacing β_{0k} and $G_k(\cdot|z)$ in (B.2) with $\hat\beta_k^*$ and $\hat G_k(\cdot|z)$, respectively.

References

Andersen, P. K. and Gill, R. D. (1982). Cox's regression model for counting processes: A large sample study (Com: p1121-1124). *Ann. Statist.* **10**, 1100–1120.

Bennett, S. (1983). Analysis of survival data by the proportional odds model. *Statist. Med.* **2**, 273–277.

Byar, D. P. (1980). The Veterans Administration Study of Chemoprophylaxis for Recurrent Stage I Bladder Tumors: Comparisons of Placebo, Pyridoxine, and Topical Thiotepa. *Bladder Tumors and Other Topics in Urological*

Oncology, eds Pavone-Macaluso, M., Smith, P. H. and Edsmyn, F., New York: Plenum, 363–370.

Cheng, S. C., Wei, L. J. and Ying, Z. (1995a). Analysis of transformation models with censored data. *Biometrika* **82**, 835–845.

Cheng, S. C., Wei, L. J. and Ying, Z. (1995b). Predicting survival probability with non-proportional hazards models, Technical Report, University of Texas, M. D. Anderson Cancer Center, Dept. of Biomathematics.

Cheng, S. C., Wei, L. J. and Ying, Z. (1997). Predicting survival probability with semi-parametric transformation models. To appear in *J. Am. Statist. Assoc.*

Cox, D. R. (1972). Regression models and life-tables (with discussion). *J. R. Statist. Soc. B* **34**, 187–220.

Cuzick, J. (1988). Rank regression. *Ann. Statist.* **16**, 1369–1389.

Dabrowska, D. M. and Doksum, K. A. (1988a). Estimation and testing in a two-sample generalized odds-rate model. *J. Am. Statist. Assoc.* **83**, 744–749.

Dabrowska, D. M. and Doksum, K. A. (1988b). Partial likelihood in transformation models with censored data. *Scand. J. Statist.* **15**, 1–23.

Fischl, M. A., Stanley, K. et al. (1995). Combination and monotherapy with zidovudine and zalcitabine in patients with advanced HIV disease. *Annals of Internal Medicine* **122**, 24–32.

Gill, R. D. (1980). *Censoring and Stochastic Integrals* (Mathematical Centre Tract No. 124). Amsterdam: Mathematisch Centrum.

Gray, R. J. (1988). A class of K-sample tests for comparing the cumulative incidence of a competing risk. *Ann. Statist.* **16**, 1140–1154.

Lin, J. S. and Wei, L. J. (1992). Linear regression analysis for multivariate failure time observations. *J. Am. Statist. Assoc.* **87**, 1091–1097.

Pettitt, A. N. (1982). Inference for the linear model using a likelihood based on ranks. *J. R. Statist. Soc. C* **33**, 169–175.

Wei, L. J. (1992). The accelerated failure time model: A useful alternative to the Cox regression model in survival analysis (Disc: p1881-1885). *Statist. Med.* **11**, 1871–1879.

Wei, L. J. and Johnson, W. E. (1985). Combining dependent tests with incomplete repeated measurements. *Biometrika* **72**, 359–364.

Wei, L. J., Lin, D. Y. and Weissfeld, L. (1989). Regression analysis of multivariate incomplete failure time data by modeling marginal distributions. *J. Am. Statist. Assoc.* **84**, 1065–1073.

Analysis of Mean and Rate Functions for Recurrent Events

J. F. Lawless
C. Nadeau
R. J. Cook

ABSTRACT Methods for the analysis of mean and rate functions for recurrent events are reviewed, in the context of longitudinal studies involving multiple subjects. Semiparametric models involving an arbitrary baseline mean or rate function multiplied by a parametrically specified function of covariates are emphasized. Methods of dealing with censoring or terminal events which are related to the recurrent events are considered. An example involving rejection episodes for kidney transplant recipients is discussed.

1 Introduction

In many longitudinal studies subjects experience recurrent or repeated events. For example, patients at risk for cerebrovascular events may experience transient ischemic attacks (Hobson et al. 1993), cancer patients may experience recurrent superficial tumors (Byar 1980), and individuals with intractable epilepsy may experience seizures (Albert 1991). Several types of recurrent events may be of interest in any one study but we assume there is just one type for most of this paper. Broad objectives of analysis for recurrent events include (i) understanding and characterizing event occurrence processes for individual subjects, (ii) explaining subject to subject variability in terms of treatments, fixed covariates, or other factors, and (iii) assessing the relationship of time varying covariates or other processes to event occurrence.

To motivate and illustrate later developments we introduce an example.

Example. Rejection Episodes in Kidney Transplant Recipients

Cole et al. (1994) describe a study in which kidney transplant recipients were randomly assigned to receive either rabbit anti-thymocyte serum (RATS) or a drug, OKT3, for 10 to 14 days after transplant as prophylaxis against graft rejection. There were 158 patients with fully functioning kidney grafts, of which 75 were in the RATS group and 83 were in the

OKT3 group. The main objective of the study was to compare the occurrence of rejection "episodes" for patients in the two treatment groups; a rejection episode was defined in terms of elevated serum creatinine levels for a patient.

Subjects were followed for 6 months and the times of rejection episodes noted. To simplify the discussion we will suppose that the exact event times are known, although Cole et al. (1994) report the times in intervals. A complicating factor was that 20 subjects experienced total graft rejection before 6 months: 9 in the RATS group and 11 in the OKT3 group. This is a competing risk that complicates assessment of rejection episodes, since total graft rejection is not strictly independent of the rejection episode history for a subject. The average number of rejection episodes per patient in the study was about .4 for the RATS group and .8 for the OKT3 group; the data are summarized in Table 1 below. Cook and Lawless (1996) discuss this example in some detail.

Comprehensive approaches to the modelling and analysis of recurrent events are provided by counting process (e.g. Andersen et al. 1993) or point process (e.g. Cox and Lewis 1966; Cox and Isham 1980) models. With these approaches a model that specifies the probability distribution of the event history for an individual is provided, so that maximum likelihood or partial likelihood methods of inference may be used. However, for purposes such as the assessment of treatments or fixed covariates, methods that focus on marginal mean or rate functions for the recurrent events are valuable. They do not require a complete specification of the event process and may be made robust to assumptions about the underlying processes. Pepe and Cai (1993) introduced rate functions for the analysis of recurrent events. Lawless and Nadeau (1995) and Nelson (1995) emphasize mean functions, following earlier work of Nelson (1988). A comparison of marginal methods with those based on full process models is given by Lawless (1995).

This article will review the analysis of marginal mean and rate functions and outline some new methods. The emphasis is hence on the objectives (ii) and (iii) mentioned in the first paragraph of the paper. Some preliminary notation and definitions follow.

Suppose that k subjects may experience some type of recurrent event. Let $N_i(t)$ denote the number of events over the time period $(0,t]$ for the i'th subject and $dN_i(t)$ denote the number of events over the small interval $[t, t+dt)$. We assume that subject i is observed over the interval $(0, \tau_i]$, and refer to τ_i as the termination time; we also define $\delta_i(t) = I(t \le \tau_i)$, where the indicator function $I(A)$ for event A equals 1 if A is true and 0 otherwise. Exact occurrence times for events in $(0, \tau_i]$ are assumed known, so that $\{N_i(t),\ t \ge 0\}$ is a counting process with jumps at event occurrence times $t_{i1} < t_{i2} < \cdots$. Observational schemes where subjects may be observed intermittently or where only total numbers of events for different time intervals are observed can also be addressed but will not be considered here.

Full event process models may be specified via an intensity function which specifies the probability distribution of $dN_i(t)$ conditional on past events (see Lawless 1995). In this article we focus on the marginal mean functions

$$M_i(t) = E\{N_i(t)\} \qquad t \geq 0 \tag{1.1}$$

and, if the events occur in continuous time, the corresponding rate of occurrence functions $m_i(t) = M_i'(t)$. If time is measured on a discrete scale, say $t = 0, 1, 2, \ldots$, then we define $m_i(t) = E\{dN_i(t)\}$.

The remainder of the article deals with modelling and estimation of $M_i(t)$ or $m_i(t)$. Section 2 discusses robust estimating equation methods, first proposed for rates by Pepe and Cai (1993). Section 3 addresses termination or censoring related to the recurrent events. Section 4 considers the example above as an illustration, and Section 5 concludes with some comments on extensions to the current methodology.

2 Inferences About Mean Functions

In this section we assume that the termination times τ_i are independent of the event processes $\{N_i(t),\ t \geq 0\}$. In that case the mean function (1.1) is unambiguous, and it is possible to develop robust estimating equation methodology for inference (Pepe and Cai 1993; Nadeau 1995; Lawless and Nadeau 1995).

Following Lawless and Nadeau (1995), let $x_i(t)$ be a vector of external covariates (i.e. they do not depend on the event history $\{N_i(s),\ s < t\}$) associated with subject i at time t and assume that, conditional on the covariate history $x_i^* = \{x_i(t),\ t \geq 0\}$ we have

$$E\{dN_i(t)|x_i^*\} = dM_i(t) = dM_0(t)g(x_i(t);\beta), \tag{2.1}$$

where β is a vector of regression coefficients, and g and $dM_0(t)$ are positive-valued functions. Stieltjes integral notation is used to accommodate discrete or continuous time processes; in the continuous time case $dM_i(t) = m_i(t)dt$ and $dM_0(t) = m_0(t)dt$, where $m_0(t)$ is a baseline rate function. In the discrete time case $dM_i(t) = m_i(t)$. Other types of regression specification can be considered, but the multiplicative form (2.1) is flexible and natural in many applications.

Fully parametric models for (2.1) employ a specified function g and a specified parametric function $dM_0(t;\alpha)$, where the parameter vector $\theta = (\beta, \alpha)$ is unknown. In trying to avoid strong assumptions about $M_0(t)$, we might make $dM_0(t;\alpha)$ a rather flexible family involving several parameters, as in Pepe and Cai (1993). Semiparametric models for (2.1) would make no restrictions on $M_0(t)$ other than it be non-decreasing. In this case we will also write $\theta = (\beta, M_0(\cdot))$ to represent the model "parameters" though $M_0(\cdot)$ is not finite-dimensional.

Nadeau (1995) proposes that inference about θ be based on linear estimating functions of the form

$$U(\theta) = \sum_{i=1}^{k} U_i(\theta) = \sum_{i=1}^{k} \int_0^{\infty} \delta_i(t) a_i(t)[dN_i(t) - dM_i(t)], \qquad (2.2)$$

where the $a_i(t)$'s are specified vectors of functions which may depend on both θ and any covariates in x_i^*. This is the same as Pepe and Cai's (1993) suggestion applied to the present context. It is clear that when $\{\delta_i(t),\ t \geq 0\}$'s are independent of the event processes $\{N_i(t),\ t \geq 0\}$ and (2.1) is correct, the estimating function $U(\theta)$ has expectation 0. Thus it may be shown in the parametric case (Pepe and Cai, 1993) that under mild conditions an estimator $\hat{\theta}$ obtained by solving $U(\theta) = 0$ is consistent and asymptotically normal as $k \to \infty$. In the semiparametric case (Lawless and Nadeau 1995, Nadeau 1995), it is easily established that $\hat{\beta}$ and estimators $\hat{M}_0(t_1), \ldots, \hat{M}_0(t_r)$ at a finite set of t-values are consistent and asymptotically normal. Results concerning the convergence of $\sqrt{k}\{\hat{M}_0(t) - M_0(t),\ t \geq 0\}$ to a Gaussian process under slightly more stringent conditions seem possible, but have not been investigated.

A natural question concerns what the $a_i(t)$'s should be. Pepe and Cai (1993) and Lawless and Nadeau (1995) use the maximum likelihood estimating functions that arise when $\{N_i(t),\ t \geq 0\}$ is a Poisson process with rate function (2.1). In the fully parametric case this yields

$$U(\theta) = \sum_{i=1}^{k} \int_0^{\infty} \delta_i(t) \frac{\partial \log m_i(t)}{\partial \theta} [dN_i(t) - dM_i(t)]. \qquad (2.3)$$

That is, $a_i(t) = \partial \log m_i(t)/\partial\theta$.

Pepe and Cai (1993) and Lawless and Nadeau (1995) discuss the solution of (2.3) and of (2.4) and (2.5) below, respectively and provide robust variance estimates that rely only on the correctness of (2.1) and not on strong assumptions about the underlying event processes. In the fully parametric case, the following results are given by Pepe and Cai (1993) for the general case involving (2.2). Define $p \times p$ matrices (where $p = dim(\theta)$)

$$A(\theta) = \frac{-\partial U(\theta)}{\partial \theta^T}, \quad B(\theta) \doteq \sum_{i=1}^{k} U_i(\theta) U_i(\theta)^T,$$

where all vectors are written in column form and V^T denotes the transpose of a matrix or vector V. Under mild conditions on the event processes and the $m_i(t;\theta)$ functions, $\sqrt{k}(\hat{\theta} - \theta)$ converges as $k \to \infty$ to a limiting normal random vector with mean 0 and covariance matrix that is estimated consistently by $kA(\hat{\theta})^{-1}B(\hat{\theta})A(\hat{\theta})^{-T}$.

For the semiparametric case the general equations (2.2) become

$$U_\beta(\theta) = \sum_{i=1}^{k} \int_0^\infty \delta_i(t) a_{1i}(t)[dN_i(t) - dM_i(t)] \tag{2.4}$$

$$U_{M_0}(\theta) = \sum_{i=1}^{k} \delta_i(t) a_{2i}(t)[dN_i(t) - dM_i(t)], \quad t \geq 0 \tag{2.5}$$

where $a_{1i}(t)$ is a column vector of length $dim(\beta), a_{2i}(t)$ is scalar, and (2.5) is to be solved for all $t \geq 0$. The Poisson equations analogous to (2.3) have $a_{1i}(t) = \partial \log g(x_i(t); \beta)/\partial\beta$ and $a_{2i}(t) = 1$. Setting (2.5) equal to 0 gives

$$dM_0(t) = \frac{\sum_{i=1}^{k} \delta_i(t) a_{2i}(t) dN_i(t)}{\sum_{i=1}^{k} \delta_i(t) a_{2i}(t) g_i(t)}, \tag{2.6}$$

where $g_i(t) = g(x_i(t); \beta)$. This is a generalization of the extended Nelson-Aalen or Breslow estimator (e.g. Andersen et al. 1993, Section VII.2) used with intensity-based counting process models. It may be inserted into (2.4) to yield an equation in β alone which may be solved to obtain $\hat{\beta}$. Substitution of $\hat{\beta}$ into (2.6) then gives $d\hat{M}_0(t)$ and thus $\hat{M}_0(t) = \int_0^t d\hat{M}_0(s)$.

Specifically, define

$$a_1g(t) = \sum_{i=1}^{k} \delta_i(t) a_{1i}(t) g_i(t), \quad a_2g(t) = \sum_{i=1}^{k} \delta_i(t) a_{2i}(t) g_i(t).$$

Then substitution of (2.6) into (2.4) yields, after a little rearrangement,

$$U^{(1)}(\beta) = \sum_{i=1}^{k} \int_0^\infty \delta_i(t)[a_{1i}(t) - \frac{a_1g(t)}{a_2g(t)} a_{2i}(t)] dN_i(t). \tag{2.7}$$

It is especially easy to consider β jointly with functionals of $M_0(t)$ such as $\psi = M_0(t_1)$, where t_1 is a specified value. We note that (2.6) leads to $U^{(2)}(\beta, \psi) = 0$, where

$$U^{(2)}(\beta, \psi) = \sum_{i=1}^{k} [\int_0^{t_1} \delta_i(t) \frac{a_{2i}(t)}{a_2g(t)} dN_i(t) - \frac{\psi}{k}], \tag{2.8}$$

where $\psi = M_0(t_1)$. Solving $U^{(1)}(\beta) = 0$ yields $\hat{\beta}$ and then $U^{(2)}(\hat{\beta}, \psi) = 0$ yields $\hat{\psi} = \hat{M}_0(t_1)$. Furthermore, the joint asymptotic normality of $\sqrt{k}(\hat{\beta} - \beta, \hat{\psi} - \psi)$ is easily established. (It is equally easy to deal with $\hat{\beta}$ and

$\hat{M}(t_1), ..., \hat{M}(t_r)$.) In particular, the asymptotic covariance matrix of $\sqrt{k}(\hat{\beta}-\beta, \hat{\psi}-\psi)$ follows from a simple extension of the approach in Lawless and Nadeau (1995, see Appendix), and may be estimated consistently as follows: define

$$A(\beta,\psi) = \begin{pmatrix} -\partial U^{(1)}/\partial\beta^T & -\partial U^{(1)}/\partial\psi \\ -\partial U^{(2)}/\partial\beta^T & -\partial U^{(2)}/\partial\psi \end{pmatrix} = \begin{pmatrix} -\partial U^{(1)}/\partial\beta^T & 0 \\ -\partial U^{(2)}/\partial\beta^T & 1 \end{pmatrix}$$

$$B(\beta, M_0) = \sum_{i=1}^{k} U_i(\beta, M_0)U_i(\beta, M_0)^T,$$

where $U_i(\beta, M_0) = (U_i^{(1)}(\beta, M_0)^T, U_i^{(2)}(\beta, M_0))^T$ with

$$U_i^{(1)}(\beta, M_0) = \int_0^\infty \delta_i(t)[a_{1i}(t) - \frac{a_1 g(t)}{a_2 g(t)} a_{2i}(t)][dN_i(t) - dM_i(t)]$$

$$U_i^{(2)}(\beta, M_0) = \int_0^{t_1} \frac{\delta_i(t)a_{2i}(t)}{a_2 g(t)}[dN_i(t) - dM_i(t)].$$

Here M_0 stands for the function $M_0(t)$, $t \le \max(\tau_1, ..., \tau_k)$. Observe that (2.7) and (2.8) may be rewritten as

$$U^{(1)}(\beta) = \sum_{i=1}^{k} U_i^{(1)}(\beta, M_0)$$

$$U^{(2)}(\beta, \psi) = \sum_{i=1}^{k} U_i^{(2)}(\beta, M_0).$$

Then the asymptotic variance of $\sqrt{k}(\hat{\beta}-\beta, \hat{\psi}-\psi)$ may be estimated by $kA(\hat{\beta},\hat{\psi})^{-1}B(\hat{\beta},\hat{M}_0)A(\hat{\beta},\hat{\psi})^{-T}$. Lawless and Nadeau's (1995) results (see their equations (3.10) and (A.6)) are obtained for the case of Poisson estimating functions, where $a_{1i}(t) = \partial \log g_i(t)/\partial\beta$ and $a_{2i}(t) = 1$.

It should be noted that the case where there are no covariates and $M_i(t) = M_0(t)$ for all $i = 1, ..., k$ is governed by the results above. In that case $g_i(t) = 1$ and there is only (2.6) to consider. Lawless and Nadeau (1995, Section 2) and Nelson (1995) consider this case in detail.

Nadeau (1995) and Nadeau and Lawless (1996) discuss efficiency and optimality for estimating functions in the class (2.1). If there is information about the covariance structure of the recurrent event processes it may be possible to select functions $a_i(t)$, $a_{1i}(t)$ or $a_{2i}(t)$ that give more efficient (smaller asymptotic variance) estimators than the Poisson equations. However, if a variance specification for the processes involves additional unknown parameters, they must be estimated. In practical situations where there is little prior information about variance structures, use of the Poisson equations with appropriate robust variance estimates is recommended.

3 Related Termination or Censoring

In some studies there are terminal events which preclude the occurrence of further recurrent events. For example, in the kidney transplant study described in Section 1, 20 out of 158 patients suffered total graft rejection before the end of the 6 month study period, thus precluding further rejection episodes. In other situations a subject may withdraw early from a study because of factors related to their past history of recurrent events even though future recurrent events are not strictly precluded. In this case we say that termination or censoring is "related" to the recurrent events if the risk of a subject's withdrawal at any time point depends upon their previous event history. (We do not allow termination or censoring to depend upon future events.)

A precise mathematical treatment of this type of termination or censoring is given by Andersen et al. (1993, Section III.2). This is somewhat unfortunately termed "independent" censoring in the literature, and we will refer to the case where termination or censoring times are independent of the entire event history process as "completely independent" censoring. It is well known that for intensity-based event history models with independent censoring, likelihood inference about the recurrent event process is not affected in the sense that the censoring process does not need to be modelled explicitly, and that the likelihood for the recurrent events has the same form as for completely independent censoring. The estimating function methods of Section 2, however, require completely independent censoring. The problem is not solely that the distributional properties of the procedures break down under "related" censoring; the definitions of marginal rates and mean functions also need adjustment.

For simplicity, suppose that all subjects in a study have a common maximum censoring time τ^*, defined by the length of the study; in the kidney transplant study this is 6 months. Suppose that subjects may also withdraw earlier due to some terminal event, so that $\tau_i = \min(\tau^*, T_i)$, where T_i is the time of the terminal event. It is possible to assess whether the hazard function for T_i depends upon previous recurrent event history by basing time-dependent covariates on the recurrent events and using standard survival analysis methods. If there is some relationship, then we must decide how to model and analyze the recurrent events. One approach is to model the terminal and recurrent events jointly as a multivariate counting process (Andersen et al. 1993), but this does not generally lead to tractable specifications for marginal rates or mean functions. If a marginal analysis is sought, then we propose what Little (1995) refers to as "pattern" models, where recurrent event rates are defined conditional on values of T_i. Cook and Lawless (1996) consider the issue in some detail, and we describe some of the ideas here.

Letting x_i^* denote any covariate history for subject i, we define the conditional rate

$$dM_i(s;t) = E\{dN_i(s)|T_i \geq t, x_i^*\}, \quad s \leq t, \tag{3.1}$$

and corresponding mean function

$$M_i(s;t) = \int_0^s m_i(u;t)du = E\{N_i(s)|T_i \geq t, x_i^*\}, \quad s \leq t. \tag{3.2}$$

Both (3.1) and (3.2) have simple interpretations. When there are no covariates, so that $M_i(s;t) = M(s;t)$, the trivial nonparametric estimator

$$\hat{M}(s;t) = \frac{\sum_{i=1}^{k} \delta_i(t) N_i(s)}{\sum_{i=1}^{k} \delta_i(t)} \tag{3.3}$$

is available. A slight extension of results in Section 2 of Lawless and Nadeau (1995) shows that the variance of $\hat{M}(s;t)$ is estimated consistently by

$$\hat{V}(s;t) = \sum_{i=1}^{k} \left[\frac{\delta_i(t)}{\delta.(t)}\right]^2 \{N_i(s) - \hat{M}(s;t)\}^2, \tag{3.4}$$

where $\delta.(t) = \sum_{i=1}^{k} \delta_i(t)$. Other rates or means may be obtained by combining (3.1) or (3.2) with a marginal model for T_i. Defining $\bar{F}_i(t) = Pr(T_i > t|x_i^*)$, we have for example that

$$E\{\int_0^t \delta_i(s)dN_i(s)\} = \int_0^t \bar{F}_i(s)dM_i(s;s) \tag{3.5}$$

and for $s \leq a$,

$$E\{N_i(s)|a \leq T_i < b\} = \frac{M_i(s;a)\bar{F}_i(a) - M_i(s;b)\bar{F}_i(b)}{\bar{F}_i(a) - \bar{F}_i(b)}. \tag{3.6}$$

We reiterate that although maximum likelihood methods are valid under "related" censoring, standard intensity-based models do not generally yield simple estimates of marginal means. In particular, the Nelson-Aalen estimator $\hat{\Lambda}(t)$ (Andersen et al. 1993, Section IV.1) is based on

$$d\hat{\Lambda}(s) = \frac{\sum_{i=1}^{k} \delta_i(s) dN_i(s)}{\sum_{i=1}^{k} \delta_i(s)}.$$

Thus, it estimates $dM_i(s;s)$ and hence $\hat{\Lambda}(t)$ estimates $\int_0^t dM_i(s;s)$, which is not the expected value of anything relevant. (The estimate of the weighted integral (3.5), on the other hand, is.)

When there are covariates of interest, modelling and estimation are more difficult. It is simplest then to consider models which condition on the exact value of T_i,

$$dR_i(s;t) = E\{dN_i(s)|T_i = t, x_i^*\},$$

but estimation is hampered by the fact that T_i is not observed if $T_i > \tau^*$. Some methods are discussed by Cook and Lawless (1996) but further investigation is needed. Pepe and Cai (1993) consider regression models for rate functions defined by $r_i(t)dt = dM_i(t;t)$; these do not yield interpretable mean functions, as indicated in the preceeding paragraph.

4 Example

As an example we consider the kidney transplant study described in Section 1, wherein subjects were monitored for rejection episodes over a 6 month period. As noted previously, 20 out of 158 subjects withdrew from the study before 6 months because of total graft rejection. Table 1 shows the total numbers of rejection episodes across all subjects in the two treatment groups, and the number of total rejections (withdrawals).

Table 1. Rejection Episodes and Withdrawals

Treatment	No. of Subjects	Total No. of Rejection Episodes 0	1	2	3	No. of Withdrawals
RATS	75	44	19	7	5	9
OKT3	83	28	33	17	5	11

One approach to treatment comparison in terms of rejection episodes that allows for related withdrawals is to condition on the termination time, as suggested in Section 3. A very simple approach, since relatively few subjects experienced total graft rejection and early withdrawal, is to compute separate estimates (3.3) for each treatment group, with $t = 6$ months. That is, we compare the rejection episode mean functions for subjects who have not experienced total rejection. Figure 1 shows the estimated mean functions. A formal comparison could be based on a two-sample test described by Cook and Lawless (1996), or a regression model linking the two treatment groups. Noting that the curves in Figure 1 are roughly proportional, we consider a model where x_i is a binary covariate equalling 1 for subjects on OKT3 and 0 for subjects on RATS, with $M_i(s;\tau) = M_0(s;\tau)\exp\{\beta x_i\}$, and τ equal to 6 months. The estimation procedure based on (2.4) and (2.5) in Section 2 may be applied, with $M_i(t;\tau)$ in place of $M_i(t)$. Using the Poisson estimating equations, we find $\hat{\beta} = 0.561$ with a standard error, obtained from (3.10) in Lawless and Nadeau (1995), of 0.052. This yields an approximate 95% confidence interval for $\exp(\beta)$ of (1.58, 1.94), representing the estimated ratio of either rates or mean functions for patients on OKT3 vs. patients on RATS.

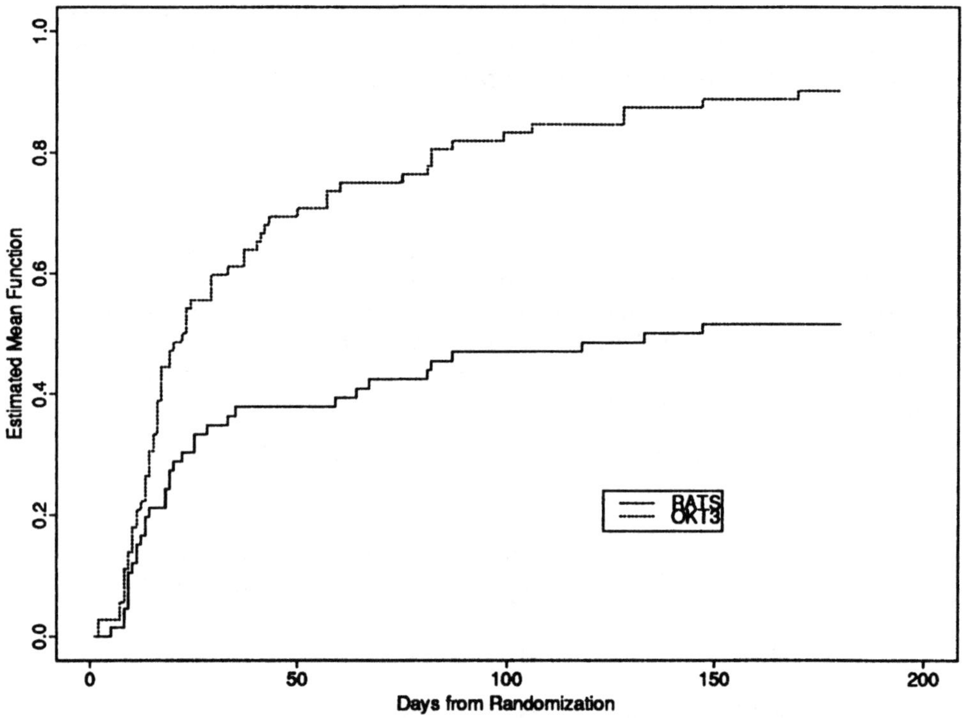

Figure 1. Estimated Mean Functions for Two Treatment Groups

More complicated models that include the 20 subjects who withdrew early may also be fitted. Cook and Lawless (1996) do so and find mild evidence of a higher rate of rejection episodes for patients with early withdrawals, and a slightly reduced treatment effect. However, the effect of early withdrawal is not significant even at the 10% level, perhaps as a consequence of there being so few withdrawals.

5 Conclusion

The analysis of marginal mean functions for recurrent events provides a robust, convenient, and easily interpreted way to assess covariates or treatment effects. It is particularly useful when there are too few events per subject to develop a full model for the recurrent event processes, or when attention is naturally focussed on average or total numbers of events. The illustration in Section 4 involved only a single treatment indicator covariate but the methods are applicable with arbitrary fixed or external time-dependent covariates.

The main ideas of the approach have been outlined here. Various extensions are possible, for example to multiple treatment comparisons (Cook et al. 1996) or multiple event types and to repeated analysis in a clinical trial context (Cook and Lawless 1996). Diagnostic checks of regression assumptions are considered by Lawless and Nadeau (1995) and Nadeau (1995).

Pepe and Cai (1993) emphasize the analysis of rate functions in a broader framework than here. However, they define rate functions without regard to whether they correspond to derivatives of mean functions. Thus, for example, in the context of Section 3 they consider models for rates such as $\Lambda_i(t) = E\{dN_i(t)|T_i \geq t, x_i\}$. These methods are useful but do not necessarily admit interpretation in terms of mean numbers of events.

Smooth estimates of rate functions may be obtained by using kernel smoothing or local weighting (Pepe and Cai 1993, Nadeau 1995). For example, in the case in Section 2 where time is continuous, there are no covariates, and we employ Poisson estimating equations, we have $g_i(t) = 1$, $a_{2i}(t) = 1$ in (2.6) and thus

$$d\hat{M}_0(t) = \frac{\sum_{i=1}^{k} \delta_i(t) dN_i(t)}{\sum_{i=1}^{k} \delta_i(t)}.$$

A smooth estimate of the rate $m_0(t) = dM_0(t)/dt$ is given by

$$\tilde{m}_0(t) = \int_0^\infty K\left(\frac{s-t}{b}\right) d\hat{M}_0(s) \Big/ \int_0^{\max(\tau_i)} K\left(\frac{s-t}{b}\right) ds$$

where $K(\cdot)$ is a bounded function which integrates to 1 on $(-\infty, \infty)$, and b is a selected window size parameter. This estimate is formally equivalent to the smoothed Nelson-Aalen estimator (Andersen et al. 1993, Section IV.2) but can be assigned a robust variance estimate that is valid for a wide variety of processes, and not just Poisson models (Nadeau 1995). Pepe and Cai (1993) discuss similar estimates of rate functions based on splines.

Acknowledgments: This research was supported by grants to all three authors from the Natural Sciences and Engineering Research Council of Canada.

References

Albert, P.S. (1991). A two-state Markov mixture model for a time series of epileptic seizure counts. *Biometrics*, **47**, 1371-1381.

Andersen, P.K., Borgan, O., Gill, R.D. and Keiding, N. (1993). *Statistical Methods Based on Counting Processes.* New York: Springer.

Byar, D.P. (1980). The Veterans' Administration Study of Chemoprophylaxis for recurrent stage 1 bladder tumors: Comparisons of placebo, pyridoxine and topical thiopeta. In *Bladder Tumors and Other topics in Urological Oncology.* (eds. M. Pavone-Macaliso; P.H. Smith and F. Edsmyr), pp. 363-370, New York: Plenum.

Cole, E.H. et al. (1994). A Comparison of Rabbit Anti-Thymocyte Serum and OKT3 as Prophylaxis Against Renal Allograft Rejection. *Transplantation*, **57**, 60-67.

Cook, R.J. and Lawless, J.F. (1996). Interim monitoring of longitudinal comparative studies with recurrent event responses. To appear in *Biometrics.*

Cook, R.J. and Lawless, J.F. (1996). Marginal analysis of recurrent events and a terminating event. To appear in *Statistics in Medicine.*

Cook, R.J., Lawless, J.F. and Nadeau, C. (1996). Robust tests for treatment comparisons based on recurrent event responses. *Biometrics, 52*, 557-571.

Cox, D.R. and Isham, V. (1980). *Point Processes.* London: Chapman and Hall.

Cox, D.R. and Lewis, P.A.W. (1966). *The Statistical Analysis of Series of Events.* London: Methuen.

Hobson, R.W., Weiss, D.G., Fields, W.S., Goldstone, J., Moore, W.S., Towne, J.B., Wright, C.B. and the Veteran Affairs Cooperative Study Group (1993). Effect of carotid endarterectomy for asymptomatic carotid stenosis. *New England Journal of Medicine*, **328**, 221-227.

Lawless, J.F. (1995). The analysis of recurrent events for multiple subjects. *Applied Statistics*, **44**, 487-498.

Lawless, J.F. and Nadeau, C. (1995). Some simple robust methods for the analysis of recurrent events. *Technometrics*, **37**, 158-168.

Little, R.J.A. (1995). Modeling the dropout mechanism in repeated-measures studies. *Journal of the American Statistical Association*, **90**, 1112-1121.

Nadeau, C. (1995). Inference for Point Processes Through Estimating Functions. Unpublished Ph.D. Thesis, University of Waterloo.

Nadeau, C. and Lawless, J.F. (1996). Inference for means and covariances of point processes through estimating functions. Submitted for publication.

Nelson, W.B. (1988). Graphical analysis of system repair data. *Journal of Quality Technology 20*, 24-35.

Nelson, W.B. (1995). Confidence limits for recurrence data - applied to cost or number of product repairs. *Technometrics 37*, 147-157.

Pepe, M.S. and Cai, J. (1993). Some graphical displays and marginal regression analyses for recurrent failure times and time dependent covariates. *Journal of The American Statistical Association 88*, 811-820.

Extending the Cox Model

Terry M. Therneau

ABSTRACT Since its introduction, the proportional hazards model proposed by Cox has become the workhorse of regression analysis for censored data. In the last several years, the theoretical basis for the model has been solidified by connecting it to the study of counting processes and martingale theory. These developments have, in turn, led to the introduction of several new extensions to the original model. These include the analysis of residuals, time dependent coefficients, multiple/correlated observations, multiple time scales and time dependent strata.
These new techniques have not, however, been easily available to the practictioner, having as of yet not appeared as options in the usual statistical packages. The aim of this monograph is to show how many of these techniques can be approached using standard statistical software, in particular the S-Plus and SAS packages. As such, it should be a bridge between the statistical journals and actual practice.

1 Introduction

Since its introduction, the proportional hazards model proposed by Cox [8] has become the workhorse of regression analysis for censored data. In the last several years, the theoretical basis for the model has been solidified by connecting it to the study of counting processes and martingale theory, as discussed in the books of Fleming and Harrington [13] and of Andersen et. al. [2]. These developments have, in turn, led to the introduction of several new extensions to the original model. These include the analysis of residuals, time dependent coefficients, multiple/correlated observations, multiple time scales,and time dependent strata.

The aim of this monograph is to show how many of these can be approached using standard statistical software, in particular the S-Plus and SAS packages. As such, it should be a bridge between the statistical journals and actual practice. The focus on SAS and S-Plus is based largely on the author's familiarity with these two packages, and should not be taken as evidence against the abilities of other software products. The text uses the labels 'S' and 'S-Plus' interchangeably, the former is the package developed by Bell Laboratories and the latter is the commercial version of the same. Since nearly every installation of "S" consists of the latter this shorthand notation should cause no harm. All the examples given in the text actually refer to S-Plus.

Sections 2 and 3 lay a foundation for our methods. In section 2, we discuss the *counting process* formulation of a Cox model, the software implementation of this model, and the flexibility that it allows. Section 3 defines a set of residuals for the Cox model, based on the counting process and the mathematical formalism of a martingale.

Sections 4 and 5 use residuals to test the two basic assumptions of a Cox model: proportional hazards and the functional form of the covariates. The counting process formulation allows us to extend these methods to time-dependent covariate models as well.

Section 6 discusses leverage residuals and their use as a basis for robust/resistant estimates of variance. Such estimates are particularly needed for correlated data.

2 The counting process formulation of a Cox model

The Andersen-Gill formulation of the proportional hazards model as a counting process has proven very useful in theoretical development. Represent the ith subject as a counting process where

- $N_i(t)$ is the cumulative number of events to date for the subject.
- $Y_i(t)$ is an indicator function, $Y_i(t) = 1$ iff the subject is at risk and under observation at time t.

From a data analysis viewpoint, each subject is treated as an observation of a (very slow) Poisson process. A censored subject is thought of not as "incomplete data", but as one whose event count is still zero. Time dependent covariates effect the rate for upcoming events, and can depend in any way on past observation of the subject. Intervals of observation need not be contiguous.

To cast a data analysis in this framework has advantages as well. In the computer data set, each subject i is represented by a set of observations: s_{ij}, t_{ij}, δ_{ij}, x_{ij}, k_{ij}, $j = 1, \ldots, n_i$; where $(s_{ij}, t_{ij}]$ is an interval of risk, open on the left and closed on the right, $\delta_{ij} = 1$ if the subject had an event at time t_{ij} and 0 if the subject did not have an event, x_{ij} is the covariate vector over the interval, and k_{ij} is the stratum during the interval. Data sets like this are easy to construct with a package such as SAS or S-Plus.

2.1 *Particular cases*

Multiple events

The original motivation for adding the counting process form to the S-Plus coxph function was a study of a calcium channel blocker, diltiazem, in post myocardial infarction patients, where one of the events of interest was fatal or non-fatal re-infarction [34]. Several patients had multiple events, though none had more than 3. A "standard" Cox analysis was performed using first cardiac event as an endpoint, but there was a question about whether more power would be obtained if all of the events could be used. This can be incorporated by breaking any patient with multiple events into multiple intervals of risk. For instance, assume a subject has an event on days 100 and 185 and has now been followed to day 250. He would be coded as 3 observations or "lines" of data whose intervals are (0, 100], (100, 185], (185, 250] with corresponding status codes of 1, 1 and 0.

Time dependent covariates

The most common type of time dependent covariate is a repeated measurement on a subject or a change in the subject's treatment. Both of these are straightforward in the proposed formulation. As an example consider the well known Stanford heart transplant study [10], where treatment is a time dependent covariate. Select two patients whose time from enrollment to death is 102 and 343 days, respectively; the second patient had a transplant 21 days from enrollment. The data file for these two would be

Interval	Status	Transplant	Age at Entry	Prior Surgery
(0,102]	1	0	41	0
(0,21]	0	0	48	1
(21,343]	1	1	48	1

Note that static covariates such as age are simply repeated for a patient with multiple lines of data.

Multiple lab values are easily coded as well. A patient with tests on days 0, 60 and 120, and follow up to day 140 would be coded using 3 time intervals 0–60, 60–120 and 120–140. This does implicitly assume that the time dependent covariate is a step function with jumps at the measurement points. It might be more reasonable to break at the midpoints of the measurement times, or to use an interpolated value over many smaller intervals of time, but in practice these refinements appear to make little practical difference in a model's results. If a lab test varies *markedly* from visit to visit interpolation strategies may become important, but the adequacy of the study design would then also be in question.

Discontinuous intervals of risk

In a study of tumor progression and its relationship to a particular blood marker, the key time-dependent variable was the monthly measurement of this marker. Patients were expected to have a measurement every 3-6 months. One patient, however, had a significant hiatus in her visit record. One choice for the analysis was to interpolate the values over the missing 2 year period. A more conservative course, and that chosen by the investigators, was to treat the data value as missing. This effectively removes the subject from the risk set over that interval, without having to remove her entire experience from the study.

Other cases where a subject is lost from and then returns to observation can certainly be imagined. For instance, consider a situation where multiple events are possible, but the treatment for an event temporarily protects the patient from further injury. In a study of falls in the elderly, hospitalization following a fall would temporarily protect the subject from further falls. (For conditions with a low event rate, however, this refinement is likely to be insignificant.)

Alternate time scales

The usual Cox model forms risk groups based on time since entry. For some studies a more logical grouping might be based on another alignment, such as age or time since diagnosis.

Andersen et al. [2] discuss this issue in several of their examples, and then in depth in chapter 10 of the book. One example concerns nephropathy and mortality among insulin dependent diabetics. Patients can be in one of 3 states: 0-alive without diabetic nephropathy (DN), 1-alive with DN, and 3-dead. Relevant time scales for the 0-1 transition are age, calendar time, and duration of diabetes, and for the 1-2 transition the duration of DN.

As another example consider a study conducted at the Mayo Clinic on the effect of L-dopa for patients with Parkinson's disease [41]. It was felt that time from diagnosis was the most appropriate time scale for analysis. However, Mayo is a major tertiary referral center, and many of the study's patients were not seen here until well after that date. For each patient we have the date of diagnosis, the date of referral to Mayo, and the date of last follow-up or death. The patient diagnosed on 8Feb82 with referral on 28Apr85 and last contact on 18Jun90 will be represented as a single interval (1175, 3052]. It is not correct to enter them as (0, 3052] (which is equivalent to a standard non-interval Cox model with follow-up time of 3052) since the patient was not at risk for an *observable* death during that interval. Such data, where the patient enters the risk set after time 0, is said to be left truncated.

Time dependent strata

When a patient is represented as multiple lines of data or "observations", both the covariates and the strata indicator may change from line to line. Coding a time dependent strata is thus quite easy.

Time alignment within the strata may require more thought, however. As an example consider a study of Dutch patients with primary biliary cirrhosis of the liver (PBC). PBC is a rare but fatal chronic liver disease of unknown cause, with a prevalence of about 50 cases per million population. The hazard rate for a diseased patient grows over time, as does the rate of degeneration in their hepatic function as tracked by various blood tests. A portion of the patients receive a liver transplant at some point during the follow up.

One point of the study was to assess the value of covariates such as age and bilirubin in predicting patient outcome, both before and after transplantation. Transplant was treated as a time dependent stratification variable. In the post transplant strata the most "natural" hazard function is based on time since transplant. Surgical death is a major risk for such an extensive procedure, and this time scale properly aligns the patient's clock with the dominating hazard.

The "proper" alignment for time dependent strata is not always so clear. One appealing method of analysis for the diltiazem study is to place patients into new strata after their second, third, etc cardiac event (all have had one event as that was the trigger for enrollment). The baseline hazard after a second infarction may be quite different than the group as a whole. It is not obvious, however, whether time since enrollment or time since last event is the better index of an appropriate risk group.

2.2 Implementation in SAS and S-Plus

The S-Plus function `coxph` and the SAS `phreg` procedure accommodate these extensions by a simple programming artifice. The input data set is assumed to consist of observations or rows of data, each of which contains the covariate values Z, a status indicator 1=event 0=censored and an optional stratum indicator, along with the time interval $(start, stop]$ over which this information applies. In the notation above, this means that each row is treated as a separate subject whose Y_i variable is 1 on the interval $(start, stop]$ and zero otherwise. Within the program, it means that the risk set at time t only uses the applicable rows of the data. In order to avoid double-counting any subject within a risk set, both packages disallow zero length intervals, i.e. $(x, x]$. At the time of this writing S-Plus will fail with an error message if such intervals are present, and SAS silently deletes them. I slightly prefer the former behavior since the presence of such intervals is often indicative of a more general mistake in setting up the data set.

The code has no specific "hooks" to accommodate time-dependent covariates, time-dependent strata, multiple events, or any of the other special features mentioned above. Rather, it is the responsibility of the user to first construct an appropriate data set. The strategy, originally motivated by sloth, leads to a fitting program that is simpler, shorter, easier to debug, and more efficient than one with multiple specific options. A significantly more important benefit has become apparent over time, i.e., the flexibility inherent in building a data set has allowed analyses that were not considered by the original coder — left truncation is a case in point.

The S-Plus code has had the ability to fit these models since version 2.0 (1992) using the `agreg` function; from S-Plus version 3.3 (1995) onward it is easier to use the `coxph` function. The counting process form is an option within `phreg` for SAS releases 6.10 and later; the actual date of availability depends on your specific computing platform, e.g. 1995 for DOS/PC and 1996 for SUN workstations.

Stanford Heart Transplant Data

Since the Stanford data set is so well known, it is worthwhile to show the data setup in complete detail. We use the data set as it is found in the paper by Crowley and Hu [10]. A copy of the data can be obtained from statlib (http://lib.stat.cmu.edu) in the *jasa* section. Three issues arise in setting up the data:

1. The covariates in the data set are moderately colinear. Because of the presence of an interaction term, the coefficients found in table 5.2 of Kalbfleisch and Prentice [21] will be recreated only if the covariates are defined in *exactly* the correct way. (Conclusions are not changed, however). This is the reason for using fractional age (days/365.25), centered at 40 years, and for centering the enrollment date at Oct 1, 1967.

2. One subject died on the day of entry. However $(0, 0]$ is an illegal time interval for the programs. To avoid this, treat an interval from 10/21 to 10/22, say, as 2 days of risk. This leads to the "1 +" expression in several lines of the program.

3. A subject transplanted on day 10 is considered to be on medical treatment for days 1–10 and on surgical treatment for days 11–last contact. Thus if Smith died on day 10 and Jones was transplanted on day 10, we in effect treat the transplant as happening later in the day than the death; *except* for patient 38, who died during surgery on day 5. This person should certainly be counted as a treatment death rather than a medical one. The problem is resolved by moving his transplant to day 4.9.

(If there are ties between the time at which some time dependent covariate changes value and an event or censoring time, I have usually found it most accurate to explicitly resolve the ambiguity via a fractional time value, rather than try to remember and apply any tie breaking rule used by the program code. However, for those who wish to know there are two principles. Since time intervals are open on the left and closed on the right, changes in a covariate by default happen *after* the deaths and/or censoring at a time point. For ties between death and censoring times SAS and S-Plus place deaths first, in accordance with common practice.)

Here is SAS code to create the analysis data set.

```
data temp;
    infile 'data.jasa';
    input id  @6 birth_dt mmddyy8. @16 entry_dt mmddyy8.
              @26 tx_dt mmddyy8.   @37 fu_dt mmddyy8.
              fustat  prior_sx ;

    format birth_dt entry_dt tx_dt fu_dt date7.;

data stanford;
    set temp;
    drop fu_dt fustat birth_dt entry_dt tx_dt;

    age =  (entry_dt - birth_dt)/365.25  - 48;
    year = (entry_dt - mdy(10,1,67))/ 365.25; *time since 10/1/67;
    wait =  1 + (tx_dt - entry_dt);
    if (id = 38) then wait = wait - .1;

    if (tx_dt =.) then do;
        rx = 0;               * standard therapy;
        start = 0;
        stop  = 1 + fu_dt - entry_dt;
        status= fustat;
        output;
        end;

    else do;
        rx =0;          *first an interval on standard treatment;
        start = 0;
        stop  = wait;
        status= 0;
        output;

        rx =1;          *then an interval on surgical treatment;
        start = wait;
        stop  = 1 + fu_dt - entry_dt;
        status= fustat;
        output;
        end;

proc print;
```

And here is S-Plus code to fit the 6 models found in table 5.2 of Kalbfleisch and Prentice [21]. The > symbol is the package's interactive prompt. The formula language uses ~ for "is modeled as", and the right hand side symbols are simlar to the GLIM and GENSTAT programs: + for main effects, : for interaction, and * for main effects plus interaction. The contrast option that I have chosen causes it to use the first treatment as the reference category.

```
> options(contrasts="contr.treatment")

> sfit.1 <- coxph(Surv(start, stop, status)~ (age + prior.sx)*rx,
                                  data=stanford, method='breslow')
> print(sfit.1)
                coef exp(coef) se(coef)      z    p
         age  0.0139     1.014   0.0181  0.768 0.44
    prior.sx -0.5465     0.579   0.6109 -0.895 0.37
          rx  0.1195     1.127   0.3277  0.365 0.72
      age:rx  0.0346     1.035   0.0272  1.270 0.20
 prior.sx:rx -0.2929     0.746   0.7582 -0.386 0.70

Likelihood ratio test=12.5  on 5 df, p=0.0288  n= 172

> sfit.2 <- coxph(Surv(start, stop, status)~ year* rx,
                                  data=stanford, method='breslow')

> sfit.3 <- coxph(Surv(start, stop, status)~ (age + year)* rx,
                                  data=stanford, method='breslow')

> sfit.4 <- coxph(Surv(start, stop, status)~ (year +prior.sx)*rx,
                                  data=stanford, method='breslow')

> sfit.5 <- coxph(Surv(start, stop, status)~ (age + prior.sx)*rx
                        + year, data=stanford, method='breslow')

> sfit.6 <- coxph(Surv(start, stop, status)~ age* rx + prior.sx +
                        year, data=stanford, method='breslow')
```

Because of several tied death times in the data set, the Efron approximation would be preferred to the Breslow method, but the difference in this case is slight and we wished to match the printed table. (The Efron approximation is the default in S-Plus). The SAS code to analyse the data set requires explicit creation of the interaction variables. An example for the first of the models is given below

```
data temp2; set stanford;
    age_rx   = age * rx;
    prior_rx = prior_sx * rx;
proc phreg data=temp2;
   model (start stop)*status(0) = age prior_sx rx age_rx prior_rx;
```

2.3 Computational considerations

The more common way to deal with time dependent covariates is to call a computational subroutine at each death time. The counting process style has been implemented in the S-Plus package, and more recently in SAS as well. There are some advantages for the procedure proposed here.

1. It was easy to code the routine.

2. The counting process code may be more efficient. Both routines begin by sorting the data according to the end point of the interval. And in both, the time needed to accumulate the score and information matrices is identical. The difference is that counting process routine must spend time scanning the data set to select the risk set — those observations whose (start, stop] interval brackets a given death time. The other style routine must make n_l subroutine calls, where n_l is the size of the risk set. If the subroutine is at all complex the tradeoff will favor the counting process routine, particularly if only a few subjects have multiple lines of data. For example, in a recent study here with multiple lab tests, the proportion with 1, 2, ... tests was geometrically decreasing; only 2 patients had five values. Slightly over half the subjects had only one value.
   The counting process function will often run much faster when there are stratification variables in the model. When strata are introduced the program spends less time searching out whom is part of the current risk set since it need look only within the strata; without strata it has to scan the entire data set.

3. The counting process style may be easier to use. This is particularly true for the case of multiple lab tests. Assume that there is a maximum of 5 lab tests, then the older style code requires the creation of 9 ancillary covariates containing the first lab value, the time of the second test, the value of the second test, the time of the third test, the value of the third test, etc. Those with fewer than 10 tests have the remaining "time of test" values set to some number greater than that patient's follow up time. Then a code fragment similar to the following is placed into the subroutine:

```
lab = lab1;
if (TIME > t1) then lab=lab2;
if (TIME > t2) then lab=lab3;
...
```

   As well, if some patient has the effrontery to come for an 11th visit, then the subroutine and *all* ancillary data must be updated to include a new pair of variables.

The major disadvantage of the counting process style is that it is difficult to accommodate a smoothly varying covariate. One main use of this is to test the proportional hazards assumption by the addition of a product variable `x*time` or `x* log(time)`. As we will see later, there is a much better way to test for and visualize non-proportional hazards.

The above paragraph did not say one *cannot* accommodate a smooth covariate. With sufficient effort, general time dependent covariates can be managed in any package that allows stratified models. Make each unique death time a separate strata, then within the strata for each given time t^* place one observation for each subject at risk at time t^*, setting his/her covariates to the appropriate time-dependent values. Creation of such a data set would of course be a tedious process.

3 Residuals

3.1 Mathematical definitions

Several ideas for residuals based on a Cox model have been proposed on an ad hoc basis, most with only limited success. The current, and most successful, methods are all based on counting process arguments, and in particular on the subject-specific martingale process that arises from this formulation. Barlow and Prentice [3] give the original definition of "martingale residuals", and further work was done by Therneau, Grambsch and Fleming [40]. To begin, we must expand on the mathematical notation introduced in the previous section.

As before, let $Y_i(t)$ be the indicator that a given subject is at risk and under observation at time t, $N_i(t)$ be the cumulative number of "events" for the subject up to time t, and $\overline{N}(t) = \sum N_i$ be the total event process. Let $Z_{ij}(t)$ be the jth covariate of the ith person (possibly time dependent), where $i = 1, \ldots, n$ and $j = 1, \ldots, p$; and $Z_i(t)$ be the entire covariate set for a subject, represented as a $p \times 1$ column vector. Define $r_i(t)$ to be $\exp[\beta' Z_i(t)]$, i.e., the risk score for the ith subject. In actual practice β will be replaced by $\hat{\beta}$ and the subject weights r_i by $\hat{r}_i$.

If $\Lambda_i(t)$ is the true cumulative event rate function for subject i, then the (observed - expected) process

$$M_i(t) = \int Y_i(t)\{dN_i(t) - d\Lambda_i(t)\}$$

will be a subject-specific martingale.

The Cox model assumes that the hazard for subject i is

$$\lambda(t; Z_i) = \lambda_0(t) r_i(t)$$

where λ_0 is an unspecified baseline hazard. Assuming no tied death times, the log partial likelihood is defined as

$$l(\beta) = \sum_{i=1}^{n} \int_0^\infty \left[Y_i(t) r_i(t) - \log\{\sum_j Y_j(t) r_j(t)\} \right] dN_i(t)\,.$$

The first derivative is the p by 1 vector

$$U(\beta) \quad = \quad \sum_{i=1}^{n} \int_0^\infty [Z_i(t) - \bar{Z}(\beta, t)]\, dN_i(t) \qquad (1.1)$$

$$= \sum_{i=1}^{n} \int_0^\infty [Z_i(t) - \bar{Z}(\beta,t)]\, dM_i(\beta,t)\,. \tag{1.2}$$

M is defined below. Equation 1.1 is the usual form in which the Cox score is written, but 1.2 is more useful in some contexts. The transition from 1.1 to 1.2 involves simple algebraic manipulations The p by p information matrix is

$$\mathcal{I}(\beta) = \sum_{i=1}^{n} \int_0^\infty \frac{\sum_j Y_j(t) r_j(t) [Z_i(t) - \bar{Z}(t)][Z_i(t) - \bar{Z}(\beta,t)]'}{\sum_j Y_j(t) r_j(t)} dN_i(t)\,, \tag{1.3}$$

where $\bar{Z}$ is the weighted mean (vector) of those still at risk at time t:

$$\bar{Z}(\beta,t) = \frac{\sum Y_i(t) r_i(t) Z_i(t)}{\sum Y_i(t) r_i(t)}\,. \tag{1.4}$$

In practice the martingale residual is estimated by $\widehat{M}$, with $\hat{\beta}$, $\hat{r}$ and $\hat{\Lambda}$ substituted for β, r, and Λ. The most usual estimate of the baseline hazard is the Breslow or Nelson–Aalen estimate

$$\hat{\Lambda}_0(t) = \int_0^t \frac{d\overline{N}(s)}{\sum_{j=1}^{n} Y_j(s) \hat{r}_j(s)}.$$

3.2 *Martingale and deviance residuals*

The martingale residual for subject i at time t is

$$M_i(t) = N_i(t) - \int_0^t r_i(\beta,s) Y_i(s) d\hat{\Lambda}_0(\beta,s).$$

The SAS and S-Plus functions return the residual at $t = \infty, \beta = \hat{\beta}$. If there are no time-dependent covariates, then $r_i(t) = r_i$ and can be factored out of the integral, giving $\widehat{M_i} \equiv \widehat{M_i}(\infty) = N_i - \hat{r}_i \hat{\Lambda}_0(\hat{\beta}, t_i)$.

The martingale residual is really a simple difference, $O - E$, between the observed number of events for a subject and the expected number given the current model. Like Poisson data, $\text{var}(M_i) = E_i$, the expected number of events. The residuals have some properties that are reminiscent of the residuals from an ordinary linear model.

1. $E(M_i) = 0$
2. $\sum \widehat{M_i} = 0$
3. $\text{cov}(M_i, M_j) = 0$
4. $\text{cov}(\widehat{M_i}, \widehat{M_j}) < 0.$

The last property is a consequence of the fact that the observed residuals are constrained to sum to zero. The covariance is small, however; of the same order as the covariance of the residuals from a linear model.

The martingale residual is highly skewed for single-event (survival) data, since the limiting distribution (under the true model) of $\Lambda_0(t)$ is a censored exponential distribution. This suggests the application of a normalizing transform for the residual, with the idea that such plots would be "easier on the eye". The deviance residual [40] is such a normalizing transform and is very similar in form to the deviance residual for a Poisson distribution,

$$d_i = \text{sign}(\widehat{M}_i) * \sqrt{-\widehat{M}_i - N_i \log((N_i - \widehat{M}_i)/N_i)}$$

In practice it has not been very useful.

3.3 Score residuals

The score process $L_{ij}(t)$ for the ith subject and the jth variable is

$$L_{ij}(t) = \int_0^t [Z_{ij}(s) - \bar{Z}_j(s)]\, d\widehat{M}_i(s),$$

where $\bar{Z}_j(s)$ is the weighted mean of the jth covariate over the subjects still in the risk set at time s, as defined in equation 1.4.

One can think of the score process as a 3 way array with dimensions of subject, covariate, and time. Lin, Wei and Ying [30] suggest a global test of the proportional hazards model based on the maximum of the array.

The *score residual* is defined as a $L_{ij}(\infty)$, and is an n by p matrix. It is the sum of the score process over time. The first, and most obvious use of the score residual is as a measure of leverage. The *dfbeta* matrix is defined as $D = -L\mathcal{I}^{-1}$, where $\mathcal{I}^{-1}$ is the usual Cox variance matrix. The elements of D are the leverage residuals derived by Cain and Lange [5] using case weights, and by Reid and Crépeau [38] using another method. The ij element D_{ij} is an estimate of the change in $\hat{\beta}_j$ if observation i were to be removed from the sample. It is straightforward to show that the column sums $1'D$ are the Newton-Raphson increment $\Delta\hat{\beta}$ at each iteration, and thus that $1'D = 0$ at the final solution. A useful variant is *dfbetas*, the change in the coefficients scaled by the standard errors of the coefficients. It is common to plot either D_i (dfbeta) or dfbetas against $i = 1, 2, \ldots, n$ as a visual way of checking influence.

A second use of the leverage residuals matrix is to form a robust variance estimate. This will be discussed further in section 6.

3.4 Schoenfeld residuals

Another useful transform is to sum the score process up over individuals to get a total score process $L_j(t)$ which is a function of time. This yields a residual first proposed by Schoenfeld [39]. Because $\hat{\Lambda}_0$ is discreet, our estimated score process will also be discreet, having jumps at each of the unique event times. The Schoenfeld residuals are a p column matrix with one row per death which contains these increments.

3.5 General residuals

Barlow and Prentice [3] define a generalized martingale residual by

$$e_i(f_i) = \int f_i(t)\, d\widehat{M}_i(t)$$

for a predictable process f_i, where a predictable process is any function of time or the covariates that depends only on the past. They show that since the predictable covariation process for M_i and M_j is zero, $e_i(f_i)$ and $e_j(f_j)$ are asymptotically uncorrelated. The variance process is

$$\text{var}(e_i(f_i)) = \int f_i(t)[f_i(t)]'\, d\Lambda_i(t)$$

This formula can be used to motivate several residuals. The martingale residual is $e(1)$ and the score residual is $e(Z - \bar{Z})$. Other possibilities suggested by the authors are $e(t)$, "a more traditional residual giving the difference between the failure time t_i for an uncensored subject and a corresponding projected quantity under the model, and $e(\eta)$ where η is the linear predictor $Z'\beta$.

3.6 Counting process data

Assume that we have a counting process style of data set, where each subject may be represented by multiple (start, stop] intervals. One nice feature of the martingale and score residuals is that the residual for a subject is the sum of the residuals for his/her observations. This is a natural consequence of the integral representation. For example, assume that subject "Smith" is represented in the data as four observations with intervals (0,10], (10,25], (25,27] and (27,50]. The martingale residuals for the individual observations are

$$\int_0^{10} Y_i(s) d\widehat{M}_i(s)\,, \quad \int_{10}^{25} Y_i(s) d\widehat{M}_i(s)\,,$$
$$\int_{25}^{27} Y_i(s) d\widehat{M}_i(s)\,, \text{ and } \int_{27}^{50} Y_i(s) d\widehat{M}_i(s)\,.$$

The martingale residual for the subject is the integral from 0 to 50, which is obviously the sum of these four quantities.

The Schoenfeld residuals are defined at each unique death, and are unchanged by a counting process formulation.

4 Functional form

In the Cox model, we assume that the hazard function satisfies

$$\lambda_i(t) = \lambda_0(t)\exp(\beta' Z).$$

That is, a proportional hazards structure with a log-linear model for the covariates. In this section we investigate the correct functional form for the covariates. Perhaps one of the variables, Z_j, should be replaced by Z_j^2, $\ln Z_j$, $I_{\{Z_j>c\}}$ or some other transform.

We use the PBC data set as an example. The data come from a Mayo Clinic trial in primary biliary cirrhosis of the liver conducted between 1974 and 1984. PBC is a progressive disease thought to be of an auto-immune origin; the subsequent inflammatory process eventually leads to cirrhosis and destruction of the liver's bile ducts. A description of the study along with a listing of the data set can be found in Fleming and Harrington [13]. A more extended discussion can be found in Dickson, et al. [11]. Through the work done for these two analyses, the important variables and their proper transformation is already 'known'.

4.1 Simple approach

The simplest approach is one examined by Therneau, Grambsch and Fleming [40] who suggested smoothed residual plots. Consider the martingale residuals from a null model, i.e., one with $\hat{\beta} = 0$. They show that if the correct model is $\exp(\beta f(Z))$ for some smooth function f, then a plot of the smoothed martingale residuals versus Z will display the form of f. That is

$$E(M_i) \approx cf(z_i)\,,$$

where c is a constant which depends on the amount of censoring. Since c simply scales the labeling of the y axis, it has no effect on the visual appearance of the plot. In many ways, this plot is similar in spirit to the ordinary y vs. x plots used for standard data; the censoring process forces us to use this modification.

The following example shows the creation of such a plot in S, using the PBC data set.

```
> fit.pbc0 <- coxph(Surv(futime, fustat) ~ age, iter=0, data=pbc)
> rr <- resid(fit.pbc0)            #default is martingale residuals
> plot(pbc$age, rr, xlab="Age", ylab='Residual')
> lines(lowess(pbc$age, rr, iter=0), lty=2)
> title(main="PBC data", sub='Figure 1')
>
> plot(pbc$bili, rr, xlab="Bilirubin", ylab='Residual')
> lines(lowess(pbc$bili, rr, iter=0), lty=2)
> title(main="PBC data", sub='Figure 2')
```

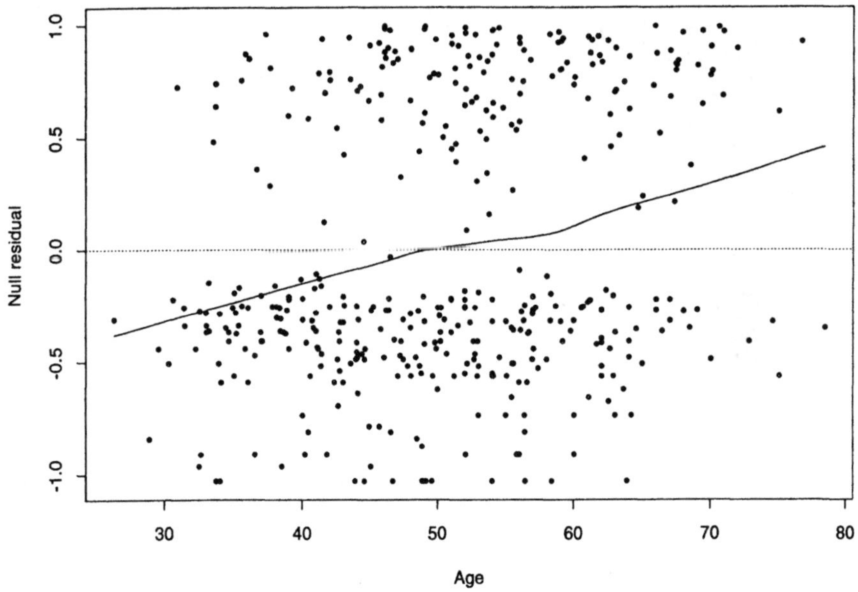

FIGURE 1. PBC Data, functional form for age

The `iter=0` option of the `coxph` function causes it to return after no iterations, i.e., the residuals and fit for a model with $\beta = 0$. The `resid` function returns martingale residuals by default. The `iter=0` option of the lowess smoother causes it to skip its outlier detection phase, and is necessary because of the extreme skewness of the martingale residuals. (A similar problem can occur with logistic regression residuals when the proportion of postive responses is small). Figures 1 and 2 show that age is reasonably linear, but that bilirubin is certainly not so.

Creation of this same plot in SAS is more subtle, as `phreg` has no direct facility to fit a model with $\hat{\beta}$ forced to a given value; a "not converged" error message is invariably produced. However, it can be done by using a single dummy variable as follows.

```
libname save 'sasdata';

data temp; set save.pbc;
    dummy = 0;

proc phreg data=temp;
    model futime*fustat(0)= dummy;
    output out=temp2 resmart=rr / order=data;
```

This method works well when the data are uncorrelated, but fails when correlations are present. The same failure occurs for ordinary scatter plots in uncensored data: if $y = 2x_1 + 0x_2$ and $\text{cor}(x_1, x_2) = .9$, then a plot of

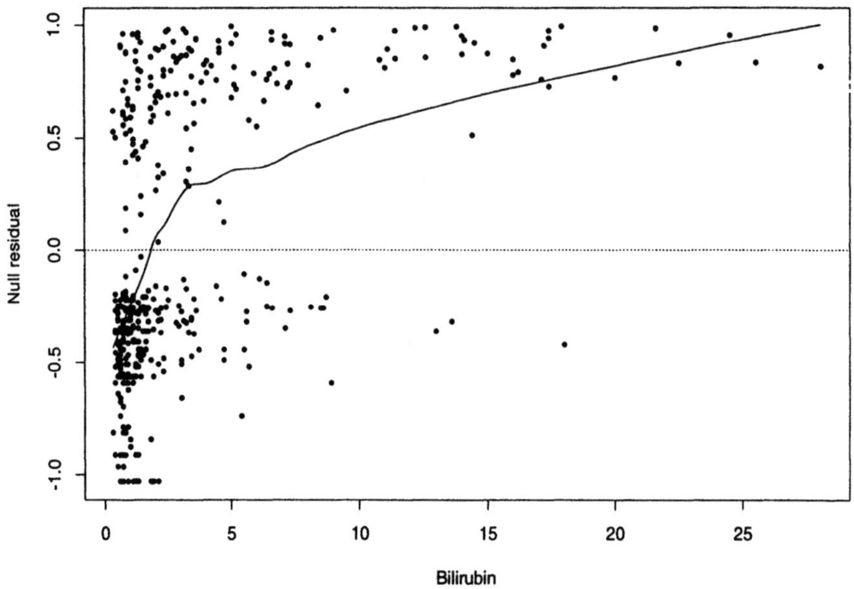

FIGURE 2. PBC Data, functional form for bilirubin

y versus x_2 will show a spurious relationship. There are other problems with the plot as well: the skewness of the data may force the fitted curve to occupy only a small region, and there are no clear methods for creating a confidence band.

4.2 Poisson approach

Grambsh, Therneau and Fleming [16] later extended the method to address these deficiencies. The basic idea is to use the residuals from a linear fit as the building block. Let M_i be the martingale residuals from a Cox model using all of the covariates, and E_i the expected number of events for each subject, based on the fitted model. (Since the residual satisfies $M_i = N_i - E_i$, the expected number of events is just the 0/1 status variable minus M_i). Two different plots are suggested. The first is based on the relationship

$$f \approx \log\left\{\frac{\text{smooth}(N)}{\text{smooth}(E)}\right\} + \hat{\beta}_j Z_j + c$$

where c is an unknown additive constant. If all of the covariates are linear except for the jth, this should reveal the functional form of that covariate, Z_j. The approach performs well, but can experience numerical difficulties since smooth(E) may not always be > 0.

Their second idea, illustrated here, is to use the residual data as input to Poisson regression, taking advantage of modeling tools already available for

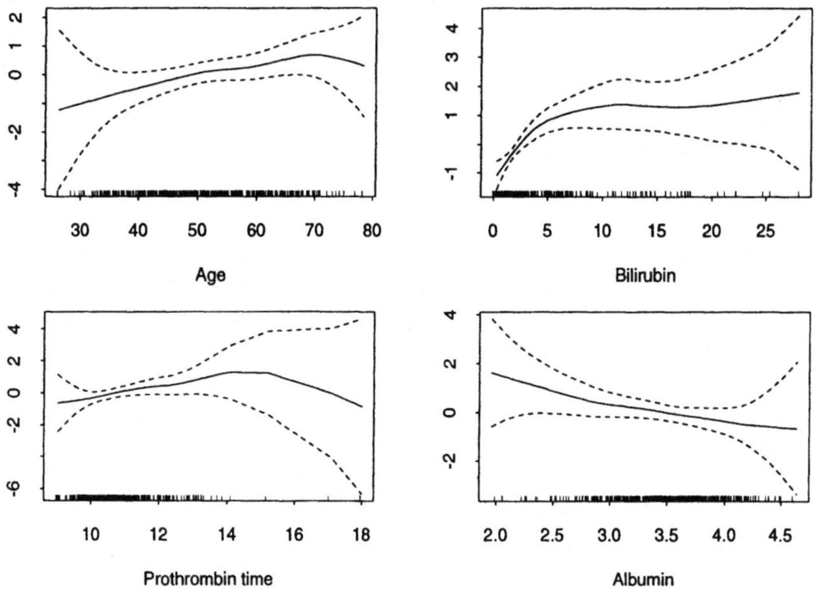

FIGURE 3. PBC Data, functional form using Poisson approach

that method. In S-Plus for instance this involves the **gam** function, which implements the Generalized Additive Models of Hastie and Tibshirani [19].

```
> fit.pbc <- coxph(Surv(futime, fustat) ~ age + edema + bili +
                      protime + albumin, data=pbc)
> print(fit.pbc)

            coef exp(coef) se(coef)     z       p
     age  0.0383      1.04  0.00806  4.75 2.0e-06
   edema  0.9351      2.55  0.28186  3.32 9.1e-04
    bili  0.1158      1.12  0.01302  8.90 0.0e+00
 protime  0.2006      1.22  0.05661  3.54 3.9e-04
 albumin -0.9682      0.38  0.20533 -4.72 2.4e-06

Likelihood ratio test=182  on 5 df, p=0  n=416
(2 observations deleted due to missing)

> exp.pbc <- predict(fit.pbc, type="expected")
> xbeta   <- predict(fit.pbc, type="lp")      #the linear predictor
> gfit <- gam(fustat ~ s(age) + edema + s(bili) + s(protime) +
                        s(albumin) + offset(log(exp.pbc) - xbeta),
                  data=pbc, family=poisson, na.action=na.omit)
> plot(gfit, se=T, rug=T)
```

For the PBC data set prior work had shown that age and edema were reasonably modeled as linear terms, and that bilirubin, prothrombin time and albumin levels fit well with a logarithmic transform. Hopefully, the plots will clearly reveal this trend.

In the **gam** function, the `s(age)` term asks for the fit of a smoothing spline in age with the default 4 degrees of freedom. The edema variable has only 3 values, 0, .5 and 1, and is fit as a linear term. The offset term includes both the log(expected) term usual to a Poisson model, but also the linear predictor $X'\hat{\beta}$. This latter causes the linear term, already fit by the Cox program, to be reflected in the plots.

The plots are shown in figure 3 (a linear plot for edema is also produced by the plot command, but is omitted). We can see the logarithmic form clearly for bilirubin. No transform appears necessary for protime and albumin. The **rug** option to the plot command produces the set of tick marks along the bottom of the plot, one at the location of each of the x-values for the data. One can see that the apparent downturn at the right extreme of the protime plot is based on only 2 data points.

A logical next step would be to replace bilirubin with its logarithm and repeat the process, however, we will jump directly to the final model.

```
> fit.pbc2 <- coxph(Surv(futime, fustat)~ age + edema + log(bili)
                                   + log(protime) + log(albumin),
                              data=pbc, method="breslow")
> print(fit.pbc2)

                  coef exp(coef) se(coef)      z       p
          age  0.0396    1.0404  0.00767   5.16 2.4e-07
        edema  0.8946    2.4463  0.27165   3.29 9.9e-04
    log(bili)  0.8630    2.3703  0.08295  10.40 0.0e+00
 log(protime)  2.3856   10.8654  0.76876   3.10 1.9e-03
 log(albumin) -2.4966    0.0824  0.65280  -3.82 1.3e-04

Likelihood ratio test=231  on 5 df, p=0  n=416
  (2 observations deleted due to missing)

> exp.pbc <- predict(fit.pbc2, type="expected")
> lbili <- log(pbc$bili)
> lpro  <- log(pbc$protime)
> lalb  <- log(pbc$albumin)
> xbeta <- c(cbind(lbili, lpro, lalb) %*% fit.pbc2$coef[3:5])

> gfit <- gam(fustat ~s(lbili) + s(lpro) + s(lalb)
                                + offset(log(exp.pbc) - xbeta)
                       data=pbc, family=poisson, na.action=na.omit)
> plot(gfit, se=T, rug=T)
```

In this run we have also made the refinement of dropping out the linear terms for age and edema, both from the **gam** fit and from the compensating variable `xbeta`. The final plots are shown in figure 4, and show that the transformations are quite satisfactory.

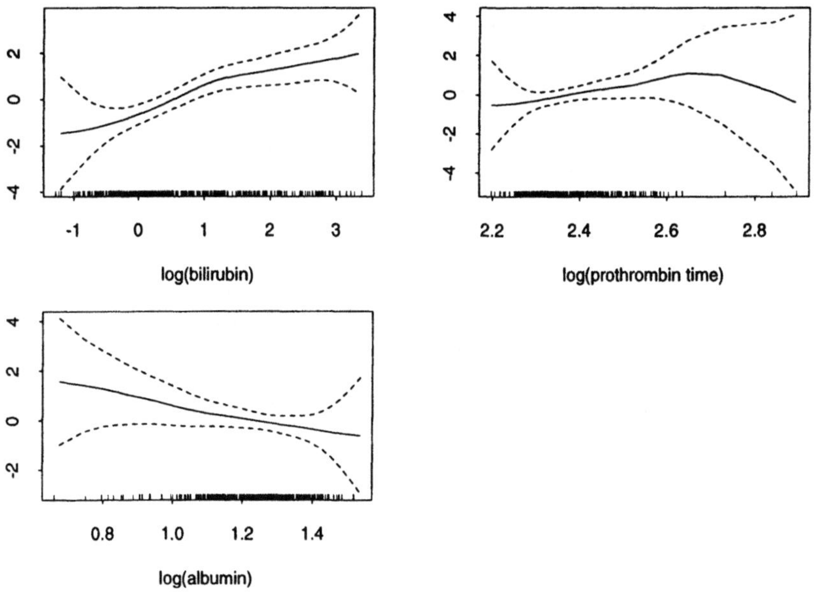

FIGURE 4. PBC Data, test of final functional form

With smaller data sets and/or a large number of variables, this method should be applied to one variable at a time rather than all-at-once, to avoid an excessive number of degrees of freedom in the gam model. Grambsch et al [16], however, show one constructed example with highly correlated predictors and nonlinear effects where one-at-a-time plots do not completely succeed.

For the SAS code, we cannot take advantage of a standard additive models procedure. Instead, for each predictor variable in turn a natural spline is fit using the genmod procedure, followed by a plot of the predicted values of the fit. The daspline macro creates a set of basis vectors **age1, age2, age3, age4** which allows a spline to be fit with standard procedures [18].

```
proc phreg data=save.pbc outest=fit;
    model futime*fustat(0) = age edema bili protime albumin
                                          /ties=efron;
    output out=temp resmart=rr/ order=data;

data temp2; merge temp save.pbc;
    keep fustat rr expected age bili protime albumin;
    expected = fustat - rr;

data temp3; set fit;
    keep beta;
```

```
    beta = age;
data temp4; merge temp2 temp3;
    xbeta = age * beta;
    off = log(expected) - xbeta;

%daspline(age, nk=5);              * spline basis with 4 df ;
data tempd; set temp4;
    &_age;

proc genmod data=tempd noprint;
    model fustat = age age1 age2 age3 / offset=off dist=poisson
                                        obstats ;
    make 'OBSTATS' out=temp5;

*plot of x=age, y=xbeta, xbeta-2*std, xbeta+2*std;
data temp6;
    merge temp4 temp5;
    lower = xbeta - 2*std;
    upper = xbeta + 2*std;
    res   = xbeta + resdev;
proc sort; by age;
proc gplot;
    plot xbeta*age=1
         lower*age=2
         upper*age=3
         res*age  =4  /overlay vaxis=axis1 haxis=axis2;

    symbol1 i=join l=1;
    symbol2 i=join l=2;
    symbol3 i=join l=2;
    symbol4 i=none v=dot h= .1 cm;

    axis1 label=(r=0 a=90 "smooth(age), df=4");
    axis2 label=("age");

*repeat the temp3 through plot process for bili, etc.
```

For the S-Plus example, the gee function's plotting and smoothing abilities make this Poisson residual technique attractive. In the SAS example, however, it would be just as easy to use the regression splines directly in the phreg procedure.

4.3 Other methods

Another method to adjust the plots for possible correlation between the predictor variables is to borrow techniques from the linear models literature. Let m' be the martingale residual from a model omitting age. The *adjusted variable* plot uses m' on the y-axis, and the residual from a linear regression of age on the remaining predictors in the model on the x-axis. If the plot is linear, then age is presumed to enter linearly into the multivariate model. Other variants on this theme are the ***partial residual*** and ***augmented partial residual*** plots. See Chambers et al [9] for an explanation and examples of the methods.

Chen and Wang [7] discuss another method, ***constructed variable*** plots, which should be useful for detecting a power transform. They are based on the Taylor series expansion

$$x^{(\lambda)} \approx x + (\lambda - 1)(x \log x - x) + (\lambda - 2)\,,$$

where

$$x^{(lambda)} \equiv \begin{cases} (x^{\lambda} - 1)/\lambda & if \lambda \neq 0 \\ \log x & if \lambda = 0 \end{cases}$$

which suggests the use of $z = x \log x - x$ as the extra variable in an adjusted variable plot. The slope of a fitted line will suggest the appropriate power transform.

In the author's experience, none of these methods work as reliably as the Poisson regression based fits of the section above.

4.4 Time dependent covariates

The martingale residual for a subject with time-dependent covariates is well defined, although the computation requires more bookkeeping. (With a time-dependent covariate defined by programming statements, `phreg` presently will not produce the martingale residual). However, it is not at all clear how to plot each observation — what should be used as the x-axis value?

If the time dependent covariate is encoded using the counting process style of data, however, the above methods can be used. In this case each subject is represented as one or more observations, each consisting of a time interval, the status, and *fixed* covariate values over that interval. The martingale residual per observation is easily computed.

One could explore functional form using the simpler method by plotting one point for each observation. Depending on their length of follow-up, different subjects may have different numbers of observations, however, and for a given subject, observations may encompass different intervals of time. This can introduce bias in the simple method by effectively giving different weights to subjects, e.g., a person with 10 observations (10 points on the scatter plot) will have a larger influence on the smooth than a subject with only one observation.

The Poisson based method, however, preserves the necessary weighting through the expected values e_i and can be used without modification in such a data set. One exercise, not shown here, is to randomly divide some of the subjects in the PBC data set into two intervals. It is easy to show that the fitted Cox model and the poisson regression estimate of functional form are unchanged by this manipulation.

5 Testing proportional hazards

5.1 Time dependent coefficients

Many possible alternatives to proportional hazards exist. One easily expressed alternative is provided by models with a *time-dependent coefficient*

$$\lambda(t) = \lambda_0(t) \exp[\beta(t)Z] .$$

The proportional hazards model, for a given covariate j, corresponds to the restriction $\beta_j(t) = \beta$, i.e., that a plot of $\beta_j(t)$ versus time will be a horizontal line.

Let $V(\beta, t)$ be the covariance of Z at time t, defined analogously to the running mean $\bar{Z}$, whose j, j' element is estimated as

$$V_{jj'}(\beta, t) = \frac{\sum_i Y_i(t) r_i(t) [Z_{ij}(t) - \bar{Z}_j(t)][Z_{ij'}(t) - \bar{Z}'_j(t)]}{\sum_i Y_i(t) r_i(t)} .$$

Let s_k be the Schoenfeld residual for the kth death in the study, and s^*_k be the rescaled Schoenfeld residual $V^{-1}(\beta, t_k)s_k$.

Grambsch and Therneau [15] show that if $\hat{\beta}$ is the coefficient from an ordinary fit of the Cox model, then

$$E(s^*_k + \hat{\beta}) \approx \beta(t_k) .$$

This suggests a plot of $s^* + \hat{\beta}$ versus time or some function of time $g(t)$ as a method for visualizing the extent of non-proportional hazards. A line can be fit to the plot followed by a test for zero slope; a non-zero slope is evidence against proportional hazards. If T is the test statistic for zero slope, it is clear that different choices of the time-scale g lead to different tests for model misspecification.

1. If $g(t)$ is a specified function of time, then T is a score test for the addition of the time-dependent variable $g(t) * Z$ to the model, a test initially suggested by Cox [8]. Chappell [6] describes the relationship between this test and the test of Gill and Schumacher [14].

2. If g is piecewise constant on non-overlapping time intervals with the intervals and constants chosen in advance, T is the score test proposed by O'Quigley and Pessione [36], which generalizes and extends goodness of fit tests proposed by Schoenfeld [39] and Moreau, O'Quigley and Mesbah [33]. As the authors point out, this test has the disadvantage that the investigator must choose a partition of the time axis, but they suggest guidelines for doing so.

3. If $g(t) = \overline{N}(t-)$ then T is the covariance between the scaled Schoenfeld residual and the rank of the event times. The resulting test is similar to one proposed by Harrell [17], who uses the correlation between the unscaled residuals and rank of the event times. This test is familiar to users of the (now discontinued) SAS phglm procedure.

4. Lin [27] suggests comparing $\hat{\beta}$ to the solution $\hat{\beta}_g$ of a weighted Cox estimating equation

$$\sum_i \int g(t)\,[Z_i(t) - \bar{Z}(t)]dN_i(t) = 0$$

with $g(t)$ one of the scalar weight functions commonly chosen for weighted log rank tests, and he showed that asymptotically $\hat{\beta} - \hat{\beta}_g$ is multivariate normal with mean 0 and a variance matrix derived from martingale counting process theory. If the estimator $\hat{\beta}_g$ were based on a one-step Newton-Raphson algorithm starting from $\hat{\beta}$, his test would be identical to T. Lin suggested a monotone weight function such as $\widehat{F}(t)$, the left-continuous version of the Kaplan-Meier estimator for the survivor function of the entire data set, to detect monotone departures from proportionality and a quadratic function such as $\widehat{F}(t)\{1 - \widehat{F}(t)\}$ for non-monotone trends.

5. Nagelkerke, Oosting and Hart [35] suggest using the serial correlation of the Schoenfeld residuals for a univariate predictor, or for multivariate covariates, the correlation of a weighted sum, $a's$. The authors standardize by using a permutational approach to estimate the variance. They suggest $a = \hat{\beta}$ as a natural choice for the weights, followed by examination of individual covariates if the test is significant. This is equivalent to using the lagged residuals as $g(t)$.

The key point is that each of the above tests can be directly visualized as a simple trend test applied to the plot of $g(t)$ versus the scaled residuals.

In reality, the individual estimates of V at each death time may be unstable, particularly near to the end of follow-up when the number of subjects in the risk set is less than the number of elements of V. For most data sets, the variance matrix of $Z(t)$ changes slowly, and is quite stable until the last few death times. Combining this with the observation that

$$\int_0^\infty V(\beta, t)d\overline{N}(t) = \mathcal{I}(\beta)$$

where $\mathcal{I}^{-1}$ is the Cox model's usual estimate for the variance of $\hat{\beta}$, suggests the use of the average value $\overline{V} = \mathcal{I}/d$, where $d = \overline{N}(\infty)$ is the total number of deaths. With this substitution, let $t_k, k = 1, \ldots d$ be the individual death times, $g_k = g(t_k)$ the chosen transformation, $\bar{g} = \sum g_k/d$ the average of the transformed time values, and S the matrix of unscaled residuals. The tests are based on a slope parameter

$$\theta = D^{-1}[S'(g - \bar{g})]$$

with variance

$$D^{-1} = d\mathcal{I}^{-1} / \sum (g - \bar{g})^2 .$$

A global test of proportional hazards, over all p covariates is

$$T = \frac{(g - \bar{g})' S \mathcal{I}^{-1} S' (g - \bar{g})}{\sum (g_k - \bar{g})^2 / d}$$

The test for an individual covariate Z_j is

$$T_j = \frac{\sum (g_k - \bar{g}) s^*_{kj}}{\sqrt{\mathcal{I}^{-1}_{jj} \sum (g_k - \bar{g})^2 / d}} \tag{1.5}$$

Since the Schoenfeld residuals sum to zero, the above equation is the usual test of a correlation coefficient, with $\mathcal{I}$ as an estimator for the variance of s^*.

To aid in detecting the possible form of departure, a smooth curve with confidence bands is added to the plot. For both the S-Plus and SAS functions this has been done using a spline fit. Let X be the matrix of basis vectors for the spline fit of the scaled residual on $g(t)$ and B the same spline functions, but evaluated at the plotting points. (B will usually be based on 30-40 points evenly spread over the range of $g(t)$). The plotted values of the spline curve will be

$$\hat{y} = 1\hat{\beta}' + B(X'X)^{-1}X'R \equiv 1\hat{\beta}' + HR$$

The variance matrix for the jth variable is

$$S_j = \mathcal{I}^{-1}_{jj} \{dHH' + (J - HJH')\}$$

where J is a matrix of 1's. For most smoothers, smooth(constant) = constant so that $HJ = J$ and the second and third terms cancel. The resultant formula is equivalent to the standard linear model's formula for a standard error for the predicted values, with the exception that $d\mathcal{I}^{-1}_{jj}$ replaces the usual estimator of σ^2. Confidence intervals can be formed by standard linear model calculations, e.g., Sheffé intervals using the rank of S_j for simultaneous confidence bands or simple z-intervals for pointwise estimates.

If the residuals are used as input to a "standard" scatterplot smoother, the resultant confidence bands will be based on $\hat{\sigma}^2 HH'$ where $\hat{\sigma}$ is based on the sum of squared residuals from the smooth. We have not done any theoretical investigation of this estimator, but note that

- based on a *very small* number of empirical cases, the resultant bands tend to be somewhat too large (anticonservative),
- the *shape* of the bands will be correct, however.

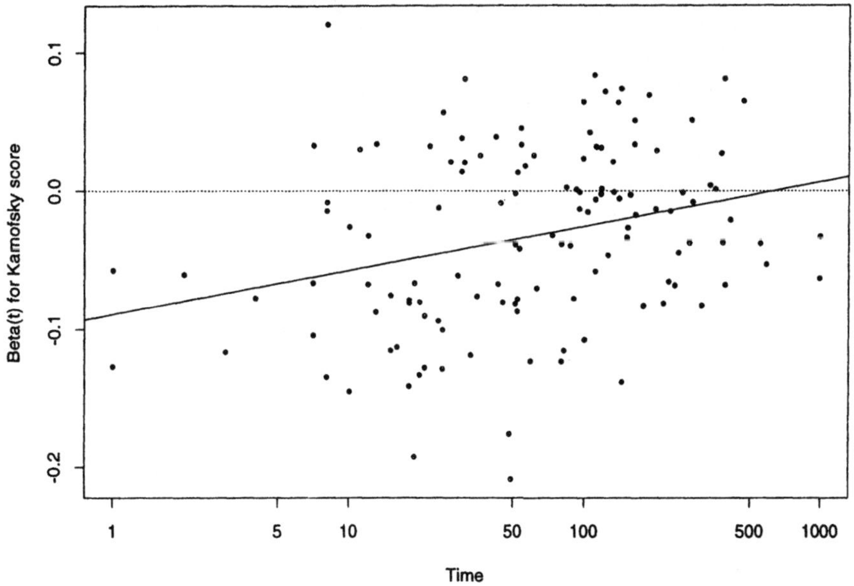

FIGURE 5. Veteran Data, tests of PH

5.2 Vetarans Administration data

As an example consider the Veterans Administration lung cancer data as found in Kalbfleisch and Prentice [21], pp. 223–224, from a clinical trial of 137 male patients with advanced inoperable lung cancer. The end point was time to death and there were six covariates measured at randomization: cell type (squamous cell, large cell, small cell, and adenocarcinoma), Karnofsky performance status, time in months from diagnosis, age in years, prior therapy (yes/no) and therapy (test chemotherapy versus standard). Lin's test [27] comparing the Cox model $\widehat{\beta}$ to a weighted estimate with the Peto-Prentice weight function found a highly significant departure ($p = 0.00002$), suggestive of non-proportionality. Figure 5 shows the scaled Schoenfeld residuals for the most significant predictor, Karnofsky score, along with a fitted least squares line. Table 1.1 shows that the test for slope=0 is highly significant.

```
> fit.vet <- coxph(Surv(futime, status) ~ rx + celltype + karno +
                     months + age + prior.rx, data=veteran)
> print(fit.vet)
                       coef exp(coef) se(coef)       z       p
               rx  0.319242     1.376  0.20949  1.5239     .13
    celltypelarge -0.799691     0.449  0.30305 -2.6388     .008
 celltypesmallcel -0.328601     0.720  0.27632 -1.1892     .23
 celltypesquamous -1.236709     0.290  0.30491 -4.0560     .0005
```

Covariate	*Chi-square*	*d.f.*	*p*
Cell type	7.39	3	0.0604
Karnofsky score	11.68	1	0.0006
Months since diagnosis	1.67	1	0.1955
Age	6.58	1	0.0103
Prior therapy	3.90	1	0.0482
Treatment	0.05	1	0.8287
Global test	27.22	8	0.0006

TABLE 1.1. Tests for the Veterans Administration data with $g = \log(\text{time})$

```
   karno -0.032886      0.968  0.00553 -5.9471  <.0001
  months -0.000269      1.000  0.00914 -0.0295    .98
     age -0.009646      0.990  0.00932 -1.0346    .30
prior.rx  0.084866      1.089  0.23312  0.3640    .72

Likelihood ratio test=62.7  on 8 df, p=1.39e-10  n= 136

> zph.vet <- cox.zph(fit.vet, transform='log')

> for (i in 1:4) {
      plot(zph.vet, var=i)
      abline(0,0, lty=3)
      title(main="Veteran data, test for PH")
      }
```

Karnofsky score and cell-type were the only significant predictors in the Cox model. Smoothed scaled Schoenfeld residuals plots for these predictors are shown in figure 6 and provide an interpretation of the non-proportionality. Because the survival times have a long-tailed distribution, $\log(t_k)$ is used for the x-axis. Use of the Kaplan-Meier values for the x-axis (Peto-Prentice scores) does a slightly better job of equi-spacing the plot points, but the figure is quite similar. Smoothed curves are shown along with pointwise 90% pointwise confidence intervals. The remaining four predictors (plots not shown) did not suggest significant nonproportionality. Table 1.1 summarizes individual predictor test statistics, using (1.5) with log(event times) for $g(t)$. (It differs from the results in [15], whose table 1 is an amalgamation of the tests from individual univariate models.)

The impact of Karnofsky score clearly changes with time. Early, a low score is protective. However, the effect diminishes over time and is effectively zero by 100 days. Another way of interpreting this would be that a 3-4 month old Karnofsky score is no longer medically useful. The down-turn at the right end of the plot is likely an artifact of small numbers and disappears if the last four points are excluded.

The effect of each cell type is less clearly marked. The plots suggest that the increased relative risk due to small cell or adenomatous as compared to large cell may not persist beyond 100 days and that the squamous cell type

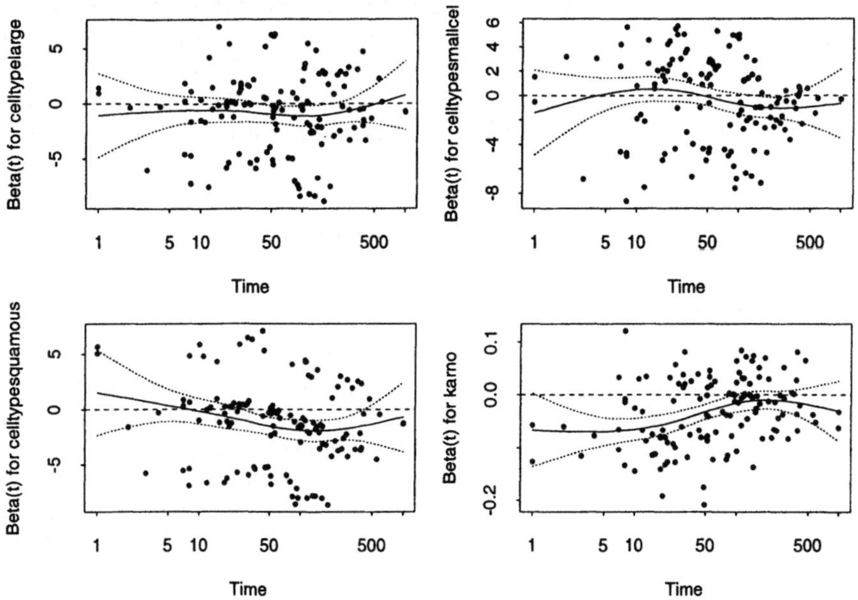

FIGURE 6. Veteran Data, test of PH for Karnofsky score

may be protective in long-term survivors (beyond 100 days) as compared to large cell.

5.3 Stratified models

Both S-Plus and SAS currently return the scaled Schoenfeld residuals based on an overall estimate of variance $V(t) \approx \mathcal{I}/d$. This average over the risk sets is appropriate if the variance matrix is fairly constant over those risk sets. One case where this may not be so is with stratified models. If there are strata by covariate interactions, the averaging is almost certainly unwise. Consider the following example: assume that `rx1` is defined as follows

$$\text{rx1} = \begin{cases} \text{treatment arm} & \text{if center} = 1 \\ 0 & \text{otherwise} \end{cases}$$

Assume `rx2`, `rx3` and `rx4` are defined similarly and that `center` has values 1, 2, 3 and 4 for four participating centers in the study. The following model is equivalent to fitting the four centers separately:

```
coxph(Surv(time, status)~ rx1 + rx2 + rx3 + rx4 +strata(center))
```

Clearly, the variable `rx1` is identically zero in strata 2, it has variance 0 within that strata, hence those data "points" can contribute no information on the appropriateness of proportional hazards for the variable `rx1`.

At present there is only a partial solution to this problem. First fit the overall model to the data. Then refit each strata separately, using the `iter=0` and `initial` options to force the same coefficients as the overall fit. Since the variance will be summed only over the individual strata, this will produce scaled Schoenfeld residuals appropriate to the subsets.

6 Leverage and Robust variance

6.1 Leverage

The most direct estimate of an observation's influence or leverage is the matrix J of jackknife residuals, where J_{ij} represents the change in $\hat{\beta}_j$ when observation i has been removed from the data set. The computation time for this residual may be excessive, however, since it involves n completely independent fits of the model. An approximate influence residual has already been encountered in section 3.3, in terms of the matrix D of *dfbeta* residuals. These residuals, as defined by Cain and Lange [5] and by Reid and Crépeau [38] are returned directly by both the SAS and S-Plus programs. One possible deficiency of D is that although it accounts for the effect of the removal of an observation on the score statistic U, it does not include a correction for the potential change in the variance matrix $\mathcal{I}^{-1}$. Since a truely large outlier will tend to inflate the variance, we might expect D to underestimate the influence of the largest outliers. Some underestimation does occur, but in our experience it is small enough to be ignorable.

D can also be used to approximate a grouped jackknife, e.g., the sum of rows 1–3 of D approximates the change in $\hat{\beta}$ if observations 1–3 were removed from the data set. (This estimate is obviously cruder than for a single subject, with respect to pairs of outliers for instance). In particular assume that the sample were formed from m groups of observations, of size $n_1, n_2, \ldots n_m$, with possible within-group correlation. Then one might form the collapsed $m \times p$ leverage matrix $\widetilde{D}$, where

$$\begin{aligned} \widetilde{D}_{1j} &= \sum_{i=1}^{n_1} D_{ij} \\ \widetilde{D}_{2j} &= \sum_{i=1+n_1}^{n1+n2} D_{ij} \\ &\vdots \end{aligned}$$

The kth row of $\widetilde{D}$ is an estimate of the leverage of the kth group.

The most common use of this estimate will arise when there are multiple observations per subject. In this case the rows of D represent the *per observation* influence and those of $\widetilde{D}$ the *per subject* influence. Plots of both of these are useful in their own right for checking a fitted model. In S-Plus $\widetilde{D}$ is obtained by using the `collapse` option of the `residuals` function. In SAS the `phreg` procedure can be followed by a call to `proc means` to add up the rows; an example of this is found below.

6.2 *Robust variance*

If one suspected that some element of the Cox model were misspecified, a natural correction would be to use the jackknife estimate of variance $(J-\bar{J})'(J-\bar{J})$, where $\bar{J} = 11'J/n$ is a matrix containing the column means of J [12]. An natural approximation to the jackknife variance is then $D'D$. (Remember that $1'D = 0$). This approximate jackknife variance estimate was noted in the appendix of Reid and Crépeau [38], although they did not persue it. The estimate $\widetilde{D}'\widetilde{D}$ corresponds to a grouped jackknife estimate.

6.3 *Relation to other methods*

These estimates are also familiar from other contexts, although the general form $D'D$ has not always been emphasised. Using the same method of derivation as Cain and Lange [5], the results for a linear model are $U_{ij} = X_{ij}(y_i - \hat{y}_i)$, $D = U(X'X)^{-1}$ and $D'D$ is the robust variance estimate proposed by White [43, 44] for linear models with heteroscedasticity or other model violations.

For a generalized linear model with log-likelihood function $l(\beta)$

$$U_{ij} = \frac{\partial l}{\partial \eta_i} \frac{\partial \eta_i}{\partial \beta_j}$$

and $\widetilde{D}'\widetilde{D}$ is the *working independence* estimate of variance proposed by Liang and Zeger [24] for generalized estimating equation (GEE) models.

Lipsitz, Laird and Harrington [26] use the Six Cities dataset to compare several estimators of variance for logistic regression with correlated data: the usual MLE estimate, the actual jackknife variance estimate, the approximate jackknife $D'D$, and a more refined one-step approximation that corrects as well for changes in the information matrix $\mathcal{I}$ due to deletion of an observation. (To derive the refined formula, frame logistic regression as an iteratively reweighted least squares problem and then apply the exact jackknife formula for a linear model to the last step of the iteration). Using a sample size of 30 and 2 covariates plus an intercept the most accurate confidence interval coverage was given by $D'D$. The refined approximation did less well and the actual jackknife did poorly; it appears to be overly sensitive to individual data points. At $n = 60$ the two approximate methods were nearly identical, with the jackknife still somewhat inferior.

Lin and Wei [28] build on the work of Huber [20], applying his methods to the Cox partial likelihood, and derive the Huber sandwich estimate

$V = \mathcal{I}^{-1}(U'U)\mathcal{I}^{-1}$. They also discuss situations in which this is estimate is preferable, including the important cases of omitted covariates and incorrect functional form for the covariate. Some further algebra reveals that their estimate is precisely $V = D'D$.

Lee, Wei and Amato [23] consider highly stratified data sets which arise from inter observation correlation. As an example they use paired eye data on visual loss due to diabetic retinopathy, where photocoagulation was randomly assigned to one eye of each patient. There are $n/2 = 1742$ clusters (patients) with 2 observations per cluster. Treating each pair of eyes as a cluster, they derive a modified sandwich estimate $V = \mathcal{I}^{-1}\sum[(U_{iL} + U_{iR})'(U_{iL} + U_{iR})]\mathcal{I}^{-1}$, where U_{iL} and U_{iR} are the contributions of subject i's left and right eyes, respectively. Some further algebra reveals that this estimate is precisely $V = \widetilde{D}'\widetilde{D}$, where $\widetilde{D}$ is obtained by summing D over each individual's pair of eyes. (A subject with only one studied eye would have one (identical) row of data in both D and $\widetilde{D}$).

Assuming a data set `eyes` with variables `subject.id`, `time`, `status` (0=censored, 1=failure) and `treatment`, and with two observations (rows of data) per subject, one S-Plus program to perform the analysis is

```
> fit    <- coxph(Surv(time, status) ~ treatment, data=eyes)
> Dtilde <- residuals(fit, type='dfbeta', collapse=subject.id)
> newvar <- t(Dtilde) %*% Dtilde
```

The first statement fits an ordinary Cox model, without regard to the clustering variable `subject.id`. The second statement retrieves a copy of the collapsed score residuals matrix $\widetilde{D}$, and the third forms the modified sandwich estimate of variance. (In S, `%*%` is matrix multiplication, `t` is the transpose function, and `Surv` is a 'packaging' function that allows both time and status to be part of the left hand side of a model formula). A second example given in Lee, Wei and Amato concerns a litter-matched experiment. In this case the number of rats/litter may vary. The resulting estimator is shown to be much more efficient than an analysis stratified by cluster.

Lin and Wei [29] use the difference between $\mathcal{I}^{-1}$ and $D'D$ as a goodness of fit test for the Cox model. The variance estimate is quite complex, however, and we are not aware of any implementations of the method.

Wei, Lin and Weissfeld [42] consider multivariate survival times, an example is the measurement of both time to progression of disease and time to death for a group of cancer patients. They suggest a marginal model, where the covariate effects $\hat{\beta}$ are estimated without accounting for the within-subject correlation between survival and progression, followed by a post-fit computation of a corrected variance. Again, it can be shown that the corrected variance is equal to $\widetilde{D}'\widetilde{D}$, so that the model is computable using standard tools.

Below is the SAS code to fit the model. The data set contains $2n$ observations, and essentially consists of the data one would have used for a time-to-progression analysis (n rows), followed by the data for a time-to-death analysis (n rows). The variables are time and status, subject id, and covariates x1-x3. There is also an indicator variable `etype` to distinguish the event type, progression vs. survival. This variable will be 1 for the first n rows (the 'progression' data set) and 2 for the next n rows.

```
proc phreg data=wlw;
    model time * status(0) = rx size number;
    strata etype;
    output out=temp dfbeta= x1 x2 x3;
    id subject;

* Form the per-subject dfbeta matrix;
proc sort data=temp; by subject;
proc means data=temp noprint;
    by subject;
    var x1 x2 x3;
    output out=temp2 sum=y1 y2 y3;

* Now use IML to compute the product D'D ;
proc iml;
    use temp2;
    read all var rx size number in x;
    v = x' *x;
    reset noname;
    vname = "rx", "size", "number";
    print , "Robust variance",,v[colname=vname rowname=vname];
```

The computational method suggested by Wei, Lin, and Weissfeld in their paper implicitly includes all of the covariate by strata interaction terms; to match their analysis we would have to create these interaction terms, and then fit a six variable model. Alternatively, the following S-Plus code fits the model directly.

```
fit <- coxph(Surv(time, status)~ (rx +size +rnumber)*strata(etype)
                            + cluster(subject), data=wlw)
print(fit)
```

Barlow [4] discusses variance estimation for the case-cohort design. He recommends the estimate $D'D$, notes the explicit connection between it and the paper of Lin and Wei [28], and provides simulation results to show that it retains the correct size for this study design.

7 Conclusion

This paper has been able to discuss only some of the possibilities for analysis that are available with modern Cox regression programs. It has shown that such analyses are relatively straightforward to perform using standard packages. It is hoped that these methods, based on the counting process model and the use of residuals, will through it become more accessable to the statistical practitioner.

References

[1] Andersen, P.K. and Gill, R.D. (1982). Cox's regression model for counting processes: A large sample study. *Ann. Stat.* **10**, 1100–20.

[2] Andersen, P.K., Borgan, O., Gill, R.D., and Keiding, N. (1992) *Statistical Models Based on Counting Processes.* Springer-Verlag, New York.

[3] Barlow, W.E. and Prentice, R.L. (1988). Residuals for relative risk regression. *Biometrika* **75**, 65–74.

[4] Barlow, W.E. (1994). Robust variance estimation for the case-cohort design. *Biometrics* **50**, 1064-1072.

[5] Cain, K.C. and Lange, N.T. (1984). Approximate case influence for the proportional hazards regression model with censored data. *Biometrics* **40**, 493–99.

[6] Chappell, R. (1992). A note on linear rank tests and Gill and Schumacher's tests of proportionality. *Biometrika* **79**, 199–201.

[7] Chen, C.H. and Wang, P.C. (1991). Diagnostic plots in Cox's regression model. *Biometrics* **47**, 841–50.

[8] Cox, D.R. (1972). Regression models and life tables (with discussion). *J. Royal Stat. Soc. B* **34**, 187–220.

[9] Chambers, J.M., Cleveland, W.S., Kleiner, B. and Tukey, P.A. (1983). *Graphical Methods for Data Analysis.* Wadsworth, Belmont.

[10] Crowley, J. and Hu, M. (1977). Covariance analysis of heart transplant survival data. *J. Am. Stat. Assoc.*, **72**, 27–36.

[11] Dickson, E.R., Grambsch, P.M., Fleming, T.R, Fisher, L.D. and Langworthy, A. (1989). Prognosis in primary biliary cirrhosis: Model for decision making. *Hepatology* **10**, 1–7.

[12] Efron, B. (1982). *The Jackknife, the Bootstrap and Other Resampling Plans*, SIAM, Philadelphia.

[13] Fleming, T.R. and Harrington, D.P. (1991), *Counting Processes and Survival Analysis*, Wiley, New York.

[14] Gill, R. and Schumacher, M. (1987). A simple test of the proportional hazards assumption. *Biometrika* **74**, 289–300.

[15] Grambsch, P.M. and Therneau, T.M. (1994). Proportional hazards tests and diagnostics based on weighted residuals. *Biometrika*, **81**, 515–26.

[16] Grambsch, P., Therneau, T.M., and Fleming T.R. (1995). Diagnostic plots to reveal functional form for covariates in multiplicative intensity models. *Biometrics* **51**, 1469-1482.

[17] Harrell, F. (1986). The PHGLM procedure. *SAS Supplemental Library User's Guide* Version 5. Cary, NC: SAS Institute Inc.

[18] Harrell, F.E., Pollock, B. G. and Lee, K. L. (1987). Graphical methods for the analysis of survival data. In *Proceedings of the Twelfth Annual SAS Users Group International Conference*, p 1107-1115. SAS Institute, Inc., Cary, NC.

[19] Hastie, T.J. and Tibshirani, R.J. (1990). *Generalized Additive Models.* Chapman and Hall, London.

[20] Huber, P.J. (1967). The behavior of maximum likelihood estimates under non-standard conditions. *Proceedings of the Fifth Berkeley Symposium on Mathematical Statistics and Probability* **1**, 221–33.

[21] Kalbfleisch, J.D. and Prentice R.L. (1980). *The Statistical Analysis of Failure Time Data,* Wiley, New York.

[22] Laird, N.M. and Olivier, D. (1981). Covariance analysis of censored survival data using log-linear analysis techniques. *J. Am. Stat. Assoc.* **76**, 231–40.

[23] Lee, E.W., Wei, L.J. and Amato D. (1992). *Cox-type regression analysis for large number of small groups of correlated failure time observations.* In Klein, J.P and Goel, P.K. (eds), Survival Analysis, State of the Art, 237–247, Kluwer Academic Publishers, Netherlands.

[24] Liang, K.Y. and Zeger, S.L. (1986). Longitudinal data analysis for discrete and continuous outcomes. *Biometrics* **42**, 121-30.

[25] Lindor, K.D., Dickson, E.R., Baldus, W.P., Jorgensen, R.A., Ludwig, J., Murtaugh, P.A., Harrison, J.M., Wiesner, R.H., Anderson, M.L., Lange, S.M., LeSage, G., Rossi, S.S and Hofman, A.F. (1994). Ursodeoxycholic acid in the treatment of primary biliary cirrhosis. *Gastroenterology* **106**, 1284–90.

[26] Lipsitz, S.R., Laird, N.M. and Harrington, D.P. (1990) Using the jackknife to estimate the variance of regression estimators from repeated measures studies. *Comm Stat Theory Meth.* **19**, 821–45.

[27] Lin, D.Y. (1991). Goodness-of-fit analysis for the Cox regression model based on a class of parameter estimators. *J. Am. Stat. Assoc.* **86**, 725–728.

[28] Lin, D.Y. and Wei, L.J. (1989). The robust inference for the Cox proportional hazards model. *J. Am. Stat. Assoc.* **84**, 1074–78.

[29] Lin, D.Y. and Wei, L.J. (1991). Goodness-of-fit tests for the general Cox regression model. *Statistica Sinica* **1**, 1–17.

[30] Lin, D.Y., Wei, L.J. and Ying, Z. (1993). Checking the Cox model with cumulative sums of martingale-based residuals. *Biometrika* **80**, 557–72.

[31] McCullagh, P. and Nelder, J.A. (1983). *Generalized Linear Models.* Chapman and Hall.

[32] Miller, R.G. (1981), *Survival Analysis,* Wiley, New York.

[33] Moreau, T., O'Quigley, J. and Mesbah, M. (1985). A global goodness-of-fit statistic for the proportional hazards model. *Applied Statistics* **34**, 212–218.

[34] Moss, A.J. and the Multicenter Diltiazem Trial Research Group (1988). The effect of diltiazem on mortality and reinfarction after myocardial infarction. *New England J. of Medicine,* **319**, 385–92.

[35] Nagelkerke, N. J. D., Oosting, J., and Hart, A. A. M. (1984). A simple test for goodness of fit of Cox's proportional hazards model. *Biometrics* **40**, 483–486.

[36] O'Quigley, J. and Pessione, F. (1989). Score tests for homogeneity of regression effects in the proportional hazards model. *Biometrics* **45**, 135–144.

[37] Prentice, R.L. and Cai, J. (1992). Covariance and survivor function estimation using censored multivariate failure time data, *Biometrika,* **79**, 495–512.

[38] Reid, N. and Crépeau, H. (1985). Influence functions for proportional hazards regression. *Biometrika* **72**: 1–9.

[39] Schoenfeld, D. (1980). Chi-squared goodness-of-fit tests for the proportional hazards regression model. *Biometrika* **67**, 145–53.

[40] Therneau, T.M., Grambsch P.M. and Fleming, T.R. (1990). Martingale based residuals for survival models. *Biometrika* **77**, 147–60.

[41] Uitti, R.J., Ahlskog, J.E., Maraganore, D.M., Muenter, M.D., Atkinson, E.J., Cha, R.H. and O'Brien, P.C. (1993). Levodopa therapy and survival in idiopathic Parkinson's disease: Olmsted County Project. *Neurology* **43**, 1918–26.

[42] Wei, L.J., Lin, D.Y. and Weissfeld, L. (1989). Regression analysis of multivariate incomplete failure time data by modeling marginal distributions. *J. Am. Stat. Assoc.* **84**, 1065–73.

[43] White, H. (1980). A heteroskedasticity-consistent covariance matrix estimator and a direct test for heteroskedasticity. *Econometrica* **48**, 817–38.

[44] White, H. (1982). Maximum likelihood estimation of misspecified models. *Econometrika* **50** 1–26.

Model-Based and/or Marginal Analysis for Multiple Event-Time Data?

David Oakes

ABSTRACT We will explore the relationship between a marginal proportional hazards formulation for ordered failure times, as proposed by Wei, Lin and Weissfeld (1989) and strongly supported by Therneau (1997), and a conditional formulation advocated by, among others, Clayton (1988), Oakes (1992) and Pepe and Cai (1993). We review the derivation given by Oakes (1992) of a conditional proportional hazards model via an underlying frailty structure. Quite generally, we show that a family of absolutely continuous bivariate survivor functions cannot simultaneously satisfy the proportional hazards model both conditionally and unconditionally. We will explore the efficiency loss in using the marginal approach compared with full likelihood procedures in two simple special cases, and find it to be quite small.

1 Introduction

This paper concerns methods for the analysis of ordered event times. That is, for each of a sample of individuals we observe the times $0 < T_1 < T_2 < \ldots < T_k$ of successive failures. Often these failures will correspond to successive milestones in a disease process, for example in AIDS studies, with the origin at the point of infection, T_1 could be the time to develop clinical AIDS, T_2 the time to first reach a specified level of CD4 counts, T_3 the time of death. These failure times may be influenced by explanatory variables x, which are measured prior to entry and do not change over time for each individual.

In the modeling of this type of data we have several choices for the primary time-scale - we may measure each failure-time from the immediately preceding event, including relevant information about the history of failures as covariates, or we may measure each failure time from the initial time-origin. Modeling of inter-event times was discussed by Prentice et al (1981) and developed by many subsequent authors. See e.g. Dabrowska et al (1992). For another line of development (modulated renewal processes) see Cox (1972), Oakes (1981, Section 7) and Oakes and Cui (1994).

When primary interest focusses on the time from entry to one of the later outcomes it is natural to model this directly. Suppose for simplicity that subjects are entered into a randomized trial at $t = 0$ and that they are randomized to one of two treatment regimes at that point. Suppose that primary interest is in comparing the distribution of T_2 among subjects assigned to active treatment ($x = 1$) and to control ($x = -1$). To preserve the validity of the randomization analysis we should not adjust for T_1. For suppose that the treatment has an effect on T_2 through delaying T_1. Then if we adjust for T_1, for example by not counting subjects as being at risk of T_2 until after they have experienced T_1, then we may be adjusting away precisely the effect we are looking for. Here T_1 plays the role of an intermediate response variable. In a subsequent analysis of mechanism, ie to describe the causal pathway, we may need to carry out such adjusted analyses.

An application that I have been much concerned with is the use of deprenyl in the treatment of Parkinson's disease (PD). One milestone in the progression of disease is the need for treatment with levodopa, an agent which relieves the symptoms of PD but does not halt the progression of the underlying disease process and may possibly hasten it. Use of deprenyl has been shown in a large double-blind randomized trial to delay the need for levodopa therapy (Parkinson Study Group, 1989, 1993) but may not have a similar effect on later milestones of the disease process, for example the development of levodopa - related side effects (Parkinson Study Group, 1996). An issue of considerable practical importance that I will not attempt to deal with here is that a treatment effect on an earlier endpoint may distort the analysis of a later endpoint by introducing differential bias and noncompliance between the treatment groups.

Consider now the opposite situation, where we think it likely that the treatment has no effect on T_1 but that it may delay T_2 through increasing the time $Y_2 = T_2 - T_1$ between T_1 and T_2. Then it is natural to analyze the Y_2 directly. However such an analysis may again lose the protection provided by randomization. An obvious answer is to change the starting point of the study, so that randomization occurs at T_1. However this may not be possible - administration of the treatment in the earlier phase (before T_1) may be critical to achieving the increase in Y_2.

A reasonable and conservative strategy is to analyze the times $T_1, T_2, \ldots$ from randomization to each endpoint separately, with an allowance for multiple comparisons if overall type - I error protection is thought to be desirable. Wei and Lachin (1984) showed how to combine test statistics for the influence of x on $T_1, T_2, \ldots$. The appropriate method of combination depends on the correlations among the test statistics under the null hypothesis, which can be estimated from the data, and on the relevant alternative hypothesis, which must be specified. Wei, Lin and Weissfeld (1989) - hereinafter WLW - discussed a similar approach to estimation, under the assumption that x influences each of $T_1, T_2, \ldots$ via proportional hazards models with possibly different baseline hazards and hazard-ratio parameters.

These methods have the counter-intuitive property that subjects are considered to be "at risk" of later events before they have experienced earlier events. This feature has been criticized by several authors including Oakes (1992) and Pepe and Cai (1993). Therneau (1997) has argued that such criticisms may be misplaced. Whether an individual is "at risk" for an event can be a rather metaphysical concept - with perfect knowledge about the state of the world the concept of being "at risk" would disappear - subjects would be at risk precisely at the times when they experience events. Intermediatc states of knowledge regarding risk factors would lead to intermediate definitions of whether subjects are at risk.

However on a more mundane level it is clear that there may be some loss of statistical efficiency (power of tests or precision of estimators) in ignoring information that is available - for example that certain subjects are definitely not at risk of an event. Surprisingly the degree of this efficiency loss does not appear to have been studied. The present paper begins this investigation, using some very simple statistical models. Hughes (1995) has carried out an extensive investigation of the gains in efficiency from using data from outcomes beyond the first event within the WLW structure - however he did not address the issue of the further gain that might be possible from a complete model specification.

In part the reason for this lack may be the difficulty of formulating models for ordered survival times in which the ordered times to events follow proportional hazards models. Of course such models for unordered bivariate data are very easy to develop - one can place a proportional hazards model on the marginals and combine this with any convenient dependence structure, for example one governed by gamma distributed frailties (Huster et al, 1989), or positive stable frailties (Manatunga and Oakes, 1995). But such models are harder to develop for ordered failure times. For example the family of alternative models used by Hughes does not apparently have any closed form analytic expression. Although the proportional hazards requirement is not essential, there will be some further loss of efficiency if it does not hold.

Section 2 attempts to clarify the relationship between two apparently similar but different models. In both models, time to each event is measured from a single origin, but in the first model, called a "conditional model" here, individuals are not considered to be at risk of a second event till they have experienced the first event. The second model, called here a marginal model, works with the marginal hazard functions as proposed by WLW. Section 3 briefly reviews work of Oakes (1992) that shows how conditional proportional hazards models can arise from a frailty structure. Section 4 shows that families of absolutely continuous bivariate survival distributions cannot satisfy proportional hazards models both conditionally and unconditionally. Section 5 discusses the consequences of a marginal analysis applied to the very simplest conditional model - an ordinary Poisson process. Section 6 introduces a marginal model with a very specific form of dependence structure and gives some explicit algebraic results for the ARE of the WLW estimates compared with those from a full likelihood approach. In Section 7 we draw some general conclusions, which might be better expressed as hypotheses requiring further investigation using more complex models.

2 Two Models for Multiple Event-Time Data

We suppose that, for a given subject, the intensity rate of the $k'th$ event at time t given that the subject has already experienced exactly $k-1$ events, before time t is

$$h_k(t) = \lim_{\Delta \to 0} \frac{1}{\Delta} \mathrm{pr}\{T_k \leq t + \Delta | T_{k-1} < t, T_k \geq t\} = h_{0k}(t) \exp(\beta_k x). \quad (1)$$

Here x is a vector of explanatory variables. Important features of this model are that (i) time is always measured from the original time origin for that subject, (ii) an individual is not viewed as being at risk of the $k'th$ event until after experiencing events $1, 2, \ldots, k-1$, (iii) the covariate vector x for a given individual is the same for all k, although its coefficient β_k may change, (iv) the model is in a literal sense both well-specified and constructive - if the quantities $x, \beta_k, h_{0k}(u)$ are known the entire process can be simulated easily and in an essentially unique way. Depending on the values of these quantities, the process may explode with infinitely many events in finite time, or (more usually) die out, with the number of events to an individual remaining bounded as $t \to \infty$. (v) the number of events $N(t) = \sup\{j : T_j < t\}$ carries all information about the history of the process in $(0, t)$ that is relevant for predicting the future of the process from t on.

Usually we will observe failure histories for each individual i only up to some censoring time c_i. Where necessary we will assume for simplicity that these censoring times are independent realizations of a random variable C with survivor function $G(c) = \mathrm{pr}(C > c)$. In particular we shall assume here that the distribution of C is not influenced by the explanatory variables x, a condition we call *equal censoring.*

Under this observation scheme, the times to a particular event, say the $k'th$, will be subject to both left truncation and right censorship. An observation on T_k will be subject to the condition $T_k > t_{k-1}$, where t_{k-1} is the time of the $k-1'th$ event, and will be right-censored at c i.e. if $T_k > c$ we observe this fact, but we do not observe the value of T_k. Observations with $T_{k-1} > c$ yield no information about $\beta_k, h_k(t)$.

In Oakes (1992) I described this as a "generalized Andersen-Gill model" by analogy with that of Andersen and Gill(1982), in which $\beta_k = \beta$ and $h_k(t) = h(t)$, for all k. The original Andersen and Gill model forces a Poisson distribution for the number of events to any individual over any finite time - this property gives the model the virtue of very easy falsifiability. Afficionados of BMDP will know how to obtain a semiparametric fit for the generalized Andersen - Gill model using the "time-dependent stratification" option. For simplicity, I use the term conditional model here to recognize that the hazard for each event conditions on the history of previous events. Clayton (1988) emphasized the importance of incorporating features such as dynamic stratification into statistical software for survival analysis, particularly for epidemiologic applications.

Wei, Lin and Weissfeld (1989) proposed an essentially different though superficially similar approach to the analysis of multiple event times, based on *marginal* models. They postulated a proportional hazards form for the marginal hazards

$$\tilde{h}_k(t) = \lim_{\Delta \to 0} \frac{1}{\Delta} \mathrm{pr}\{T_k \le t + \Delta | T_k \ge t\} = \tilde{h}_{0k}(t) \exp(\tilde{\beta}_k x). \tag{2}$$

An advantage of this formulation is that it leads to simple semi-parametric estimation procedures for the parameters using the "generalized estimating approach" of Liang and Zeger (1986). One writes down separate partial likelihoods in the $\tilde{\beta}_k$, maximizes them by solving the score equations as if the likelihoods in the $\tilde{\beta}_k$ were independent, and then forms a direct estimate of the correlation matrix of the $\{\tilde{\beta}_k\}$ from the sample correlations of the corresponding score statistics. One can test hypotheses of the form $\tilde{\beta}_j = 0 \quad (j = 1, 2, \ldots, k)$ or $\tilde{\beta}_j = \tilde{\beta}_1 \quad (j = 2, 3 \ldots, k)$, by appropriate combinations of the sample scores and their covariances. One can estimate $\tilde{\beta}_j$, individually, or subject to conditions such as $\tilde{\beta}_j = \tilde{\beta}_1$. See Wei and Lachin (1984) for the results on hypothesis testing, and Wei, Lin and Weissfeld (1989), for estimation.

It is a tribute to the popularity of counting process methods in survival analysis that considerable work on both a theoretical level (proving asymptotic results) and practical level (software development and implementation) has been devoted to developing semiparametric estimates of the parameters β and $\tilde{\beta}$ in these two models. Corresponding parametric formulations have scarcely been considered, although (because?) they require much less in the way of technical development. We focus here on on parametric formulations. In view of results such as those Efron (1977) and Oakes (1977) on the generally excellent asymptotic relative efficiency (ARE) of the semiparametric estimates relative to the parametric estimates we can take the more easily derived parametric results as a good guide to the corresponding results for the semiparametric model. In particular under the equal censoring condition, and with the covariate vector centered so that $\sum x = 0$, the ARE of the parametric test of the overall null hypothesis (either $\beta_k = 0$ for all k or $\tilde{\beta}_k = 0$ for all k) with $h_{0k}(t)$ or $\tilde{h}_{0k}(t)$ completely known, and semiparametric tests, with these quantities completely unknown, will be the same. Unlike (1), model (2) is not well-specified, in the sense that further information is needed regarding the structure of the dependence between the T_j before the model can be simulated.

3 An Illustration Involving a Frailty Model

Oakes (1992) described how a special conditional model could arise from a frailty structure, in which an unobserved random effect introduces heterogeneity among subjects. In essence the idea is due to Greenwood and Yule (1920). Suppose that the intensity of an event at t does not depend explicitly on the number of previous events, but does depend on an unobserved random quantity W, called a frailty, so that, for some relative risk parameter θ,

$$h_j(t, w, \theta) = \lim_{\Delta \to 0} \frac{1}{\Delta} \text{pr}\{T_j \leq t + \Delta | T_{j-1} < t, T_j \geq t, W = w\} = wb(t, \theta).$$

We then find that

$$\begin{aligned} h_j(t, \theta) &= E\{h_j(t, W, \theta) | N(t) = j - 1\} \\ &= b(t, \theta) E\{W | N(t) = j - 1\} \\ &= b(t, \theta) \frac{p^{(j)}\{B(t, \theta)\}}{p^{(j-1)}\{B(t, \theta)\}}, \end{aligned}$$

where $p^{(j)}(s)$ is the $j'th$ derivative of the Laplace transform $p(s) = E\{\exp(-sW)\}$ of the distribution of W and $B(t, \theta) = \int_0^t b(u, \theta) du$ is the integrated hazard function from $b(t, \theta)$. The simplest model gives W a gamma distribution with unit mean and variance κ, so that the Laplace transform is

$$p(s) = \left(\frac{1}{1 + \kappa s} \right)^{\frac{1}{\kappa}}.$$

We then find that

$$h_j(t, \theta) = \{1 + (j - 1)\kappa\} h_1(t, \theta),$$

the various conditional hazards are proportional to each other, though not, of course, to $b(t, \theta)$.

So far, we have left the form of the dependence of $B(t, \theta)$ on θ unspecified. If we now set

$$B(t, \theta) = \frac{\{1 + \kappa B(t, 1)\}^\theta - 1}{\kappa},$$

we ensure proportionality of the conditional hazards in θ as well as in j. The model then becomes

$$h_j(t, \theta) = \theta\{1 + (j - 1)\kappa\} h_{01}(t). \tag{3}$$

A different simple model with proportional marginal hazards is constructed in Section 6.

4 Incompatibility of Marginal and Conditional Proportional Hazards

We show that there is no family of absolutely continuous family of bivariate density functions that satisfies *both* the conditional proportional hazards model (1) and the the marginal model (2). The marginal hazard for T_1 is clearly the same as the conditional hazard, $\tilde{h}_1(t) = h_1(t)$. However the marginal survivor function $\tilde{S}_2(t)$ of the time T_2 to the second event is given in terms of the density $f_2(t)$ and survivor function $S_2(t)$ corresponding to the conditional hazard $h_2(t)$ by the equation

$$\tilde{S}_2(t) = S_1(t) + S_2(t) \int_0^t \frac{f_1(u)}{S_2(u)} du.$$

Differentiation in t and substitution yields

$$\frac{\tilde{h}_2(t)}{h_2(t)} = 1 - \frac{S_1(t)}{\tilde{S}_2(t)}.$$

Note that this equation also gives an explicit formula for the conditional hazard function $h_2(t)$ in terms of the marginal distributions of T_1 and T_2.

We are interested in parameterizations that would allow each of $h_1(t) = \tilde{h}_1(t)$, $h_2(t)$ and $\tilde{h}_2(t)$ to have proportional hazards forms, with proportionality factors $\theta_1(x)$, $\theta_2(x)$ and $\tilde{\theta}_2(x)$, depending on the covariate x, for some baseline hazards $h_{10}(t)$, $h_{20}(t)$ and $\tilde{h}_{20}(t)$ respectively. The ratio of $\tilde{h}_2(t)/h_2(t)$ to $\tilde{h}_{20}(t)/h_{20}(t)$ is then, in an obvious notation,

$$\frac{\tilde{\theta}_2(x)}{\theta_2(x)} = \frac{1 - S_{10}(t)^{\theta_1(x)}/\tilde{S}_{20}(t)^{\tilde{\theta}_2(x)}}{1 - S_{10}(t)/\tilde{S}_{20}(t)}.$$

This is impossible, at least if $\theta(x) = \{\theta_1(x), \theta_2(x)\}$ takes more than two distinct values, because the right side of this equation cannot be free of t.

5 A Simple Poisson Model

The simplest possible model for two ordered events is that they represent the first two events of a Poisson process of rate λ (say). The time to the first event is Exponential with parameter λ,. The marginal distribution of the time to the second event is $\mathcal{G}(2, \lambda)$ (Gamma, with index 2 and scale parameter λ). The density, survivor function and hazard are

$$\tilde{f}_2(t) = \lambda^2 t e^{-\lambda t}, \quad \tilde{S}_2(t) = e^{-\lambda t} + \lambda t e^{-\lambda t}, \quad \tilde{h}_2(t) = \frac{\lambda^2 t}{1 + \lambda t}.$$

This process can be viewed as a conditional model with $h_1(t) = h_2(t) = \lambda$. In this special case it is well known that the conditional distribution of T_1

given $T_2 = t$ is uniform over $(0, t)$ and that $\sum T_2$ is sufficient for λ - given the time to the second event, the time to the first event is irrelevant for inference about λ.

The optimal test statistic for the hypothesis $\lambda = 1$ is based on $U = \sum T_2$. Under the null hypothesis this has mean and variance $2n$. We can of course achieve the same test by forming the appropriate weighted sum - actually just the simple sum - of the test statistics $U_1 = \sum T_1$ and $Y_2 = \sum (T_2 - T_1)$, of the independent test statistics coresponding to the times to the first event and between the first and second events.

Suppose we incorrectly assume that the distribution of T_2 follows a proportional hazards model, with proportionality parameter ϕ, but that we correctly specify the null value of the survivor function as $S_{20}(t) = (1+t)e^{-t}$, i.e. we assume the model $\tilde{S}_2(t) = S_{20}(t)^{\phi}$. The optimal test statistic for the hypothesis $\phi = 1$ is the sum $\tilde{U} = \sum \{T_2 - \log(1 + T_2)\}$ of cumulative hazard functions. Under the null hypothesis this has unit mean and variance, since the cumulative hazard function transformation converts any continuous distribution to unit exponential. The ARE of $\tilde{U}$ to U when the Poisson model is correct is just the squared correlation between U and $\tilde{U}$. Routine calculation shows that with

$$J = \int_0^{\infty} \frac{e^{-u}}{1+u} du = 0.59633$$

the ARE equals

$$\frac{(2-J)^2}{2} = 0.9851.$$

The nonproportionality does not hurt very much here!

We can even improve on this a smidgen, by forming the appropriate weighted combination with the test statistic $U_1 = \sum T_1$ based on the times of the first events. In view of the aformentioned sufficiency of $\sum T_2$, there is no benefit to forming a combination of U_1 with the optimal statistic U, but there is a benefit to combining U_1 with $\tilde{U}$. The squared correlation of the weighted average $\alpha U_1 + (1 - \alpha)\tilde{U}$ with U is

$$\frac{\{2 - J - \alpha(1 - J)\}^2}{2(1 - \alpha J + \alpha^2 J)},$$

which is maximized when

$$\alpha = \frac{J^2 - 4J + 2}{J^2 - 3J} = 0.0207.$$

The improvement in ARE from introducing U_1 is approximately 0.00025 .

By contrast use of the simple unweighted average ($\alpha = 0.5$) gives an ARE of 0.849, a noticeable loss of efficiency.

6 A Simple Model with Exponential Marginals

A simple class of distributions whose ordered marginals are exponential can be obtained as follows. Let $\mathcal{E}(\lambda)$ denote the exponential distribution with parameter λ, density $\lambda \exp(-\lambda t)$ and survivor function $\exp(-\lambda t)$.

With $\lambda_1 > \lambda_2$ suppose that $Y_1 \sim \mathcal{E}(\lambda_1 - \lambda_2)$ and $Y_2 \sim \mathcal{E}(\lambda_2)$ are independent and let $T_1 = \min(Y_1, Y_2)$ and $T_2 = Y_2$. Equivalently, let $T_1 \sim \mathcal{E}(\lambda_1)$ and $U \sim \mathcal{E}(\lambda_2)$ be independent. Conditionally on $T_1 = t$ and $U = u$, take $T_2 = t$ with probability ρ, where $\rho = \lambda_2/\lambda_1$ and $T_2 = t + u$ with probability $1 - \rho$.

It is easily seen that, $T_1 \sim \mathcal{E}(\lambda_1)$ and $T_2 \sim \mathcal{E}(\lambda_2)$, and that the correlation between them is $\rho = \lambda_2/\lambda_1$. The joint distribution of T_1 and T_2 is not absolutely continuous, there being positive probability that $T_1 = T_2$. In many medical studies ascertainment of clinical outcome could be delayed until the patient is seen at a follow-up visit, so the possibility of two (or more) outcome events occurring (or rather, being recorded) simultaneously is not necessarily unrealistic.

The log-likelihood from a sample of size n is

$$n \log(\lambda_1) - \lambda_1 s_1 + m \log(1 - \frac{\lambda_2}{\lambda_1}) + (n - m) \log(\frac{\lambda_2}{\lambda_1}) + m \log(\lambda_2) - \lambda_2 s_2$$

$$= n \log(\lambda_2) + m \log(\lambda_1 - \lambda_2) - \lambda_1 s_1 - \lambda_2 s_2.$$

Here $(s_1 = \sum t_1, s_2 = \sum(t_2 - t_1), m = \#\{t_2 > t_1\})$ forms a three dimensional sufficient statistic for the two-dimensional parameter (λ_1, λ_2). The maximum likelihood estimates of (λ_1, λ_2) can be derived explicitly as

$$\hat{\lambda}_1 = \frac{n}{s_1 + s_2} + \frac{m}{s_1}, \quad \hat{\lambda}_2 = \frac{n}{s_1 + s_2}.$$

The asymptotic variances can be calculated from the inverse of the expected information matrix as

$$\lim n\mathrm{var}(\hat{\lambda}_1) = \lambda_1^2 + \lambda_2^2 - \lambda_1\lambda_2 = \lambda_1^2\{1 - \rho(1 - \rho)\},$$

$$\lim n\mathrm{var}(\hat{\lambda}_2) = \lim n\mathrm{cov}(\hat{\lambda}_1, \hat{\lambda}_2) = \lambda_2^2.$$

The singular nature of the distribution allows the full sufficient statistic to be reconstructed from the marginal data. However we shall assume that only the sufficient statistics s_1 for λ_1 and $s_1 + s_2$ for λ_2 are available for estimation in the marginal model, as would be the case if the form of the joint distribution were unknown. The marginal estimates are just $\tilde{\lambda}_1 = n/s_1$, $\tilde{\lambda}_2 = n/(s_1 + s_2) = \hat{\lambda}_2$. The marginal estimate of λ_1 differs from the maximum likelihood estimate (MLE) and has greater limiting variance. Its ARE varies from 0.75 to 1.0, depending on ρ, with minimum when $\rho = 0.5$.

Suppose now that for $j = 1, 2$, $\lambda_j = \theta\lambda_{j0}$, where the "baseline hazards" λ_{j0} are assumed known but the relative risk parameter θ, assumed common to the two types of failure, is to be estimated. The "marginal" estimate $\tilde{\theta}$ of θ is the best linear combination of the estimates $\tilde{\theta}_1 = \tilde{\lambda}_1/\lambda_{10}$ and $\tilde{\theta}_2 = \tilde{\lambda}_2/\lambda_{20}$. It can be shown that the MLE $\hat{\theta}$ and marginal estimate $\tilde{\theta}$ have limiting variances $\lim n\text{var}(\hat{\theta}) = \theta^2/(2-\rho)$ and $\lim n\text{var}(\tilde{\theta}) = \theta^2(1+\rho)/2$. Again the marginal estimate is fully efficient only at the extremes $\rho = 0, 1$. The minimum ARE, again achieved when $\rho = 0.5$, is 8/9. This U-shaped curve for the ARE of marginal estimates against full likelihood estimates occurs also in applications to unordered failure time data - see e.g. Huster et al (1989) and Manatunga and Oakes (1995).

Suppose that a treatment affects the time between T_1 and T_2 but does not affect the time to T_1. Does one gain by analyzing just the time $Y_2 = T_2 - T_1$ instead of T_2 which includes the irrelevant T_1? The answer depends on whether λ_1 is known. If it is, then the asymptotic variance of $\hat{\lambda}_2$ from the data $(m, \sum y_2)$ is

$$\lim n\text{var}(\hat{\lambda}_2) = \lambda_2^2(1-\rho) = (1-\rho)\lim n\text{var}(\tilde{\lambda}_2).$$

However if λ_1 is unknown, one should use the full log-likelihood to estimate it, and there is no benefit. If only the data $(m, \sum y_2)$ are available for estimation of both λ_1 and λ_2 then the MLE of λ_2 has limiting variance

$$\lim n\text{var}(\hat{\lambda}_2') = \frac{\lambda_2^2}{(1-\rho)}.$$

showing a substantial loss of efficiency compared with the use of all the marginal data on T_2 to estimate λ_2.

Now let us consider the effect that censoring might have on these calculations. Suppose that the data are subject to independent censoring by potential censoring variables that are themselves independent and exponentially distributed with scale parameter μ. We may assume that μ is known: it is orthogonal to (λ_1, λ_2). Let d_1 and d_2 be the number of observed 'first' failures (T_1) and observed 'second' failures distinct from 'first' failures $(T_2 > T_1)$ and let r_2 be the number of individuals at risk of a 'second' failure at some time after an observed 'first' failure. Let s_1 and s_2 denote the total times at risk up to the 'first' failure, or a censoring that precedes the 'first' failure, and after a 'first' failure, respectively. Then the total number of observed 'second' failures is $d_1 + d_2 - r_2$ and the corresponding total time at risk is $s_1 + s_2$. We write $\rho = \lambda_2/\lambda_1$ as before and $\tau = \mu/\lambda_1$. The following distributional properties hold: (i) $d_1 \sim \mathcal{B}(n, 1/(1+\tau))$, (i.e. Binomial with index n and success probability $1/(1+\tau)$), (ii) $s_1|d_1 \sim \mathcal{G}(n, \lambda_1 + \mu)$, (iii) $r_2|d_1, s_1 \sim \mathcal{B}(d_1, 1-\rho)$, (iv) $s_2|d_1, s_1, r_2 \sim \mathcal{G}(r_2, \lambda_2 + \mu)$ (v) $d_2|d_1, s_1, r_2, s_2 \sim \mathcal{B}(r_2, \rho/(\rho+\tau))$.

The log-likelihood in (λ_1, λ_2) is

$$d_1 \log \lambda_1 - s_1\lambda_1 + r_2 \log(1-\rho) + (d_1 - r_2)\log(\rho) + d_2 \log(\lambda_2) - s_2\lambda_2$$
$$= (d_1 + d_2 - r_2)\log(\lambda_2) + r_2 \log(\lambda_1 - \lambda_2) - s_1\lambda_1 - s_2\lambda_2.$$

We find that the Fisher information for (λ_1, λ_2) has elements $\mathcal{I}_{11} = -\mathcal{I}_{12} = -\mathcal{I}_{21} = \{(1-\rho)(1+\tau)\}^{-1}$, $\mathcal{I}_{22} = \mathcal{I}_{11} + \{\rho(\rho+\tau)\}^{-1}$. On inverting we find that the asymptotic variances of the MLE's of λ_1 and λ_2 are given by

$$\lim n\text{var}(\hat{\lambda}_1) = \lambda_1^2\{1 - \rho(1-\rho) + \tau\},$$

$$\lim n\text{var}(\hat{\lambda}_2) = \lambda_1^2\rho(\rho + \tau).$$

The estimates from the marginal distributions are $\tilde{\lambda}_1 = d_1/s_1$ and $\tilde{\lambda}_2 = (d_1 + d_2 - r_2)/(s_1 + s_2)$ respectively. Their asymptotic variances are now

$$\lim n\text{var}(\tilde{\lambda}_1) = \lambda_1^2(1+\tau), \quad \lim n\text{var}(\tilde{\lambda}_2) = \lambda_1^2\rho(\rho+\tau).$$

As before the marginal estimate of λ_2 is fully efficient - in fact $\hat{\lambda}_2$ and $\tilde{\lambda}_2$ are identical. The ARE of $\tilde{\lambda}_1$ relative to $\hat{\lambda}_1$ is $1 - \rho(1-\rho)/(1+\tau)$, an increasing function of τ which again attains its minimum in ρ when $\rho = 0.5$.

In the case that $\lambda_1 = \theta\lambda_{10}$, and $\lambda_2 = \theta\lambda_{20}$ with λ_{10} and λ_{20} both known we find that the MLE $\hat{\theta}$ has limiting variance

$$\lim n\text{var}(\hat{\theta}) = \theta^2 \frac{(1+\tau)(\rho+\tau)}{2\rho + \tau - \rho^2}.$$

Note that the information for θ is proportional to the expected number of observed distinct failure times.

The estimates $\tilde{\theta}_j = \tilde{\lambda}_j/\lambda_{j0}$ $(j = 1, 2)$ and $\tilde{\theta}_2$ of θ from the marginal survival times have limiting variances and covariances respectively:

$$v_1 = \lim n\text{var}(\tilde{\theta}_1) = \theta^2(1+\tau),$$

$$v_2 = \lim n\text{var}(\tilde{\theta}_2) = \theta^2\frac{\rho+\tau}{\rho},$$

and

$$c = \lim n\text{cov}(\tilde{\theta}_1, \tilde{\theta}_2) = \theta^2(\rho+\tau).$$

This first two expressions are easily derived - the information from each marginal is proportional to the expected numbers of failures from that marginal. The last expression was derived by repeated use of the delta method combined with distributional properties (i)-(v). The best linear combination $\tilde{\theta}$ of $\tilde{\theta}_1$ and $\tilde{\theta}_2$ has limiting variance

$$\lim n\text{var}(\tilde{\theta}) = \frac{v_1 v_2 - c^2}{v_1 + v_2 - 2c} = \frac{\theta^2(\rho+\tau)(1+\rho+\tau)}{2\rho+\tau}.$$

The ARE of $\tilde{\theta}$ relative to $\hat{\theta}$ is

$$\frac{\lim n\text{var}(\hat{\theta})}{\lim n\text{var}(\tilde{\theta}}) = \frac{(1+\tau)(2\rho+\tau)}{(1+\rho+\tau)(2\rho - \rho^2 + \tau)}.$$

This is always lies between 8/9, its minimum value in the uncensored case and unity.

7 Conclusions

What have we learned from these exercises - or more precisely what hypotheses can we formulate for more detailed investigation?

First, modeling of ordered event times is intrinsically more difficult than modeling unordered times. We do not have true independence to fall back on as a special case.

Second, Weissfeld (1989) does well in the situations we have considered, provided the tests and estimates are correctly focussed.

Third - Estimation of the distribution of the time to the first event is often helped by consideration of the time to the second event, but the reverse is apparently untrue.

Fourth - in estimation problems, where we deal with fixed "true" values of parameters, it is assumed that we know what we are estimating, e.g. a constant hazard ratio across ordered event times in two groups. If the hazard ratio is not constant, we will be able to test for that, and adjust our analyses appropriately, with no loss of asymptotic efficiency. In testing problems, we do not always know what we are testing against: if we test an overall null hypothesis against a (local) alternative of a constant hazard ratio across groups we cannot, even asymptotically, adjust our test for the possibility that hazard ratio is not constant, without some cost in power.

Fifth - censoring has little effect on our results concerning the relative efficiency of marginal to full likelihood estimates. If anything, censoring appears to improve the performance of marginal estimates relative to MLE's.

Acknowledgements

I thank Bo Yang and two anonymous reviewers for careful reading of this manuscript. This work was supported by grant R01 CA 52572 from the National Cancer Institute.

References

Andersen, P.K. and Gill, R.D. (1982). Cox's regression model for counting processes, a large sample study. *Ann. Statist.*, **10**, 1100-1120.

Clayton, D.G. (1988). The analysis of event history data: a review of progress and outstanding problems. *Statist. Med.*, **7**, 819-841.

Cox, D.R. (1972). The statistical analysis of dependencies in point processes. In Lewis, P.A.W. (ed). *Symposium on Point Processes,* New York, Wiley. pp 55-66.

Dabrowska, D.M., Sun, G.W. and Horowitz, M.H. (1992). Cox regression in a Markov renewal model with an application to the analysis of bone marrow transplant data. *J. Amer. Statist. Assoc.* **89**, 867-877.

Efron, B. (1977). The efficiency of Cox's likelihood function for censored data. *J. Amer. Statist. Assoc.*, **72**, 557-565.

Greenwood, M. and Yule, G.U. (1920). An enquiry into the nature of frequency distributions representative of multiple happenings with particular reference to the occurrence of multiple attacks of disease as repeated accidents. *J. R. Statist. Soc.*, **83**, 255-279.

Hughes, M.D. (1995). Power Considerations for Clinical Trials Using Multivariate Time-to-Event Data. *Unpublished Manuscript.*

Huster, W.J., Brookmeyer,R. and Self, S.G. (1989). Modelling paired survival data with covariates. *Biometrics*, **45**, 145- 156.

Liang, K.Y. and Zeger, S.L. (1986). Longitudinal data analysis using generalized linear models. *Biometrika*, **73**, 13-22.

Manatunga, A.K. and Oakes, D. (1995). Parametric analysis for matched pair survival data. *Unpublished Manuscript.*

Oakes, D. (1977). The asymptotic information in censored survival data. *Biometrika*, **64**, 441-448.

Oakes, D. (1981). Survival times, aspects of partial likelihood. *Int. Statist. Rev.*, **49**, 199-233.

Oakes, D. (1992). Frailty models for multiple event-time data. In Klein, J.P. and Goel P.K. (eds). *Survival Analysis: State of the Art.* Kluwer, pp 371-380.

Oakes, D. and Cui, L. (1994). On semiparametric inference for modulated renewal processes. *Biometrika*, **81**, 83-90.

Parkinson Study Group (1989). Effect of deprenyl on the progression of disability in early Parkinson's disease. *New Engl. J. Med.*, **320**, 1364-1371.

Parkinson Study Group (1993). Effects of tocopherol and deprenyl on the progression of disability in early Parkinson's disease. *New Engl. J. Med.*, **328**, 176-183.

Parkinson Study Group (1996). Impact of deprenyl and tocopherol treatment on Parkinson's disease in DATATOP patients requiring levodopa. *Ann. Neurol.*, **39**, 37-45.

Pepe, M.S. and Cai, J. (1993). Some graphical displays and marginal regression analyses for recurrent failure times and time-dependent covariates. *J. Amer. Statist. Assoc.*,**88**, 811-820.

Prentice, R.L., Williams, B.J. and Peterson, A.V. (1981). On the regression analysis of multivariate failure time data. *Biometrika*, **68**, 373-37.

Therneau, T. (1997). Extending the Cox model. In *Proceedings of the First Seattle Symposium in Biostatistics: Survival Analysis*, Eds. D. Y. Lin and T. R. Fleming, pp. xxx-xxx. New York: Springer-Verlag.

Wei, L.J. and Lachin, J.M. (1984). Two-sample asymptotically distribution-free tests for incomplete multivariate observations. *J. Amer. Statist. Assoc.* **79**, 653-661.

Wei, L.J., Lin, D.Y., and Weissfeld, L. (1989). Regression analysis of multivariate incomplete failure time data by modelling of marginal distributions. *J. Amer. Statist. Assoc.* **84**, 1065-1073.

Artificial Insemination by Donor: Discrete time survival data with crossed and nested random effects

David Clayton
René Ecochard

ABSTRACT A discrete time survival problem arising in studies of artificial insemination by donor is described. This problem involves two levels of " frailty" effect to model heterogeneity of female fecundability, together with a further two nested sets of random effects for sperm donor and donation. Parametric and non–parametric approaches to modelling such data are discussed, and computational difficulties highlighted. Attention is also drawn to the relationship of such problems to the extensive literature on generalised linear mixed models.

1 Introduction

This paper will attempt to demonstrate, via an example, that random effect or "frailty" models can be the most appropriate approach to some scientific problems, and will identify some obstacles, both theoretical and computational, to their wider use. We shall demonstrate the main manifestations of frailty in observed data and explore some techniques for estimating the heterogeneity of frailty and reconciling these manifestations. We shall also identify avenues for further methodological research.

The example is drawn from the study of human fecundability. In common with many studies in that field, it presents a problem in discrete time survival analysis, the discrete time scale being a series of ovulatory cycles at each one of which conception may take place. The dataset concerns a study of artificial insemination by donor (AID) and is unusual in that there is potentially a very complex correlation structure in the data resulting from repeated attempts to conceive by the same woman, and repeated use of sperm donors.

Attempt	Women	Cycles	Successes
1	1 901	9 740	909
2	432	1 902	244
3	90	378	54
4	13	69	6
5	1	11	0
All	2437	12 100	1 213

TABLE 1.1. The size of the AID study

The data

The data set concerns the treatments delivered between January 1985 and March 1994 at the Centre for Study and Preservation of Eggs and Sperm, Department of Biological Reproduction, HCL France. Only *couples* who have been waiting for some time for conception and in whom male infertility is the probable primary cause of sterility are accepted for AID treatment by the centre.

The data concern a hierarchically nested set of Bernoulli trials. At each *attempt*, a woman is inseminated at each of a consecutive series of ovulatory *cycles*. This continues at successive cycles until success or right censoring. No attempt was prolonged past 12 cycles in these data. There is also some left truncation since some women were midway through an attempt when data collection commenced. A total of 12 100 cycles observed in 2 437 attempts by 1 901 women (Table 1.1). There is already a multivariate aspect to this problem — we might expect correlation between waiting times to conception in different attempts by the same woman. A further source of complexity is that insemination at each cycle was linked to 1328 sperm *donations* by 279 different *donors*. This might give rise to correlations between the outcomes of insemination which use the same sperm donation or, more widely, between inseminations traced to the same donor. Since the donation and donor effects are crossed with woman effects, correlation is induced throughout the data. It should perhaps be noted that there was no systematic assignment of donor to recipient, although there is a small degree of association between donor and recipient as a result of the passage of calendar time — recipients undergoing AID earlier in the study could only receive sperm from donors who donated early in the study. However, the fact that there are repeated attempts within women and repeated donations within donor reduces this association.

A range of covariates are available, and the most important are set out below together with level at which they operate.

- *Cycle*
  - Insler score, a measure of timeliness of insemination,
  - Pharmacological stimulation of ovulation using clomiphene cit-

rate.

- *Woman*
  - Age,
  - Evidence of *azoospermia* of husband, a marker which indirectly suggests that the woman has normal fecundability.
- *Donation*
  - Sperm count,
  - Sperm mobility,
  - Sperm quality. This last measure is a composite measure taken *after* rethawing an aliquot of the donation. The other two measures were taken before freezing.

There is also a theoretical possibility of covariates at *donor* level. We only had one such candidate, donor's age, but this turned out to be unrelated to outcome. A further possibility is the use of "compositional covariates" [BR92]. For example, the mean sperm count for all donations from one donor can be entered as a covariate at donor level. The use of compositional covariates in these data will be explored elsewhere.

The analysis

Heterogeneity of fecundability of both donor and recipient induces correlation throughout the data. For the most part we shall use random effects models rather than marginal models such as GEE[LZ86] for two reasons:

1. Although it would be straightforward to devise estimating equations to allow estimation of regression models for marginal conception rates, the correlation in the data must be taken into account when estimating standard errors of parameter estimates. However in this application the correlation structure is too complicated to be taken account of in the usual ways. Since random effects are crossed, there is no repeated block structure which can be exploited in the "information sandwich" formula[Hub67] [Whi80]. For the same reason, the related approaches of the "jackknife" and "bootstrap" do not have an obvious application.

2. The magnitude of "unexplained" variance components is of at least as much interest as the covariate effects, most of which are already well understood.

We shall start, in Section 2, with a univariate discrete time survival analysis which ignores the donor effects in order to briefly review the discrete time survival data models which have been used in this field of study and to introduce some methods which we shall generalize in later sections.

Cycle	Number of Ovulations	Successes	% Successful
1	1 538	194	12.6
2	1 332	136	10.2
3	1 176	119	10.1
4	1 022	101	9.9
5	914	76	8.3
6	806	66	8.2
7	660	49	7.4
8	584	58	9.9
9	503	32	6.4
10	454	33	7.3
11	402	25	6.2
12	349	20	5.7

TABLE 1.2. Marginal hazard estimates

2 Univariate analysis

In this section we shall ignore the multivariate nature of the data and consider for the most part only the first attempt at pregnancy by each woman. The discrete hazard at cycle t, λ_t, is defined as the probability of conception at cycle t, conditional upon no conception in previous cycles. Our data is subject to both right censoring and left truncation since some women left the programme before conception occurred and some women had already partially completed a course of treatment when data collection started. Assuming that censoring and truncation are non–informative, the usual likelihood factorization allows estimation of each λ_t from the binomial likelihood obtained by considering the number of successful inseminations at the t–th cycle in relation to the total number of women observed at that cycle. These calculations are shown in Table 1.2. The clear implication of this table is that the hazard falls over time. Of course, this cannot be interpreted as suggesting that fecundability of individual women falls with repeated insemination. Biologically it is much more plausible that the fecundability of each woman remains approximately constant and that the pattern observed in Table 1.2 is attributable to heterogeneity of fecundability — the women with the highest fecundability conceive early, leaving a population of less fecund women to be observed in later cycles. Thus, the behaviour of *woman–specific* hazard may be quite different from that of *marginal* or *population–averaged* hazard.

If these were the only data available, it would not be possible to discriminate between explanations which attributed the falling hazard to heterogeneity of fecundability or to falling fecundability at the individual level. Nevertheless the biology tends to suggest the former, and a frequently used approach it to assume constant hazard for each woman. With this model, the waiting time to conception for a woman with hazard λ will follow a geometric distribution, the probability of conception at cycle t being $(1-\lambda)^{t-1}\lambda$. This is analogous to the exponential distribution of failure times in continuous time. The model goes on to relate λ to observed explanatory variables and to unobserved heterogeneity or frailty. The distribution of frailty can be modelled either parametrically or non-parametrically. The literature on frailty models for time-to-conception data has a long history. The work of Sheps[Shc64, SM73] was particularly influential. A recent review is given by Heckmann and Walker[HW90a].

The beta-geometric model

For a woman with (constant) hazard of conception λ observed from entry to the study at cycle t_0 to exit at cycle t, either due to successful conception $(d = 1)$ or censoring $(d = 0)$, is

$$\lambda^d(1-\lambda)^{t-t_0+1-d},$$

which is algebraically equivalent to that for a binomial observation. An obvious model for heterogeneity is to assume λ to take the conjugate distribution — the beta distribution with density

$$\frac{1}{\mathrm{B}(\nu,\tau)}\lambda^{\nu-1}(1-\lambda)^{\tau-1}.$$

This distribution has mean $\nu/(\nu+\tau)$ and variance

$$\frac{\nu\tau}{(\nu+\tau)^2(\nu+\tau+1)}.$$

This model was discussed by Weinberg and Gladen[WG86]. The marginal probability of survival until cycle t without conception is $\mathrm{B}(\nu,\tau+t)/\mathrm{B}(\nu,\tau)$ and the marginal hazard of conception at cycle t, μ_t say, is $\nu/(\nu+\tau+t-1)$. Thus

$$\frac{1}{\mu_t} = \frac{1}{\mu_1} + \frac{1}{\nu}(t-1)$$

so that the model implies a linear regression model for the reciprocal of the marginal hazard against cycle, t. Unfortunately further regression generalization of the model along these lines does not seem attractive.

Although the beta-geometric model has been widely used in demographic applications in which little data is available on individual women, it has been less used in applications in which covariates are available. This is probably attributable to the absence of a regression generalization capable of dealing with covariates measured both at *woman* and *cycle* level. This has led some researchers to abandon the parametric modelling approach in favour of semiparametric regression modelling of the marginal hazards. If x_{it} represents the value of the covariate vector for woman i at cycle t, such models have the general form

$$g(\mu_{t,x_{it}}) = \alpha_t + x_{it}^T\beta.$$

With logit link, $g(\mu) = \log \mu/(1-\mu)$, this leads to Cox's regression model for life tables in discrete time[Cox72, WBW94]. Note, however, that covariate "effects" in this model are expressed only in terms of contrasts of marginal hazards; they have no interpretation in terms of the probability of conception of one specific woman.

Regression models with frailty

In this section we introduce regression models whose effects are interpretable at the level of the individual woman. The main idea is to specify the model at the woman level, as a regression model involving both fixed and random effects. Random effects may be envisaged as the effects of *unobserved* covariates, and it is therefore most natural to assume that fixed and random effects act additively on the same scale. Rather than a logistic regression model we prefer a complementary log–log link function. The model relates cycle– and woman–specific hazard for woman i, λ_{it}, to possibly time dependent covariates x_{it} and random woman effects ξ_i as follows

$$\log - \log(1 - \lambda_{it}) = x_{it}^T\beta + \log(\xi_i).$$

Note that this model is multiplicative in $-\log(1-\lambda)$ so that ξ_i represents a random multiplier, or "frailty", for the i–th woman operating upon $-\log(1-\lambda) \approx \lambda$ for small λ. The ξ_i are assumed to be iid, drawn from an unknown distribution — the *frailty distribution.*

Dropping the woman subscript, i, and writing $\eta_t = x_t^T\beta$, the conditional probability of surviving until cycle t without conception given frailty ξ is

$$\prod_{u=1}^{t}(1-\lambda_u) = \exp\left(-\xi \sum_{u=1}^{t} \exp \eta_u\right).$$

The marginal survival probability is given by

$$\int \exp\left(-\xi \sum_{u=1}^{t} \exp \eta_u\right) dF(\xi) = \mathrm{M}\left(-\sum_{u=1}^{t} \exp \eta_u\right)$$

where $M()$ is the moment generating function of the frailty distribution.

The likelihood contribution for a woman observed from cycle 1 to cycle t is given by the marginal distribution function for uncensored observations

($d = 1$) or the marginal survivor function for right–censored observations ($d = 0$):

$$\mathrm{M}\left(-\sum_{u=1}^{t} \exp \eta_u\right) - d\mathrm{M}\left(-\sum_{u=1}^{t-1} \exp \eta_u\right).$$

It should be noted that this likelihood is correct only if we extend the usual definition of non–informative right censoring. For the full likelihood to factorize into a term based only upon the model for conception (the above function) and a term based only on the censoring process requires the assumption of

1. conditional independence of censoring and conception given covariates *and frailty*, and

2. independence of frailty from covariates.

In the presence of left truncation at t_0, the likelihood contribution becomes

$$\mathrm{M}\left(-\sum_{u=t_0}^{t} \exp \eta_u\right) - d\mathrm{M}\left(-\sum_{u=t_0}^{t-1} \exp \eta_u\right).$$

In deriving this expression, we have additionally assumed that left truncation and conception are conditionally independent given covariates and frailty.

This brief discussion shows that maximum likelihood estimation can be implemented quite easily for any mixed model in this class in which the frailty distribution has a closed form MGF. This closely parallels results concerning heterogeneous frailty models in continuous time survival analysis[Hou84]. The most widely used model for the frailty distribution is the gamma distribution with unit mean and variance γ, for which

$$\mathrm{M}(\phi) = (1 - \gamma\phi)^{-1/\gamma}.$$

We shall call this the *gamma–geometric* model. It closely parallels the gamma–binomial model for overdispersed binomial data[Con90], although maximum likelihood estimation is made considerably easier by the fact that no more than one "success" occurs in each cluster of trials.

In the absence of covariates the model reduces to a gamma–geometric discrete survival time distribution. This assumes that woman–specific constant hazards, λ are distributed such that $-\log(1 - \lambda)$ is a gamma variate. When the majority of the distribution extends over small values of λ, the beta–geometric and gamma–geometric models agree closely, and this is the case in the present application. Indeed they can scarcely be differentiated. In Fig 1, showing the fit of the models to the marginal hazards for first attempts (Table 1.2), the two fitted curves actually coincide. In Fig 2, showing the fitted distributions of the woman–specific hazards, they are

FIGURE 1. Fitted marginal hazards in first attempts to conceive

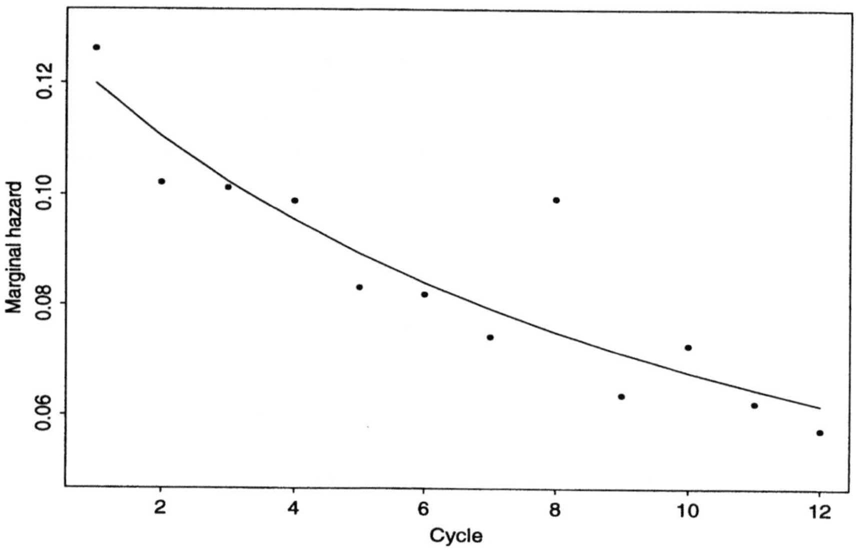

Attempt	Mean (λ)	Variance (λ)	logit(λ)
First	0.12	0.0083	0.74
Second	0.15	0.0080	0.49
Subsequent	0.18	0.0188	0.89

TABLE 1.3. Fitted means and variances for the woman–specific hazard, λ

only just discernibly different. Note that hazard, λ, is plotted on a logit scale and, on this scale, both distributions are skewed to the left. Table 1.3 shows means and variance of λ and the variance of logit(λ) for beta– and gamma–geometric models fitted to data for first, second, and subsequent attempts. The two models could not be differentiated in any of these datasets. It may be seen that women going on to further attempts, having already demonstrated their fecundability, have higher mean hazard.

3 Bivariate models

We have encountered two manifestations of heterogeneity of fecundability:

FIGURE 2. Fitted distributions for woman–specific hazards

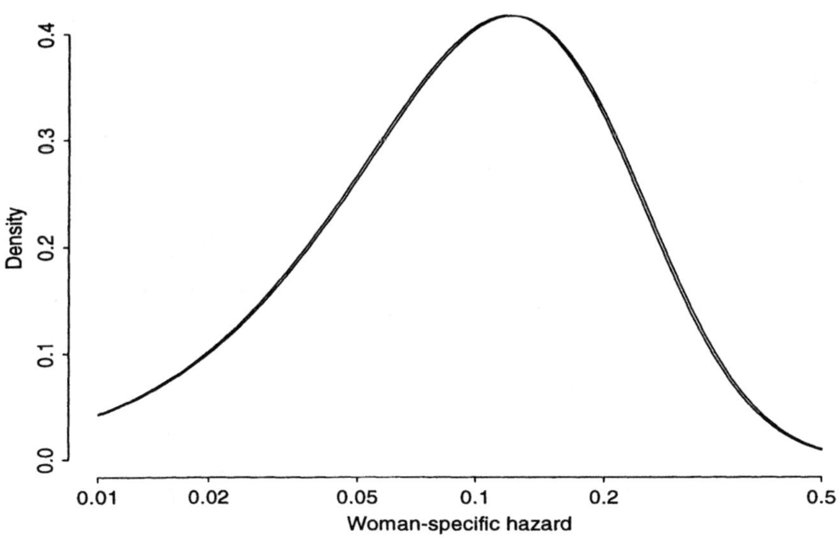

1. a decline of the marginal hazard of conception with cycle within each attempt to conceive, and
2. an increase in the marginal hazard in second and subsequent attempts.

A further consequence is *association* between waiting times to conception in different attempts.

A convenient way of studying such association follows from work on continuous bivariate survival time data[Cla78, Oak86]. Denoting the waiting time to conception in one attempt by S and the waiting time in a subsequent attempt by T, we measure association by the array of odds ratios

$$\theta_{st} = \frac{\Pr(S = s, T = t)\Pr(S > s, T > t)}{\Pr(S = s, T > t)\Pr(S > s, T = t)}$$

These are convenient statistics since θ_{st} may be consistently estimated by the sample odds ratios in the 2×2 table formed by cross–classifying all women observed at cycle s in the first attempt and at cycle t in the second attempt by whether or not they conceived at each of those cycles. Of course, there are many such statistics and there will be little data for estimation of most of them.

Random effect models provide more parsimonious models for association. The simplest such model assumes that woman–specific frailties are

Attempt:	First		Second		Subsequent	
$(s+t)$	Beta	Gamma	Beta	Gamma	Beta	Gamma
2	1.85	1.88	1.54	1.56	2.08	2.17
3	1.84	1.86	1.54	1.55	2.05	2.12
4	1.83	1.85	1.53	1.54	2.03	2.08
5	1.82	1.84	1.53	1.54	2.01	2.05
...						
10	1.80	1.81	1.51	1.51	1.96	1.97
...						
15	1.78	1.79	1.50	1.50	1.93	1.94
...						
20	1.77	1.77	1.49	1.49	1.91	1.92

TABLE 1.4. Predicted values of θ_{st} using estimates of the frailty distribution from first, second and from subsequent attempts.

	Mantel–Haenszel		Regression		
		X^2	log	SE($\log\theta$)	
Attempts	$\hat{\theta}$	(1 df)	$\hat{\theta}$	Naive	Huber
1/2	1.261	7.53	0.234	0.085	0.087
1/3	1.871	10.71	0.670	0.200	0.185
2/3	1.244	1.30	0.230	0.195	0.200
All	1.329	16.23	0.287	0.071	0.074

TABLE 1.5. Empirical estimates of θ

unchanged from one attempt to the next. With this assumption, and for small λ, the beta– and gamma–geometric models approximate the bivariate continuous model proposed by Clayton[Cla78, Oak82], and the θ_{st} are approximately constant and equal to one plus the variance of logit(λ). For example, Table 1.4 shows the values of θ_{st} suggested by the beta and gamma frailty distributions described in Table 1.3. Note that θ_{st} depends only on the sum $(s+t)$. For practical purposes, θ_{st} may be regarded as constant in these models.

Empirical estimates of a constant θ may be obtained using the method proposed by Clayton (1978). If s and t can take values from 1 to 12, each pair of attempts contributes 144 different 2×2 tables in which the odds ratios are (approximately) constant. The data from these tables may be combined to give a single estimate of the common odds ratio using the Mantel–Haenszel procedure. These estimates are shown in the second column of Table 1.5. The third column of the table shows the conventional Mantel–Haenszel one degree of freedom test of the null hypothesis $\theta = 1$. Despite the fact that one woman could contribute to all 144 tables this test remains valid — a result which follows from the work of Irwin and Lancaster

[Irw49, Lan49] who showed, in the absence of censoring or truncation, that the $(O - E)$ values for diferent 2×2 tables are uncorrelated.

This quasi–independence of tables cannot be assumed when the null hypothesis does not hold, so a little more trouble must be taken in calculating confidence intervals for θ. Jackknife or bootstrap methods could be used, but a simpler possibility is to estimate the odds ratios by logistic regression, using Huber's formula (the "sandwich estimate") to estimate standard errors allowing for the fact that the data are clustered, with many observations deriving from the same woman. These results are shown in columns four to six of Table 1.5.

Since we estimate the variance of logit(λ) as 0.49 for women who have at least two attempts to conceive and 0.89 for women who go on to have three or more attempts, the beta– and gamma–geometric models would lead us to expect values for θ of 1.5, 1.9 and 1.9 for the 1/2, 1/3, and 2/3 associations respectively. Our observed values are rather smaller than this, but the confidence intervals are wide. However it remains possible that the model is over–simplistic or that the distributions of frailties are misspecified.

4 Non–parametric mixture models

An alternative approach to parametric frailty models is non–parametric modelling of the frailty distribution. In general this approach is attractive in that it allows for a bimodal frailty distribution caused by a group of women with very low fecundability due to pathological conditions. Although such allowance must frequently be made in fecundability studies, we did not expect to see this here since clinicians would be unlikely to refer such women for AID.

It is well known that the non–parametric maximum likelihood estimate of the distribution of random effects in such models collapses onto a discrete set of mass points, or latent classes. A slow but easily programmed computational method for fitting such models model is the EM algorithm[HW90b].

To illustrate this approach, a simple heterogeneity model was fitted to the data for first attempts. We used GLIM macros of Aitkin and Francis[AF95], and the results are summarized in Table 1.6 and Fig 3. The convergence difficulties of the EM algorithm are immediately apparent since the deviance did not decrease uniformly with the number of masses. Indeed there was little to choose between any of the "solutions" with three or more masses. Given sufficient computer time, it is highly likely that convergence would be obtained to a three or four mass solution.

This analysis gives a clear impression that the frailty distribution is fairly symmetric on the logistic scale with a variance of around 0.7. This agrees quite closely with the variance estimate of 0.74 obtained from the beta– and

FIGURE 3. Non–parametric estimates of the distribution of hazard (on a logistic scale)

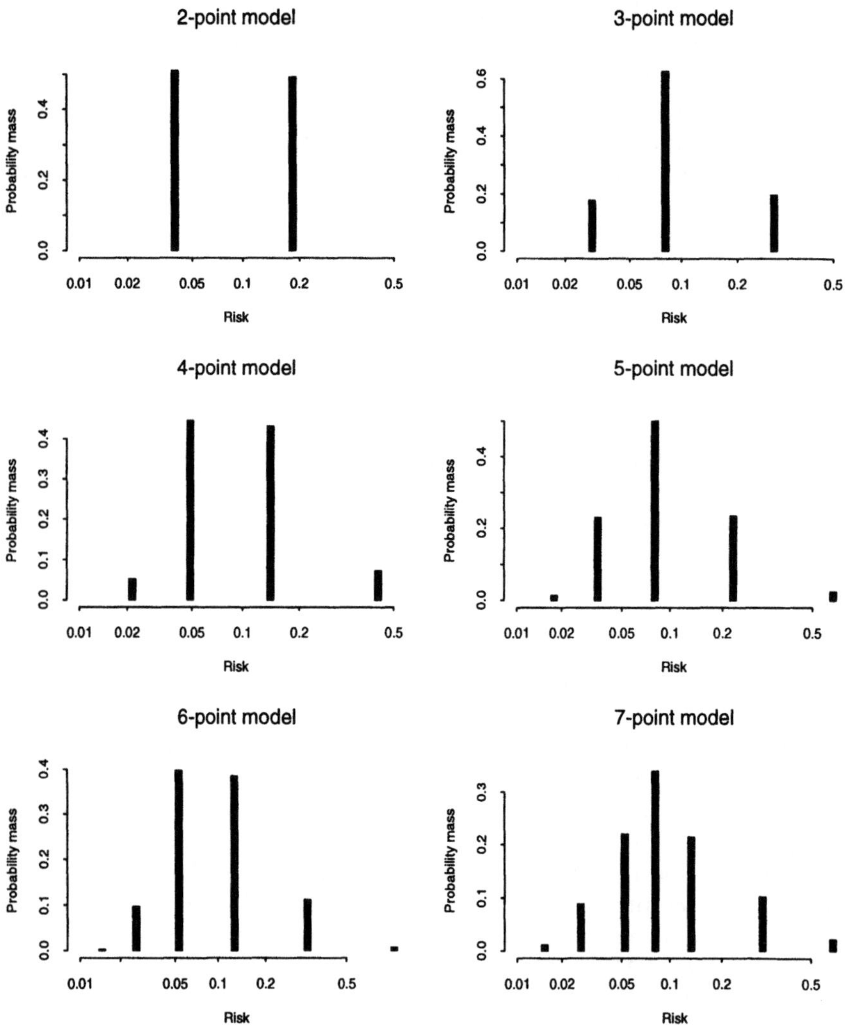

Number of masses	Deviance	Variance logit(λ)
1	6042.06	0.0
2	6016.31	0.735
3	6015.52	0.671
4	6015.09	0.737
5	6015.16	0.725
6	6015.18	0.735
7	6015.28	0.701

TABLE 1.6. Fitting non–parametric frailty distributions to the first attempt data

$(s+t)$	θ_{st}
2	2.33
3	2.05
4	1.86
5	1.71
...	
10	1.39
...	
15	1.31
...	
20	1.29

TABLE 1.7. Association predicted by the 7–point model

gamma–geometric models. As expected, there is no suggestion of an outlying infertile group. The predicted relationship between marginal hazard and cycle is shown in Fig 4. The 7–point model fitted to the first attempt data predicts a rather different pattern of association than the corresponding parametric frailty model (Table 1.7); association is stronger for shorter waiting times. The fact that here the pattern of association differs somewhat from the constant θ pattern exhibited in the beta– and gamma–frailty models is attributable to the different distribution shape. On the logit scale the non–parametric estimate of the distribution is slightly skewed to the right, while Fig 2 shows that the fitted beta and gamma distributions had a pronounced skew to the left. Thus, we might expect θ_{st} to decrease with $(s+t)$ in our data, but extending the logistic regression procedure for estimating θ_{st} yielded no evidence of such a trend. However, as stated earlier, the data for estimating patterns of association are sparse, even in this large dataset.

FIGURE 4. Marginal hazards predicted by non–parametric mixture models

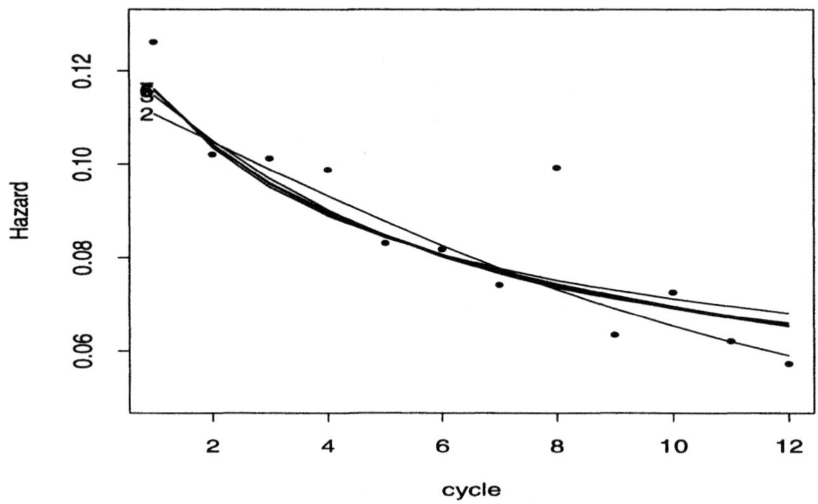

5 Generalized linear mixed models

As we have seen, a potential problem with the simple frailty models discussed above is the fact that the model accounts for two rather different aspects of the observed data — the fall of hazard with cycle within each attempt to conceive, and the association between waiting times to conception at different attempts. These two aspects of the data may be mutually consistent, but they need not be. In particular, "frailties" in first and second attempts may not be the same, although they will probably be highly correlated. This can be dealt with by extended frailty models in which frailty is decomposed into two random effects — one specific to the woman and remaining unchanged through all attempts to conceive, and another effect specific to each attempt. Such random effects can be thought of as modelling two types of unobserved covariates,

1. covariates that are constant (or at least very slowly varying) within each woman, and
2. covariates which vary from one attempt to the next.

In principle one might also envisage unobserved covariates which vary from cycle to cycle within attempts, but since each cycle is a Bernoulli trial with only two possible outcomes, any attempt to model such influences will lead to problems of identifiability.

Multiplicative models with two or more random effects are not easily dealt with. For example, beta and gamma frailty models do not readily extend since products of beta or gamma variates do not, in general, have tractable distributions. Non–parametric frailty models are more promising, but here it is necessary to incorporate constraints on the location of the random effect distributions for identifiability — not a straightforward matter. A third possibility in continuous time survival models, is to generate a multivariate gamma frailty distribution by *addition* of gamma–distributed frailty components. The resultant frailty then acts *multiplicatively* with observed covariate effects upon the hazard[YI95]. However, viewing frailty models as models for the effects of unobserved covariates, models in which fixed and random effects act additively on different scales are not attractive.

A further possibility is to use mixed models in which Gaussian random effects act additively with fixed effects on a linear predictor scale. A link function then relates the linear predictor to the woman/attempt specific hazard of conception. For example, using the logit link, the hazard for woman i at cycle t of her j–th attempt to conceive is given by

$$\log \frac{\lambda_{ijt}}{1-\lambda_{ijt}} = x_{ijt}^T\beta + \xi_i + \epsilon_{ij}$$

where ξ_i, ϵ_{ij} are between– and within–woman random effects. This is a *generalized linear mixed model* (GLMM). Maximum likelihood estimation is not tractable, but approximate methods are available. The simplest approximations use conventional mixed model computations within an iterative scheme in which, at each step, working weights and "y–values" are computed given the observed responses and current fitted values[BC93]. This is easily implemented using available mixed model software. Two variants of the iteration are available

1. MQL, which fits an approximate marginal model using quasi likelihood, and

2. PQL which fits the random effect model by maximizing a *penalized likelihood* in which the (log) random effect distributions enter the log likelihood as penalty terms. This has also been termed a *hierarchical* or *h*–likelihood by Lee and Nelder[LN96].

In both approaches, variance components are estimated using a REML correction to a pseudo–likelihood ([CR82]). The MQL approach is not suitable for survival data; the fact that failure curtails further study means that the correct model for marginal hazards is very different from the approximate model obtained by simply omitting the random terms from the conditional model. PQL, however, may be a poor approximation when, as here, little information is available concerning each random effect In such situations PQL can provide a preliminary analysis, but should be backed up by more accurate approaches, such as Markov Chain Monte

Carlo methods[ZK91, Cla94]. The use of this method for frailty models for survival data in continuous time have been discussed by McGilchrist and Aisbett[MA91, McG93]. Applying the PQL approximation in the data for first attempts with no covariates gave an estimate for the variance of logit(λ) of 0.432. This compares with estimates of between 0.7 and 0.75 obtained from the beta/gamma–geometric and non–parametric approaches. This discrepancy is probably attributable to inaccuracy of the PQL approximation.

An efficient computational method for fitting nested random effect models has been described by Goldstein[Gol86] and is implemented in the computer program *MLn*. Macros are available which implement PQL. Using this program, again without covariates, we obtained estimates of 0.522 and zero for the between– and within–woman random effects respectively. It would seem that frailty operates at the level of *woman* rather than at the level of *attempt*. The extension of this analysis to include covariate effects will be discussed after we have considered the problem of donor heterogeneity.

Before going on to consider more complicated analysis, some discussion is necessary concerning the assumptions we must make concerning missing data mechanisms. Here it is perhaps more appropriate to address this problem in relation to the *presence* of data rather than its absence, since only some women have more than one attempt to conceive. For analyses based on maximum likelihood, it is only necessary that we assume presence of data is *ignorable* in the sense of Little and Rubin[LR87]. That is, the presence of data concerning conception may depend on any *observed* data, including covariates and past conception history. Under these circumstances, the likelihood factorizes and inference concerning the model for conception is unaffected. Note that the effect of women becoming discouraged from further attempts by long waiting times to conception would be ignorable in this sense.

For models involving random effects, the likelihood factorization also requires that presence of data is unrelated to the unobserved values of random effects. If, for example, there is some relationship between the tendency to repeat attempts at conception and *fecundability* itself (as distinct from past conception history), then the missing data mechanism would no longer be ignorable. This might occur if there is a relationship between fecundability and the outcome of pregnancy. If less fecund women tended to experience greater rates of miscarriage, then this would introduce a selection bias in which the distribution of fecundability in later attempts would differ from that predicted by the model. To investigate this we related outcome to the cycle at which conception took place (Table 1.8). There is some suggestion that miscarriage or still birth is more likely when the waiting time to conception is longer, but most pregnancies have a favourable outcome and the main reason for repeated attempts is to achieve a larger family size. It is unlikely, therefore, that selection will substantially affect the analysis.

Cycle conceived	Miscarriage N	%	Live birth N	%	Other N
1	28	10.8	230	88.5	2
2–3	38	12.2	271	87.1	2
4–6	49	18.4	215	80.8	2
7 +	36	15.7	189	82.5	4

"Miscarriage" includes a small number (7) of stillbirths. The "other" category is made up of fallopian conceptions and one elective termination.

TABLE 1.8. Pregnancy outcome by cycle of conception

6 Donor heterogeneity

The most elementary analysis demonstrates that there is considerable heterogeneity of *donors*. Fig 5 plots the number of successful fertilizations against the number of attempts using sperm from each donor. The success rates varied considerably. However, it is not clear whether the bulk of the heterogeneity is at the *donor* or the *donation* level, or the extent to which it is accounted for by covariates — in particular by the measures of sperm quality.

We first fitted a mixed logit regression model. The model was marginal with respect to women, so that the fall in conception rate with increasing cycle and the increased rate in second and subsequent attempts — both manifestations of heterogeneity of fecundability — were modelled by fixed effects. We included random effects at both donor and donation level. Using the penalized/pseudo– likelihood approximation these variance components were estimated at 0.377 and 0.002 respectively, thus indicating that heterogeneity is predominantly at the donor level. Interestingly the between–donor variance component remained substantial even after inclusion of the sperm quality measurements as covariates.

Further insight into donor hetereogeneity is provided by a non–parametric random effect analysis. Fig 6 shows maximum likelihood estimates of the distribution of donor–specific conception probabilities (on the logit scale). As in the previous analysis, the effects of heterogeneity of women are modelled as fixed effects of cycle and attempt; the figure shows the fitted distributions of random donor effects in terms of the marginal conception probabilities for inseminations at cycle 1 and attempt 1. In this case there is some evidence for there being two distinct populations of donors. The estimates of between–donor variance (again on the logit scale) were substantially larger in these analyses — around 1.0 rather than the 0.4 found in the PQL analysis. Here the discrepancy between the variance estimate from the non–parametric model and the PQL analysis is rather more surprising than in the the case of heterogeneity between women. Since there is much more data on each donor than there is on each woman, the PQL approximation should be better in this case. Perhaps it is the suggestion

FIGURE 5. Donor heterogeneity

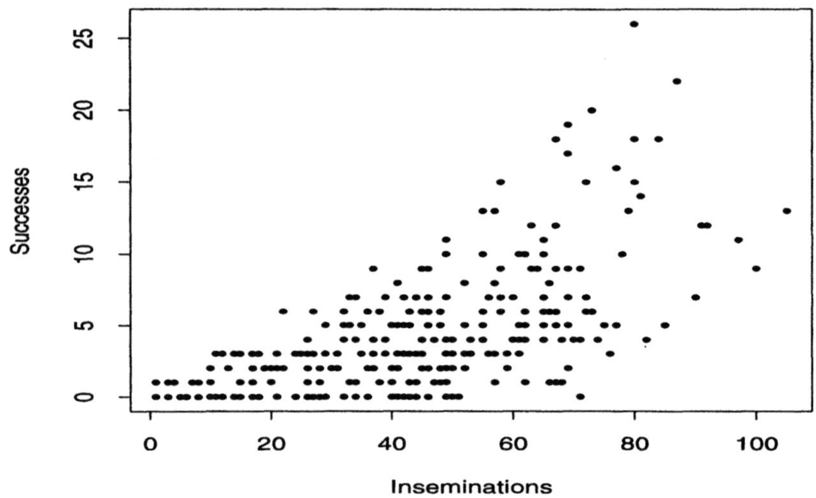

of non–normality of the donor effects that accounts for the discrepancy.

7 The full model: crossed and nested random effects

A comprehensive analysis of these data could be achieved by fitting, by maximum likelihood, a mixed model for the probability of conception occurring at each insemination. The random effects are crossed and nested — *donation* within *donor* crossed with *attempt* within *woman*. Preferably we would wish to avoid parametric assumptions concerning the distributions of random effects.

Non–parametric random effect models of such complexity present difficulties, owing to the need for constraints to preserve identifiability. Parametric models of such complexity present difficulties in that the likelihoods tend to be intractable. In principle, the PQL approximation can be used to fit the crossed model with all four random effects, using a mixed model program such as *SAS PROC MIXED* at the heart of the iteratively reweighted least squares algorithm. In practice, however, even this is not directly feasible since, for a problem of this size the mixed model equations are extremely computer intensive.

FIGURE 6. Non-parametric model for donor heterogeneity

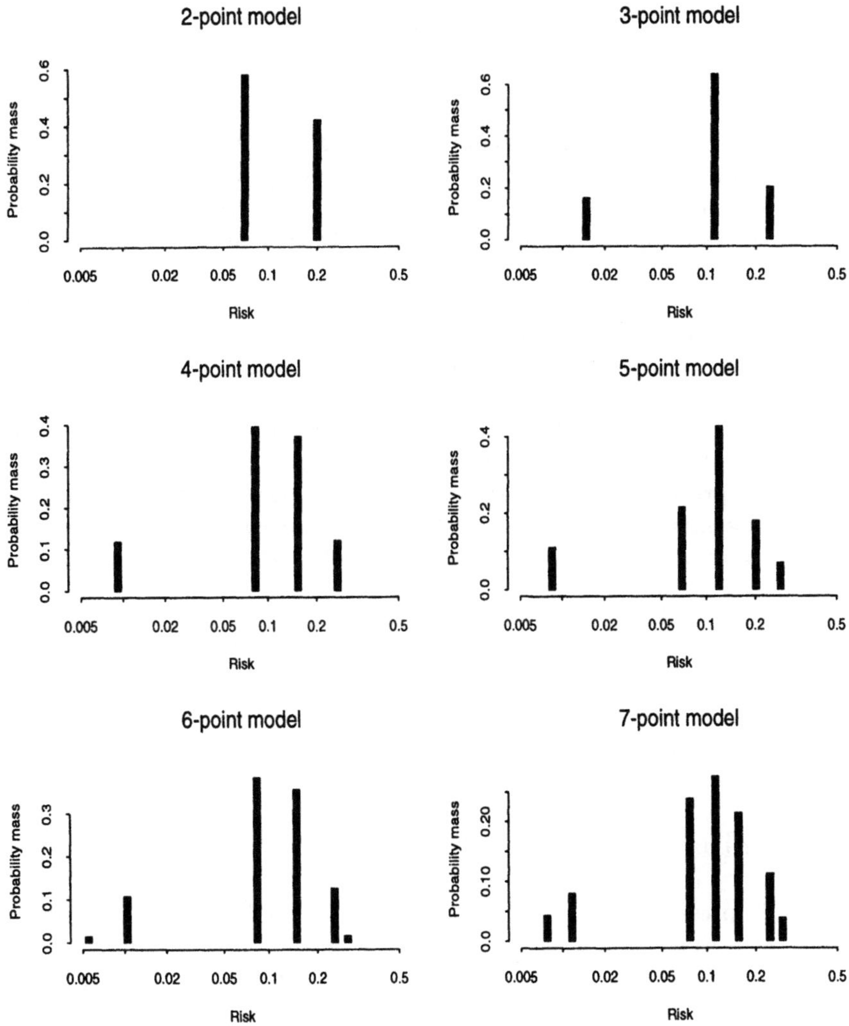

We have found that a rapid implementation of PQL in this situation is to use an efficient implementation of nested random effects (*MLn*) in an "alternating EM algorithm":

1. fit fixed effects and random *donor/donation* effects, while including empirical Bayes estimates of *woman* and *attempt* effects as constant "offsets" in the linear predictor, and

2. fit fixed effects and random *woman/attempt* random effects, this time including empirical Bayes estimates of *donor* and *donation* random effects as offsets.

Since there is a considerable degree of orthogonality of the two crossed sets of effects, convergence of this algorithm is very rapid. Unfortunately, though, estimates of standard errors are incorrect and tests and confidence intervals for parameters of interest must be based upon conmparisons of the approximate likelihood[LN96].

The estimates of the variance components for attempts within women and for donations within donor were zero, since the approximate likelihood was monotone with respect to both these parameters. Before inclusion of covariates, the variances for women and donors were 0.446 and 0.376 respectively. The covariate effects are shown in Table 1.9. As explained above, these standard errors are slightly incorrect but are presented to give some idea of the precision of estimation. The table also shows that, the unexplained variance attributable to donors is reduced somewhat, but the variance component for women is not — indeed, paradoxically, it is slightly increased!

We have tested the accuracy of this PQL analysis using an MCMC method — specifically Gibbs sampling. The results are broadly comparable. In particular the suggestion that *attempt* and *donation* random effects are negligible is supported, even though the Markov chains for these variance component parameters show considerable serial correlation.

8 Conclusions and further work

Our aim in presenting these data at this symposium has been primarily to demonstrate that random effect (frailty) models have important scientific applications, and that they are capable of providing insights which would not be available from approaches based upon marginal models. For example, the fact that considerable donor heterogeneity exists, even after we have taken account of measured sperm quality, is of some scientific interest and clinical importance.

This example also amply demonstrates several outstanding problems and difficulties. PQL provides a useful and practical way of carrying out preliminary data analysis and is easily implemented in readily available software

Parameter	Estimate	SE
Intercept	-2.303	0.038
Woman:		
Age (woman)	-0.106	0.036
Azoospermia (husband)	0.080	0.037
Cycle:		
Insler score	0.264	0.039
Early insemination	-0.137	0.038
Late insemination	-0.082	0.034
Stimulation of ovulation	-0.103	0.036
Donation:		
Sperm count	0.130	0.030
Sperm mobility	0.179	0.034
Sperm quality	0.217	0.036
Heterogeneity:		
Between women	0.500	0.072
Between donors	0.222	0.043

(Coefficients for metric covariates have been standardized and represent the effect of one standard deviation change in the variable.)

TABLE 1.9. Standardized regression coefficients

packages. However, it is not very accurate and requires validation with more accurate but computer intensive methods such as MCMC. The assumption of parametric distributions of random effects would also be better avoided if possible, and there is a need for further work on non-parametric random effect models involving crossed and nested random effects.

Acknowledgements

We would like to thank Prof J.C. Czyba and D. Cottinet for providing the data. The work of the first author was partially supported by NIH/NCI grant R01/CA61042.

9 References

[AF95] M. Aitkin and B.J. Francis. GLIM4 macros for nonparametric maximum likelihood estimation with overdispersion and variance component models in the exponential family. *The GLIM Newsletter (submitted)*, 1995.

[BC93] N.E. Breslow and D. Clayton. Approximate inference in generalized linear mixed models. *Journal of the American Statistical Association*, 88:9–25, 1993.

[BR92] A.S. Bryke and S.W. Raudenbush. *Hierarchical Linear Models: Applications and Data Analysis Methods.* Sage publications, Newbury Park, 1992.

[Cla78] D. Clayton. A model for association in bivariate life tables and its application in epidemiological studies of familial tendency in chronic disease incidence. *Biometrika*, 65:141–151, 1978.

[Cla94] D. Clayton. Generalized linear mixed models. In W.R. Gilks, S. Richardson, and D. Spiegelhalter, editors, *Markov Chain Monte Carlo in Practice.* Chapman and Hall, London, 1994.

[Con90] M.R. Conoway. A random effects model for binary data. *Biometrics*, 46:317–328, 1990.

[Cox72] D.R. Cox. Regression models and life tables (with discussion). *Journal of the Royal Statistical Society, Series B*, 34:187–220, 1972.

[CR82] R.J. Carroll and D. Ruppert. Robust estimation in heteroscedastic linear models. *Annals of Statistics*, 10:429–441, 1982.

[Gol86] H. Goldstein. Multilevel mixed linear model analysis using iterative generalized least squares. *Biometrika*, 73:43–56, 1986.

[Hou84] P. Hougaard. Life table methods for heterogeneous populations: distributions describing the heterogeneity. *Biometrika*, 71:75–83, 1984.

[Hub67] P.J. Huber. The behaviour of maximum likelihood estimates under non– standard conditions. In *Proceedings of the Fifth Berkely Symposium on Mathematical Statistics and Probability, I*, pages 221–223. 1967.

[HW90a] J.J. Heckmann and J.R. Walker. Estimating fecundability from data on waiting times to first conception. *Journal of the American Statistical Association*, 85:283–295, 1990.

[HW90b] J.P. Hinde and A.T.A. Wood. Binomial variance component models with a non–parametric assumption concerning random effects. In R. Crouchley, editor, *Longitudinal data analysis*. Avebury, Aldershot, Hants, 1990.

[Irw49] J.O. Irwin. A note on the subdivision of χ^2 into componenents. *Biometrika*, 36:130, 1949.

[Lan49] H.O. Lancaster. The derivation and partition of χ^2 in certain discrete distributions. *Biometrika*, 36:117, 1949.

[LN96] Y. Lee and J.A. Nelder. Hierarchical generalized linear models. *Journal of the Royal Statistical Society, Series B*, 58:?, 196.

[LR87] J.A. Little and D.B. Rubin. *Statistical Analysis with Missing Data.* Wiley Series in Probability and Mathematical Statistics. Wiley, New York, 1987.

[LZ86] K.Y. Liang and S.L. Zeger. Longitudinal data analysis using generalized linear models. *Biometrika*, 73:13–22, 1986.

[MA91] C.A. McGilchrist and C.W. Aisbett. Regression with frailty in survival analysis. *Biometrics*, 47:461–466, 1991.

[McG93] C.A. McGilchrist. REML estimation for survival models with frailty. *Biometrics*, 49:221–225, 1993.

[Oak82] D. Oakes. A model for association in bivariate survival data. *Journal of the Royal Statistical Society, Series B*, 44:414–422, 1982.

[Oak86] D. Oakes. Semi–parametric inference in a model for association in bivariate survival time data. *Biometrika*, 73:353–361, 1986.

[She64] M.C. Sheps. On the time required for conception. *Population Studies*, 17:85–97, 1964.

[SM73] M.C. Sheps and J.A. Menken. *Mathematical models of conception and birth.* University of Chicago Press, Chicago, 1973.

[WBW94] C.R. Weinberg, D.D. Baird, and A.J. Wilcox. Sources of bias in studies of time to pregnancy. *Statistics in Medicine*, 13:671–681, 1994.

[WG86] C.R. Weinberg and B.C. Gladen. The beta-geometric distribution applied to comparatitive fecundability studies. *Biometrics*, 42:547–560, 1986.

[Whi80] H. White. A heteroskedasticity–consistent covariance matrix estimate and a direct test for heteroskedasticity. *Econometrika*, 48:817–830, 1980.

[YI95] A.I. Yashin and I.A. Iachine. Genetic analysis of durations: correlated frailty model applied to survival of Danish twins. *Genetic Epidemiology*, 12:529–538, 1995.

[ZK91] S.L. Zeger and M.R. Karim. Generalized linear models with random effects; a gibbs sampling approach. *Journal of the American Statistical Association*, 86:79–86, 1991.

Interval Censored Survival Data: A Review of Recent Progress

Jian Huang
Jon A. Wellner

ABSTRACT We review estimation in interval censoring models, including nonparametric estimation of a distribution function and estimation of regression models. In the nonparametric setting, we describe computational procedures and asymptotic properties of the nonparametric maximum likelihood estimators. In the regression setting, we focus on the proportional hazards, the proportional odds and the accelerated failure time semiparametric regression models. Particular emphasis is given to calculation of the Fisher information for the regression parameters. We also discuss computation of the regression parameter estimators via profile likelihood or maximization of the semiparametric likelihood, distributional results for the maximum likelihood estimators, and estimation of (asymptotic) variances. Some further problems and open questions are also reviewed.

1. Introduction: interval censoring models.

Interval censored data arises when a failure time T can not be observed, but can only be determined to lie in an interval obtained from a sequence of examination times. Kongerud and Samuelsen (1991) and Samuelsen and Kongerud (1993) report two studies on respiratory symptoms and asthmatic symptoms among Norwegian aluminum workers. In these studies, the time to the development of respiratory symptoms or asthmatic symptoms is only known to be between two health examinations. Other examples of interval censored data in animal carcinogenicity and epidemiology studies can be found in Hoel and Walburg (1972), Finkelstein and Wolfe (1985), Finkelstein (1986), and Self and Grossman (1986). Diamond, McDonald and Shah (1986), and Diamond and McDonald (1991) contain examples of interval censored data from demography studies. Closely related censoring schemes also arise in AIDS studies, see for example, De Gruttola and Lagakos (1989), Shiboski and Jewell (1992), Jewell, Malani, and Vittinghoff (1994), etc.

We now briefly describe the three types of interval-censored data considered in this review. Let T (or T_i) be the unobservable failure time.

1.1. "Case 1" interval censoring or current status data.

Suppose that U is an "examination" or "observation" time. Then suppose that an observation consists of the random vector (δ, U) where $\delta = 1_{[T \le U]}$, or (δ, U, Z) if a vector Z of covariates is also available. The only knowledge about the "failure time" T is whether it has occurred before U or not. Such data is substantially different from right-censored data. In a right censorship model, the observed data is $(\min(T,Y), 1_{\{T \le Y\}}, Z)$ where Y is a "censoring time" and the probability $P\{T \le Y\}$ of observing the survival time T exactly is positive. But with current status data, we are not able to observe the exact value of the survival time at all, just $1_{\{T \le Y\}}$ (or $1_{\{T \le U\}}$).

The earliest work on nonparametric likelihood estimation (NPMLE) with current status data goes back to Ayer, Brunk, Ewing, Reid and Silverman (1955) and Van Eeden (1956, 1957). These authors introduced the pool-adjacent-violators algorithm to compute the NPMLE of a distribution function. Groeneboom (1987) established asymptotic properties of the NPMLE. See also Groeneboom and Wellner (1992) and Huang and Wellner (1995a).

In a semiparametric setting, Huang (1994, 1995, 1996) showed that the maximum likelihood estimator (MLE) for the regression parameter of the proportional hazards or proportional odds regression model with "case 1" interval censoring is asymptotically normal and efficient. Rossini and Tsiatis (1996) proved asymptotic normality and efficiency of sieve estimators for the proportional odds regression model.

There is an enormous amount of literature in econometrics on the binary choice model. This model can be thought as a linear regression model with unknown error distribution under "case 1" interval censoring. The first consistent estimator, the maximum score estimator, was introduced by Manski (1975, 1985). Kim and Pollard (1990) derived its asymptotic distribution. See also Cosslett (1983, 1987), Chamberlain (1986), Han (1987), Horowitz (1992), and Klein and Spady (1993) for related work. In particular, the estimator proposed by Klein and Spady (1993) is asymptotically efficient under appropriate conditions.

1.2. "Case 2" and "case k" interval censoring.

With "case 2" interval censored data, we only know that T has occurred either within some random time interval, or before the left end point of the time interval, or after the right end point of the time interval. More precisely, suppose that there are two examination (or observation) times U and V, the data observed is:

$$(\delta_1, \delta_2, U, V, Z) = (1_{\{T \le U\}}, 1_{\{U < T \le V\}}, U, V, Z).$$

"Case k" interval censoring arises when there are k examination times per subject. This is a generalization of "case 2" interval censoring; see e.g.

Wellner (1996).

Estimation of the proportional hazards model (Cox, 1972) and the proportional odds regression model was considered by Huang and Wellner (1995b) and Huang and Rossini (1995), respectively. They showed that the MLEs of the regression parameters in both models are asymptotically normal and efficient, even though the MLEs of the baseline cumulative hazard function or odds function only have $n^{1/3}$-rates of convergence.

1.3. A general interval censoring scheme.

Suppose that

$$0 < Y_{i,1} < Y_{i,2} < \cdots < Y_{i,n_i} < \infty$$

are ordered examination times for the ith patient, $i = 1, \ldots, n$. Denote $Y_i = (Y_{i,1}, Y_{i,2}, \cdots, Y_{i,n_i})$. Let T_i be the ith patient's unobservable failure time. Computationally, it is convenient to reduce the general interval censoring to "case 2" interval censoring by considering three possibilities: (i) the failure occurred before the first examination time. Denote $U_i = Y_{i,1}$ and let $V_i = Y_{i,2}$. Let $\delta_{1i} = 1_{[T_i \le U_i]}$ and $\delta_{2i} = 1_{[U_i < T_i \le V_i]}$. Then $\delta_{1i} = 1$ and $\delta_{2i} = 0$. (ii) T_i is known to be bracketed between a pair of examination times $(Y_{i,L}, Y_{i,R})$, where $Y_{i,L}$ is the last examination time preceding T_i and $Y_{i,R}$ is the first examination time following T_i. Denote $U_i = Y_{i,L}$ and $V_i = Y_{i,R}$. Define δ_{1i} and δ_{2i} as in (i). Then $\delta_{1i} = 0$ and $\delta_{2i} = 1$. (iii) At the last examination, the failure did not occur. Then $\delta_{1i} = 0$ and $\delta_{2i} = 0$. The effective observations are

$$(\delta_{1i}, \delta_{2i}, U_i, V_i), \quad i = 1, \ldots, n.$$

Turnbull (1976) derived self-consistency equations for a very general censoring scheme which includes interval censoring as a special case. This yields an EM algorithm for computing the NPMLE. Groeneboom's Iterative Convex Minorant algorithm can be used for computing the NPMLE. As suggested in Groeneboom and Wellner (1992), the Iterative Convex Minorant (ICM) algorithm is considerably faster than the EM algorithm, especially when the sample size is large. Finkelstein (1986) and Rabinowitz, Tsiatis and Aragon (1995) considered estimation in the proportional hazards model (Cox, 1972), and in the linear regression model, with general interval censoring, respectively. Large sample properties of their estimators are still unknown.

Computationally, the general interval censoring scheme can be reduced to "case 2" interval censoring. The estimation approach described in section 4 for "case 2" interval censoring works for general interval censoring. The distributional results do *not* carry over to the general case although they can be easily extended to "case k" interval censoring.

The organization of this paper is as follows. In section 2, computation and distributional results for the nonparametric maximum likelihood estimator $\widehat{F}_n$ (NPMLE) of a distribution function with interval censored are discussed. Section 3 contains recent results on maximum likelihood estimation of three widely used regression models (the proportional hazards, the proportional odds, and the accelerated failure time models) with "case 1" interval censored data. Section 4 contains results on estimation in these three regression models with "case 2" interval censored data We emphasize that, although the distributional results do not carry over to general interval censored data, the estimation procedures *do* carry over. In this section, consistency of the maximum likelihood estimators and information calculation in the accelerated failure time model are new results as far as we know. Section 5 contains brief discussions on further problems. Some proofs, including most of the information calculations, are included in section 6. Throughout, we emphasize the importance of information calculations in the (semiparametric) regression models we consider. These calculations not only provide benchmarks for comparison of estimators, but also help to understand the structure of the models which can facilitate proofs of the distributional results for maximum likelihood estimators and can lead to natural families of estimating equations.

2. Nonparametric likelihood estimation.

We assume that the examination times are independent of the failure time and that their distribution is independent of the distribution function of the failure time. With these conditions, the joint densities and the likelihood functions are:

(1) "Case 1" interval censoring: the joint density of a single observation $X = (\delta, U)$ is

$$p(x) = F(u)^{\delta}(1 - F(u))^{1-\delta}h(u),$$

where $h(u)$ is the density of U. The log-likelihood function of a random sample of size n is (up to an additive term not involving F)

$$l_n(F) = \sum_{i=1}^{n}\{\delta_i \log F(U_i) + (1 - \delta_i)\log(1 - F(U_i)\}.$$

(2) "Case 2", and general interval censoring: the joint density of a single observation $X = (\delta_1, \delta_2, U, V)$ of a random sample of size n is

$$p(x) = F(u)^{\delta_1}[F(v) - F(u)]^{\delta_2}(1 - F(v))^{\delta_3}h(u, v), \tag{2.1}$$

where $\delta_3 = 1 - \delta_1 - \delta_2$ and $h(u, v)$ is the joint density of (U, V). The log-likelihood function is

$$l_n(F) = \sum_{i=1}^{n}\{\delta_{1i}\log F(U_i) + \delta_{2i}\log(F(V_i) - F(U_i)) + \delta_{3i}\log(1 - F(V_i))\},$$

where $\delta_{3i} = 1 - \delta_{1i} - \delta_{2i}$. Notice that for general interval censoring, the meaning of $(\delta_{1i}, \delta_{2i}, U_i, V_i)$ is described in section 1. Since the likelihood functions $l_n(F)$ in both (1) and (2) depend on F only through its values at the observation times U_i (or (U_i, V_i) in the case of (2)), our convention is that the NPMLE $\widehat{F}_n$ of F in either (1) or (2) is a piecewise constant function with jumps only at the observation points. Although "case 1" interval censoring can be regarded as a special case of "case 2" interval censoring by taking $U_i = 0$ (or $V_i = \infty$), maximization of $l_n(F)$ given in (2) is substantially more difficult.

2.1. Computation of $\widehat{F}_n$.

For "case 1" interval-censored data, the NPMLE $\widehat{F}_n$ that maximizes $l_n(F)$ of (1) can be obtained as follows:

(i) Order the examination times: $U_{(1)} \le U_{(2)} \le \dots \le U_{(n)}$ and relabel δ_i accordingly to obtain $\delta_{(1)}, \dots, \delta_{(n)}$.

(ii) Plot $(i, \sum_{j=1}^{i} \delta_{(j)})$, $i = 1, \dots, n$.

(iii) Form the Greatest Convex Minorant (GCM) G^* of the points in step (ii).

(iv) $\widehat{F}_n(U_{(i)}) =$ left-derivative of G^* at i, $i = 1, \dots, n$.

Equivalently, $\widehat{F}_n$ can be expressed by the max-min formula:

$$\widehat{F}_n(U_{(i)}) = \max_{j \le i} \min_{k \ge i} \frac{\sum_{m=j}^{k} \delta_{(m)}}{k - j + 1}. \tag{2.2}$$

However, for "case 2" or the general interval censoring, there is no closed form expression for $\widehat{F}_n$. We describe the iterative convex minorant algorithm and the EM algorithm for computing $\widehat{F}_n$. Let P_n denote the empirical measure of a sample of size n from the "case 2" density (2.1);

$$P_n = \frac{1}{n} \sum_{i=1}^{n} \delta_{(\underline{\delta}_i, U_i, V_i)}.$$

For any distribution function F and $t \ge 0$, denote

$$\begin{aligned} W_F(t) &= \int_{u \le t} \left\{ \frac{\delta_1}{F(u)} - \frac{\delta_2}{F(v) - F(u)} \right\} d\,P_n \\ &\quad + \int_{v \le t} \left\{ \frac{\delta_2}{F(v) - F(u)} - \frac{\delta_3}{1 - F(v)} \right\} d\,P_n, \end{aligned} \tag{2.3}$$

$$\begin{aligned} G_F(t) &= \int_{u \le t} \left\{ \frac{\delta_1}{F(u)^2} + \frac{\delta_2}{(F(v) - F(u))^2} \right\} d\,P_n \\ &\quad + \int_{v \le t} \left\{ \frac{\delta_2}{(F(v) - F(u))^2} + \frac{\delta_3}{(1 - F(v))^2} \right\} d\,P_n, \end{aligned} \tag{2.4}$$

and

$$V_F(t) = W_F(t) + \int_{[0,\ t]} F(s)dG(s)\,. \tag{2.5}$$

Let J_{1n} be the set of examination times U_i such that T_i either belongs to $[0, U_i]$ or $(U_i, V_i]$, and let J_{2n} be the set of examination times V_i such that T_i either belongs to $(U_i, V_i]$ or (V_i, ∞). Furthermore, let $J_n = J_{1n} \cup J_{2n}$ and let $T_{(j)}$ be the jth order statistic of the set J_n.

PROPOSITION 2.1. *(Groeneboom, 1991) Let $T_{(1)}$ correspond to an observation U_i such that $\delta_{1i} = 1$, and let the largest order statistic $T_{(m)}$ correspond to an observation V_i such that $\delta_{1i} = \delta_{2i} = 0$. Then $\widehat{F}_n$ is the NPMLE of F_0 if and only if $\widehat{F}_n$ is the left derivative of the convex minorant of the self-induced cumulative sum diagram formed by the points*

$$P_{(j)} = (G_{\widehat{F}_n}(T_{(j)}),\ V_{\widehat{F}_n}(T_{(j)})), \qquad j = 1, 2, \ldots, m$$

and $P_{(0)} = (0,0)$.

With this characterization, Groeneboom introduced the *iterative convex minorant algorithm* to compute $\widehat{F}_n$. See Jongbloed (1995a,b) for modifications of the iterative convex minorant algorithm which always converge. Zhan and Wellner (1995) have adapted Jongbloed's argument to the related double censoring model; see also Wellner and Zhan (1996).

It can be verified that the NPMLE $\widehat{F}_n$ satisfies the self-consistency equation:

$$\begin{aligned}\widehat{F}_n(t) &= E_{\widehat{F}_n}[F_n(t) \mid \delta_i, \gamma_i, U_i, V_i, i = 1, \ldots, n] \qquad (2.6)\\ &= \frac{1}{n}\sum_{i=1}^{n}\left\{\delta_{1i}\frac{\widehat{F}_n(U_i \wedge t)}{\widehat{F}_n(U_i)} + \delta_{2i}\frac{\widehat{F}_n(V_i \wedge t) - \widehat{F}_n(U_i \wedge t)}{\widehat{F}_n(V_i) - \widehat{F}_n(U_i)}\right.\\ &\qquad\qquad \left. + \delta_{3i}\frac{\widehat{F}_n(t) - \widehat{F}_n(V_i \wedge t)}{1 - \widehat{F}_n(V_i)}\right\}\end{aligned}$$

where F_n is the (unobservable) empirical distribution function of $T_1, \ldots, T_n$. This was first derived by Turnbull (1976). However, this equation does not characterize the NPMLE $\widehat{F}_n$. That is, there may exist other distribution functions different from $\widehat{F}_n$ that satisfy (2.6).

In general, $\widehat{F}_n$ has no closed form expression. In spite of this, Groeneboom (1991) characterized the NPMLE for case 2 interval censored data and developed a fast algorithm (the iterative convex minorant algorithm) for computing the NPMLE. Aragón and Eberly (1992) and Jongbloed (1995a,b) proposed modifications of the iterative convex minorant algorithm, and Jongbloed (1995a,b) shows that his modified algorithm always converges. See Groeneboom and Wellner (1992), pages 69 - 73 and Jongbloed (1995a,b).

2.2. Distributional results.

The asymptotic distribution of $\widehat{F}_n$ with "case 1" interval censoring was established by Groeneboom (1987). See Groeneboom and Wellner (1992), page 89.

THEOREM 2.1. *(Groeneboom, 1987) Let G be the distribution function of U and $g = G'$ be the corresponding density function. Let t_0 be such that $0 < F(t_0), G(t_0) < 1$, and let F_0 and G be differentiable at t_0 with strictly positive derivatives $f(t_0)$ and $g(t_0)$, respectively. Furthermore, let $\widehat{F}_n$ be the NPMLE of F_0. Then we have, as $n \to \infty$,*

$$n^{1/3}\{\widehat{F}_n(t_0) - F_0(t_0)\} \to_d 2c_1(t_0)Z,$$

where Z is the last time where standard two-sided Brownian motion minus the parabola $y(t) = t^2$ reaches its maximum and

$$c_1(t_0) = \{[f(t_0)F_0(t_0)(1 - F_0(t_0)]/2g(t_0)\}^{1/3}.$$

With "case 2" interval-censored data, consistency of $\widehat{F}_n$ was proved by Groeneboom and Wellner (1992) and by Van de Geer (1993). Finding the asymptotic distribution of $\widehat{F}_n$ with "case 2" interval-censored data appears to be a much harder problem. Groeneboom (1991) conjectured that, with a different rate of convergence $(n \log n)^{-1/3}$ and a different scaling constant, the limiting distribution of the NPMLE takes the same form under the assumption that the joint density h of U and V is strictly positive along the diagonal $u = v$. He also showed that an approximation $\widehat{F}_n^{(1)}$ to the NPMLE $\widehat{F}_n$ has this limiting distribution (marginally at a fixed point t_0). However, this conjecture is still not proved. See also Groeneboom and Wellner (1992), page 100. Wellner (1995) proposed to explore alternative hypotheses under which the joint density h of U and V converges to zero as $u \leq v$ approach the diagonal $u = v$; in particular, the hypothesis (C2) below holds if the joint distribution H of (U, V) puts zero mass on some small strip bordering the diagonal $u = v$. Under such hypotheses, he proved a result about the "one-step" approximation to the NPMLE analogous to Groeneboom's theorem (see theorem 2.2 below), and made a corresponding conjecture about the asymptotic distribution of the NPMLE. Although conjecture of Groeneboom (1991) is still unproved, Groeneboom (1996) has succeeded in verifying Wellner's (1995) conjecture in the strict separation case.

Here we will state the version of the result for the "one-step approximation" $\widehat{F}_n^{(1)}$ at a point t_0 under the hypotheses of Wellner (1995). The hypotheses needed are as follows:

Assumption (C1): the support of F_0 is an interval $[0, \tau]$ where $\tau < \infty$, and $t_0 \in (0, \tau)$.

Define functions k_1 and k_2 by

$$k_1(u) = \int_u^{\tau} \frac{h(u,v)}{F_0(v) - F_0(u)} dv, \quad \text{and} \quad k_2(v) = \int_0^{v} \frac{h(u,v)}{F_0(v) - F_0(u)} du.$$

We also define functions $k_1(u,\epsilon)$ and $k_2(v,\epsilon)$ which isolate the contributions to k_1 and k_2 by values of v and u respectively near the diagonal $u = v$:

$$k_1(u;\epsilon) \equiv \int_u^{\tau} \frac{h(u,v)}{F_0(v) - F_0(u)} 1_{[1/(F_0(v)-F_0(u))>\epsilon]} dv$$

and

$$k_2(v;\epsilon) \equiv \int_0^{v} \frac{h(u,v)}{F_0(v) - F_0(u)} 1_{[1/(F_0(v)-F_0(u))>\epsilon]} du\,.$$

Assumption (C2): For each $\epsilon > 0$ and $i = 1, 2$

$$\alpha \int_{(t_0, t_0+t/\alpha]} k_i(u;\epsilon\alpha) du \to 0 \qquad \text{as} \qquad \alpha \to \infty. \tag{2.7}$$

Assumption (C3): (a) $F_0(t)$ and $H(u,v)$ have densities $f_0(t)$ and $h(u,v)$ with respect to Lebesgue measure on R and R^2, respectively; (b) $h(u,v)$ has bounded partial derivatives on the support of (U,V). Let $h_1(u)$ and $h_2(v)$ be the marginal densities of U and V, respectively.

Assumption (C4): $0 < F_0(t_0) < 1$ and $0 < H(t_0,t_0) < 1$.

THEOREM 2.2. *(Wellner, 1995). Suppose that assumptions (C1) to (C4) hold. Suppose that f_0, h_1, h_2, k_1 and k_2 are continuous at t_0 and $f_0(t_0) > 0$. Then,*

$$n^{1/3}\{\widehat{F}_n^{(1)}(t_0) - F_0(t_0)\} \to_d 2c_2(t_0)Z,$$

where Z is the last time where standard two-sided Brownian motion $W(t)$ minus the parabola $y(t) = t^2$ reaches its maximum,

$$c_2(t_0) = \{f_0(t_0)/2\xi(t_0)\}^{1/3},$$

and

$$\xi(t_0) = \frac{h_1(t_0)}{F_0(t_0)} + k_1(t_0) + k_2(t_0) + \frac{h_2(t_0)}{1 - F_0(t_0)}. \tag{2.8}$$

From the preceding discussion it is clear that for case 1 interval censored failure time data and examination times U with positive density h, the NPMLE $\widehat{F}_n$ does *not* satisfy the central limit theorem with the usual $n^{1/2}$-rate of convergence, and the use of observed information to construct confidence intervals or confidence bands for F_0 do not have large sample justifications. On the other hand, the general bootstrap theory of Politis and Romano (1994) does apply to yield confidence intervals for $F_0(t_0)$ as a consequence of theorem 2.1; see their example 2.1.1, page 2035, for a problem also involving a convergence rate of $n^{1/3}$. In view of the conjectures by Groeneboom and Wellner concerning $\widehat{F}_n$ with case 2 interval censored data, these remarks very likely apply to case 2 as well.

It should also be noted that with case 1 data the sequence of stochastic processes $\{n^{1/3}(\widehat{F}_n(t) - F_0(t)) : 0 \le t \le \tau\}$ (or the corresponding processes with scaling $(n \log n)^{1/3}$ for case 2 data under Groeneboom's hypotheses) is *not tight* in $D[0, \tau]$ and do *not converge weakly as processes*. If $0 < t_1 < t_2 < \tau$ are fixed points with the hypotheses of theorem 2.1 holding at both t_1 and t_2, then, much as in density estimation problems, it can be shown that

$$n^{1/3}(\widehat{F}_n(t_1) - F_0(t_1), \widehat{F}_n(t_2) - F_0(t_2)) \rightarrow_d (c_1 Z_1, c_2 Z_2)$$

where Z_1 and Z_2 are independent, and this precludes the possibility of a tight limit process. See Groeneboom (1985) for a study of the (local) dependence structure of this type of process in a closely related problem.

To complete the picture concerning the NPMLE $\widehat{F}_n$, consider case 1 interval censored data when the observation times U *fail to have a continuous density* h. Instead, suppose that U has a discrete distribution H with

$$h_j \equiv P(U = u_j), \qquad j = 1, \ldots, d$$

where we suppose $0 < u_1 < \ldots < u_d < \tau$. Then it quickly becomes clear that we can only estimate the distribution function F at the points u_i, $i = 1, \ldots, d$: note that the U_i's in the formula for the NPMLE given by (2.2) only take values in the set $\{u_1, \ldots, u_d\}$. The NPMLE $\widehat{F}_n(u_j)$ is a consistent estimator of $F_0(u_j)$, and in fact it is $\sqrt{n}$–consistent. This follows by re-expressing the NPMLE $\widehat{F}_n(u_j)$ as a monotonization of the simple binomial estimators obtained via (unconstrained) maximization of the resulting likelihood (see Ayer et al. (1955), Van Eeden (1956), or Robertson, Wright, and Dykstra (1988), example 1.5.1, page 32), and then using the continuous mapping theorem as in Chernoff (1954). It seems that similar remarks also apply to the NPMLE $\widehat{F}_n$ with case 2 or even "case k" type data (the binomial problems become trinomial or multinomial with $k+1$ cells, and the "weights" in the monotonization problem depend on the solution itself, but the continuous mapping theorem still carries the argument).

3. Regression models with interval censoring, case 1.

Several regression models can be viewed as special cases of the transformation model. This model postulates that the conditional distribution $F(t|z)$ of T given the covariate $Z = z$ satisfies

$$g(F(t|z)) = h(t) + \theta' z, \tag{3.1}$$

where g is a specified function, $h(t)$ is an unknown increasing function and θ is a d-dimensional regression parameter. If we take $g(s) = \log[-\log(1-s)], 0 < s < 1$, then (3.1) results in the famous proportional hazards model proposed by Cox (1972); in this case the model is more commonly written in terms of the cumulative hazard function as

$$\Lambda(t|z) = \Lambda(t)e^{\theta' z}, \tag{3.2}$$

where Λ is the unknown baseline cumulative hazard function.

If we take $g(s) = \text{logit}(s) \equiv \log[s/(1-s)], 0 < s < 1$, then we get the proportional odds regression model:

$$\text{logit}[F(t|z)] = \text{logit}[F(t)] + \theta' z, \tag{3.3}$$

where $F(t) \equiv F(t|0)$ is the baseline distribution function. Let $\alpha(t) = \text{logit}\, F(t)$, the baseline monotone increasing log-odds function. The proportional odds regression model is an interesting alternative to the proportional hazards model, and might be appropriate when the proportional hazards assumption is not satisfied. This model has been used by several authors in analyzing survival data; for right-censored data, see Bennett (1983), Pettitt (1984) and Parzen (1993), while for "case 1" interval censored data, see Dinse and Lagakos (1983).

Another important model which is closely related to (3.1) is the accelerated failure time regression model:

$$\log T = Z'\theta + \varepsilon, \tag{3.4}$$

where the distribution function F of ε is completely unspecified and where $\log T$ can be replaced by other more appropriate known monotone functions. In terms of conditional distributions, this model can be written as

$$F(\log T|Z) = F(\log T - Z'\theta).$$

There is an enormous amount of literature on statistical inference for the accelerated failure time model with right-censored data, but much less for this model with interval censoring. Much of the existing literature seems to be in connection with the "binary choice" model in econometrics: see e.g. Cosslett (1983, 1987), Chamberlain (1986), and Manski (1985). For further review and the relationship with interval censoring see Huang and Wellner (1996), and Huang (1993).

Throughout we assume the following basic assumptions:

(A1) The (unobservable) failure time is independent of the examination times given the covariates.

(A2) The joint distribution of the examination times and the covariates are independent of the parameters of interest.

For consistency of the maximum likelihood estimators, the following identifiability condition is needed.

(A3) (a) The distribution of Z is not concentrated on any proper affine subspace of R^d (i.e. of dimension $d-1$ or smaller). (b) Z is bounded, that is, there exists $Z_0 > 0$ such that $P(|Z| \leq Z_0) = 1$, where $|\cdot|$ denotes the Euclidean norm in R^d.

For distributional results of the maximum likelihood estimators, we need the following further regularity conditions.

(A4) $F_0(0) = 0$. Let $\tau_{F_0} = \inf\{t : F_0(t) = 1\}$. The support of U is an interval $S[U] = [\tau_0, \tau_1]$, and $0 < \tau_0 \leq \tau_1 < \tau_{F_0}$.

(A5) F_0 has a strictly positive and continuous density on $S[U]$, and the joint distribution function $H(u, z)$ of (U, Z) has bounded second order (partial) derivative with respect to u.

Under assumptions (A1) and (A2), the joint density of $X = (\delta, U, Z)$ is

$$p(x) = F(u|z)^{\delta}(1 - F(u|z))^{1-\delta} h(u, z),$$

where $h(u, z)$ is the joint density of (U, Z). So for an independent sample $(\delta_i, U_i, Z_i), i = 1, \ldots, n$ with the same distribution as (δ, U, Z), the general form of the log-likelihood function is, up to an additive constant,

$$l_n = \sum_{i=1}^{n} \{\delta_i \log F(U_i|Z_i) + (1 - \delta_i) \log(1 - F(U_i|Z_i))\}.$$

A distinct feature of "case 1" interval censoring is that, in the regression problems we discuss below, the efficient score function and the Fisher information for the regression parameter have explicit expressions. However, there are no explicit expressions for these quantities with "case 2" interval censoring in general. Consequently, properties of various estimators (in particular, maximum likelihood estimators) are better understood for regression models with "case 1" interval censoring than the corresponding problems with "case 2" interval censoring.

3.1. Proportional hazards model.

In this case, it is convenient to parametrize the model in terms of the regression parameter and the baseline cumulative hazard function. The joint density function for the proportional hazards model with case 1 data is

$$p_{\theta,\Lambda}(x) = p_{\theta,\Lambda}(\delta, u, z) = (1 - e^{-\Lambda(u)e^{\theta' z}})^{\delta} e^{-(1-\delta)\Lambda(u)e^{\theta' z}} h(u, z).$$

It follows that the log-likelihood function is (up to an additive term not involving (θ, Λ)) is:

$$l_n(\theta, \Lambda) = \sum_{i=1}^{n} \left\{\delta_i \log[1 - \exp(-\Lambda(U_i)e^{\theta' Z_i})] - (1 - \delta_i)\Lambda(U_i)e^{\theta' Z_i}\right\}.$$

3.1.1. Information for θ.

It is well known that in most parametric models and many semiparametric models (such as the Cox model with right censoring), we can estimate the finite-dimensional parameter at $\sqrt{n}$-convergence rate and asymptotically efficiently. A necessary condition is that we must have positive Fisher information. For "case 1" interval censored data, it is not clear a priori that the information is in fact positive. Actually, from the results on estimation of the distribution in the nonparametric setting, the Fisher information for the baseline cumulative hazard function is zero. Therefore, it is useful to calculate the information for the regression parameter. Positive information will suggest that it is possible to estimate θ at the $n^{1/2}$-rate of convergence as in a regular parametric model, even though it is impossible to achieve the same thing for the baseline cumulative hazard function. These comments also apply to the corresponding estimation problem with "case 2" or general interval censoring.

Define the functions

$$Q(u,\delta,z) = \delta \frac{\overline{F}(u \mid z)}{1-\overline{F}(u \mid z)} - (1-\delta), \tag{3.5}$$

and

$$O(u \mid z) = E[Q^2(U,\delta,Z) \mid U = u, Z = z] = \frac{\overline{F}(u \mid z)}{1-\overline{F}(u \mid z)}. \tag{3.6}$$

THEOREM 3.1. *Suppose that assumptions (A1) to (A5) are satisfied. Then:*
(a) The efficient score function for θ is

$$l_\theta^*(x) \quad = \quad e^{\theta' Z} Q(u,\delta,z)\Lambda(u)\left\{z - \frac{E[(Ze^{2\theta' Z})O(U \mid Z) \mid U = u]}{E[(e^{2\theta' Z})O(U \mid Z) \mid U = u]}\right\}.$$

(b) The information for θ is

$$I(\theta) = E[l_\theta^*(X)]^{\otimes 2} = E\left\{R(U,Z)\left[Z - \frac{E(ZR(U,Z) \mid U)}{E(R(U,Z) \mid U)}\right]^{\otimes 2}\right\}, \tag{3.7}$$

where $a^{\otimes 2} = aa'$ for any column vector $a \in R^d$, and $R(u,z) = \Lambda^2(u \mid Z)O(u \mid z)$.

The proof of theorem 3.1 will be given in section 6.

3.1.2. Distributional results.

Let $S_0(t)$ be the baseline survival function and let $\widehat{S}_n(t) = \exp(-\widehat{\Lambda}_n(t))$ be the corresponding estimator. Let G be the distribution function of U.

THEOREM 3.2. *(Consistency) Suppose that either (i) conditions (A1), (A2) and (A3) hold, or, (ii) condition (A1), (A2) and (A3a) holds and that the parameter space Θ of θ is bounded. Then*

$$\widehat{\theta}_n \rightarrow_{a.s} \theta_0, \quad \textit{and} \quad \widehat{S}_n(t) \rightarrow_{a.s} S_0(t) \quad G- \textit{ almost surely} .$$

Two interesting special cases of this theorem are: (i) If U is discrete, then $\widehat{S}_n(t) \rightarrow_{a.s} S_0(t)$ at all the mass points of U. (ii) If U has a continuous distribution function whose support contains the support of S_0, then

$$\sup_{0 \le t < \infty} |\widehat{S}_n(t) - S_0(t)| \rightarrow_{a.s} 0.$$

This implies for any finite $M > 0$,

$$\sup_{0 \le t \le M} |\widehat{\Lambda}_n(t) - \Lambda_0(t)| \rightarrow_{a.s} 0.$$

Although consistency is obtained with minimal assumptions, to obtain rate of convergence and asymptotic normality, further conditions (A4) and (A5) are needed.

THEOREM 3.3. *(Asymptotic normality and efficiency with "case 1" interval censoring) Suppose that θ_0 is an interior point of Θ and that assumptions (A1) to (A5) are satisfied. Then*

$$\sqrt{n}(\widehat{\theta}_n - \theta_0) \rightarrow_d N(0, I(\theta_0)^{-1}),$$

where $I(\theta_0)$ is the generalized Fisher information given in (3.7) which takes into account that the baseline cumulative hazard function Λ_0 is unknown. However,

$$\left\{ \int_{\tau_0}^{\tau_1} (\widehat{\Lambda}_n(u) - \Lambda_0(u))^2 dG(u) \right\}^{1/2} = O_p(n^{-1/3}).$$

Proofs of theorems 3.2 and 3.3 are given in Huang (1996).

3.1.3. Variance estimation.

By Theorem 3.3, $I^{-1}(\theta_0)/n$ is the asymptotic variance-covariance matrix for $\widehat{\theta}_n$. This provides one way to obtain a consistent estimator of the variance of $\widehat{\theta}_n$. Recall in the expression for $I(\theta_0)$, $R(u, z)$ is defined to be

$$R(u, z) = \Lambda^2(u \mid z) O(u \mid z) = e^{2z'\theta_0} \Lambda_0^2(u) \frac{\exp(-\Lambda_0(u) e^{z'\theta_0})}{1 - \exp(-\Lambda_0(u) e^{z'\theta_0})}.$$

Denote

$$\widehat{R}_n(u, z) = e^{2z'\widehat{\theta}_n} \widehat{\Lambda}_n{}^2(u) \frac{\exp(-\widehat{\Lambda}_n(u) e^{z'\widehat{\theta}_n})}{1 - \exp(-\widehat{\Lambda}_n(u) e^{z'\widehat{\theta}_n})}. \tag{3.8}$$

Let

$$\mu_1(u) = E(R(u, Z) | U = u), \quad \text{and} \quad \mu_2(u) = E(Z R(u, Z) | U = u).$$

Then when we have obtained reasonable estimators $\mu_{1n}(u)$ and $\mu_{2n}(u)$ for $\mu_1(u)$ and $\mu_2(u)$, we can estimate $I(\theta_0)$ by

$$\widehat{I}_n(\widehat{\theta}_n) = \frac{1}{n}\sum_{i=1}^{n}\left\{\widehat{R}_n(U_i, Z_i)\left[Z_i - \frac{\mu_{2n}(U_i)}{\mu_{1n}(U_i)}\right]^{\otimes 2}\right\}. \tag{3.9}$$

The hard work is to estimate $\mu_1(u)$ and $\mu_2(u)$. Also, θ_0 and the rest of R need to be estimated. When Z is a continuous covariate vector, $\mu_1(u) \equiv E(R(u,Z)|U = u)$ can be approximated by $E(\widehat{R}_n(u,Z)|U = u)$. Then we can estimate $E(\widehat{R}_n(u,Z)|U = u)$ by nonparametric regression approaches, see, e.g. Stone (1977).

When Z is a categorical covariate, the above nonparametric smoothing procedure does not work well because of the discrete nature of the values of $\widehat{R}_n(u,z)$. Here we consider the simplest case when Z is a dichotomous variable indicating two treatment groups, that is, Z only takes values 0 or 1 with $P\{Z = 1\} = \gamma$ and $P\{Z = 0\} = 1 - \gamma$. Thus $E[ZR(Z,U)|U = u] = R(1,u)P\{Z = 1|U = u\} = R(1,u)f_1(u)\gamma/f(u)$, and

$$\begin{aligned} E[R(Z,U)|U = u] &= R(1,u)P\{Z = 1|U = u\} + R(0,u)P\{Z = 0|U = u\} \\ &= \frac{R(1,u)f_1(u)\gamma}{f(u)} + \frac{R(0,u)f_0(u)(1-\gamma)}{f(u)}, \end{aligned}$$

where $f_1(u)$ is the conditional density of Y given $Z = 1$, $f_0(u)$ is the conditional density of Y given $Z = 0$, and $f(u)$ is the marginal density of Y. Notice that we only need to estimate the ratio the ratio of the two conditional expectations. First we can estimate γ by the total number of subjects in the treatment group with $Z = 1$ divided by the sample size. Let $\widehat{f}_{1n}(u)$ be a kernel density estimator of $f_1(u)$, and $\widehat{f}_{0n}(u)$ be a kernel density estimator of $f_0(u)$. Then a natural estimator of $E[ZR(Z,U)|U = u]/E[R(Z,U)|U = u]$ is

$$\widehat{\mu}_n(u) = \frac{\widehat{R}_n(1,u)\widehat{f}_{1n}(u)\widehat{\gamma}_n}{\widehat{R}_n(1,u)\widehat{f}_{1n}(u)\widehat{\gamma}_n + \widehat{R}_n(0,u)\widehat{f}_{0n}(u)(1-\widehat{\gamma}_n)}.$$

Here $\widehat{R}_n(u,z)$ is defined in (3.8). With a proper choice of the bandwidth and kernel in estimation of $f_1(u)$ and $f_0(u)$, the above estimator is consistent, see e.g., Silverman (1986). Hence a reasonable estimator of $I(\theta_0)$ is:

$$\widehat{I}_n(\widehat{\theta}_n) = \frac{1}{n}\sum_{i=1}^{n}\left\{\widehat{R}_n(U_i, Z_i)(Z_i - \widehat{\mu}_n(U_i))^2\right\}. \tag{3.10}$$

In the special case when Y and Z are independent, the above nonparametric smoothing is not necessary. In this case, we have

$$I(\theta_0) = E\left\{R(U,Z)\left[Z - \frac{E_Z(ZR(U,Z))}{E_Z R(U,Z)}\right]^{\otimes 2}\right\},$$

where E_Z means expectation with respect to Z. We can simply estimate $I(\theta_0)$ by

$$\widehat{I}_n(\widehat{\theta}_n) = \frac{1}{n}\sum_{i=1}^{n}\left\{\widehat{R}_n(U_i, Z_i)\left[Z_i - \frac{\sum_{j=1}^{n} Z_j \widehat{R}_n(U_i, Z_j))}{\sum_{j=1}^{n} \widehat{R}_n(U_i, Z_j)}\right]^{\otimes 2}\right\}. \qquad (3.11)$$

Two alternative approaches based on the observed Fisher information or the curvature of the profile likelihood function can also be used. These will be discussed in section 4.1.3.

3.2. Proportional odds model.

We parametrize the model in terms of the regression parameter and the baseline log-odds function $\alpha(t) = \text{logit}[F(t)]$. Then with conditions (A1) and (A2), the joint density function of the observations is

$$p_{\theta,\alpha}(x) = p_{\theta,\alpha}(\delta, u, z) = e^{\delta(\alpha(t)+\theta' z)}(1 + e^{\alpha(t)+\theta' z})^{-1} h(u, z).$$

It follows that the log-likelihood function is (up to an additive term not depending on (θ, α))

$$l_n(\theta, \alpha) = \sum_{i=1}^{n} \{\delta_i(\alpha(U_i) + \theta' Z_i) - \log(1 + \exp(\alpha(U_i) + \theta' Z_i))\}. \qquad (3.12)$$

The maximum likelihood estimator is the $(\widehat{\theta}_n, \widehat{\alpha}_n)$ that maximizes $l_n(\theta, \alpha)$ with the constraint that $\widehat{\alpha}_n$ is a nondecreasing function.

3.2.1. Information for θ.

The following result on the information bound for estimation of θ is given in Rossini and Tsiatis (1996).

THEOREM 3.4. *Suppose that conditions (A1) to (A5) are satisfied. Then:*
(a) The efficient score function for θ is

$$\dot{l}^*_\theta(x) = (\delta - E(\delta|U = u, Z = z))\left(z - \frac{E(Z\, Var(\delta|U, Z)|U = u)}{E(\,Var(\delta|U, Z)|U = u)}\right)$$

(b) The information for θ is

$$I(\theta) = E\left[(\delta - E(\delta|U, Z))^2\left(Z - \frac{E(Z\, Var(\delta|U, Z)|U)}{E(\,Var(\delta|U, Z)|U)}\right)^{\otimes 2}\right]. \qquad (3.13)$$

3.2.2. Distribution results.

THEOREM 3.5. *(Asymptotic normality) Suppose that conditions (A1)-(A5) stated earlier hold and that θ_0 is an interior point of Θ. Then*

$$\sqrt{n}(\widehat{\theta}_n - \theta_0) \to_d N(0, I(\theta_0)^{-1}),$$

where $I(\theta_0)$ is the Fisher information given in (3.13). However,

$$\left\{\int [\widehat{\alpha}_n(u) - \alpha_0(u)]^2 dG(u)\right\}^{1/2} = O_p(n^{-1/3}).$$

The proof of this result is given in Huang (1995). Although a boundedness restriction is imposed on $\widehat{\alpha}_n$ in that paper, it can be shown that this restriction can be removed; the first author intends to do this elsewhere.

3.2.3. Variance estimation.
This can be done similarly as in section 3.1.3 by using the explicit form of the asymptotic variance of $\widehat{\theta}_n$ given in terms of $I(\theta)$ in (3.13). Two alternative estimators are based on observed Fisher information or the curvature of the profile likelihood function in a neighborhood of $\widehat{\theta}_n$. We will discuss this further in section 4.1.3.

3.3. The accelerated failure time model.

For simplicity, let T be the logarithm or other appropriate transformation of the failure time and assume that the same transformation has been made for the examination time. So the model is

$$T = Z'\theta + \varepsilon,$$

where $\varepsilon \sim F_0$ unspecified. Suppose that ε is independent of Z and the examination times. Then the density of one observation is

$$p_{\theta,F}(x) = p_{\theta,F}(\delta, u, z) = F(u - \theta' z)^{\delta}(1 - F(u - \theta' z))^{1-\delta} h(u, z);$$

hence the log-likelihood function for (θ, F) is (up to an additive term)

$$l_n(\theta, F) = \sum_{i=1}^{n} \log\{\delta_i F(U_i - \theta' Z_i) + (1 - \delta_i)(1 - F(V_i - \theta' Z_i))\}.$$

The maximum (profile) likelihood estimator of (θ_0, F_0) is the $(\widehat{\theta}_n, \widehat{F}_n)$ that maximizes $l_n(\theta, F)$ with the restriction that $\widehat{F}_n$ is a (sub)distribution function. As proposed by Cosslett (1983), $(\widehat{\theta}_n, \widehat{F}_n)$ can be computed via the maximum profile likelihood approach:

(i) For any fixed θ, maximize $l_n(\theta; F)$ with respect to F under the constraint that F is a distribution function. Denote the resulting maximizer by $F_n(\cdot;\theta)$.

(ii) Substituting $F_n(\cdot;\theta)$ back into $l_n(\theta; F)$, yields the profile likelihood $l_n(\theta; F_n(\cdot;\theta))$. Then the maximum profile likelihood estimator $\widehat{\theta}_n$ is any value of θ that maximizes $l_n(\theta; F_n(\cdot;\theta))$ (assuming that it exists).

(iii) A natural estimator of F is $F_n(\cdot;\widehat{\theta}_n)$.

3.3.1. Information for θ.

The following is the result concerning the efficient score and the information bound for θ in this model.

THEOREM 3.6. *Let $k(s) = E(Z|U - Z'\theta = s)$. Suppose that k is bounded. Then the efficient score for θ is*

$$l_\theta^*(x) = f(u-z'\theta)[z-E(Z|U-Z'\theta = y-z'\theta)]\left[\frac{1-\delta}{1-F(u-z'\theta)} - \frac{\delta}{F(u-z'\theta)}\right]. \tag{3.14}$$

Moreover, the information for estimation of θ is:

$$I(\theta) = E\left\{\frac{f(U-Z'\theta)^2}{F(U-Z'\theta)(1-F(U-Z'\theta))}[Z-E(Z|U-Z'\theta)]^{\otimes 2}\right\}. \tag{3.15}$$

The proof of theorem 3.6 will be given in section 6.

3.3.2. Distributional results.

Consistency of $(\widehat{\theta}_n, \widehat{F}_n(\cdot;\widehat{\theta}_n))$ can be shown as in Cosslett (1983); alternatively, consistency can be shown as in theorem 4.7.

We have not been able to establish distributional results for $\widehat{\theta}_n$. It is not clear whether or not $\widehat{\theta}_n$ has a normal limiting distribution with $n^{1/2}$-rate of convergence since $l_n(\theta, \widehat{F}_n(\cdot;\theta))$ is not a "smooth" function of θ. This is a challenging open problem.

3.4. Computation.

We describe in detail an approach for computing the maximum likelihood estimators in the proportional hazards model.

Unlike in the nonparametric setting, there is no closed form expression for $\widehat{\Lambda}_n$. We describe an algorithm which is also applicable to general interval censoring.

In principle, computation of $(\widehat{\theta}_n, \widehat{\Lambda}_n)$ can be accomplished by maximizing (4.2) jointly with respect to (θ, Λ), or, by using a profile likelihood approach, maximizing over Λ for all fixed values of $\theta \in \Theta$ first to obtain $\widehat{\Lambda}_n(\cdot, \theta)$, and then maximizing the profile log-likelihood function $l_n(\theta, \widehat{\Lambda}_n(\cdot, \theta))$ over θ to find $\widehat{\theta}_n$ and hence $\widehat{\Lambda}_n = \widehat{\Lambda}_n(\cdot, \widehat{\theta}_n)$. For low dimensional θ, Groeneboom's iterative convex minorant algorithm (see Groeneboom and Wellner (1992), pages 69 - 74) is sufficiently fast to implement this profile likelihood approach for computing $(\widehat{\theta}_n, \widehat{\Lambda}_n)$. For higher dimensional θ we propose the following iterative algorithm for computing $(\widehat{\theta}_n, \widehat{\Lambda}_n)$. Let $\theta^{(0)}$ be an initial guess and set $k = 0$.

Step (i). Maximize $l_n(\theta^{(k)}, \Lambda)$ with respect to Λ to obtain $\Lambda^{(k)}$.

Step (ii). Maximize $l_n(\theta, \Lambda^{(k)})$ with respect to θ. Set $k = k+1$, and let $\theta^{(k)}$ be the maximizer. Go back to (i). Repeat (i) and (ii) until convergence.

It can be verified that for any fixed θ, $l_n(\theta, \Lambda)$ is concave in Λ, and for any fixed Λ, $l_n(\theta, \Lambda)$ is concave in θ. So steps (i) and (ii) are well defined concave maximization problems. Since each iteration increases the likelihood function, the algorithm converges. A much stronger conclusion holds as stated in the following proposition.

PROPOSITION 3.1. *(1) The function $l_n(\theta, \Lambda)$ has a unique maximizer $(\widehat{\theta}_n, \widehat{\Lambda}_n)$; (2) Starting from any point $\theta^{(0)}$, the algorithm produces a sequence $(\theta^{(k)}, \Lambda^{(k)})$ converging to $(\widehat{\theta}_n, \widehat{\Lambda}_n)$.*

PROOF. Reparametrize $l_n(\theta, \Lambda)$ in terms of θ and $\phi = \log \Lambda$. Let

$$\begin{aligned} l_n(\theta, \phi) &= \sum_{i=1}^{n} \{\delta_i \log[1 - \exp(-\exp(\theta' Z_i + \phi(U_i)))] \\ &\qquad - (1 - \delta_i) \exp(\theta' Z_i + \phi(U_i))\}. \end{aligned}$$

Then maximizing $l_n(\theta, \Lambda)$ with respect to (θ, Λ) with monotonicity constraints on Λ is equivalent to maximizing $l_n(\theta, \phi)$ with respect to (θ, ϕ) with monotonicity constraints on ϕ. Since functions $\log[1 - \exp(-\exp(x))]$ and $-\exp(x)$ are concave, $l_n(\theta, \phi)$ is concave in (θ, ϕ) (This follows from Theorem 5.7 of Rockafellar (1970), page 38). Thus a local maximizer of $l_n(\theta, \phi)$ will also be a global maximizer. This implies a local maximizer of $l_n(\theta, \Lambda)$ will also be a global maximizer because of the one-to-one correspondence between (θ, Λ) and (θ, ϕ). The uniqueness of the global maximizer follows since $l_n(\theta, \phi)$ is strictly concave on the product of the space of θ and the reduced space of ϕ. Here the reduced space of ϕ means the space of the distinct values of $\phi(U_{(1)}), \ldots, \phi(U_{(n)})$, since the monotonicity constraints will force $\phi(U_{(i)}), i = 1, \ldots, n$ to be blockwise constant. Λ

To implement step (i), we use the iterative convex minorant algorithm introduced by Groeneboom. To describe it in the present setting, we first introduce some notation. Let

$$q(u', z, \theta, \Lambda) = \frac{\exp(-\exp(\theta' z)\Lambda(u'))}{1 - \exp(-\exp(\theta' z)\Lambda(u'))}.$$

Define the processes W_Λ, G_Λ and V_Λ by

$$W_\Lambda(u) = \int_{u' \in [0,u]} \{1_{\{t \le u'\}} q(u', z, \theta, \Lambda) - 1_{\{t > u'\}} \exp(\theta' z)\}\, dQ_n(t, u', z),$$

$$G_\Lambda(u) = \int_{u' \in [0,u]} \left\{ 1_{\{t \le u'\}} q(u', z, \theta, \Lambda) - 1_{\{t > u'\}} \exp(\theta' z) \right\}^2 dQ_n(t, u', z),$$

and

$$V_\Lambda(u) = W_\Lambda(u) + \int_{[0,u]} \Lambda(u') dG_\Lambda(u'), \qquad u \ge 0,$$

where Q_n is the empirical measure of the (unobservable) points (T_i, U_i, Z_i), $i = 1, \ldots, n$.

THEOREM 3.7. *For any fixed θ, suppose that $\delta_{(1)} = 1, \delta_{(n)} = 0$. then $\Lambda_n(\cdot;\theta)$ maximizes $l_n(\theta, \Lambda)$ if and only if $\Lambda_n(\cdot;\theta)$ is the left derivative of the greatest convex minorant of the "self-induced" cumulative sum diagram, consisting of the points*

$$P_{(j)} = \left(G_{\Lambda_n(\cdot;\theta)}(U_{(j)}), V_{\Lambda_n(\cdot;\theta)}(U_{(j)}) \right), \qquad j = 1, \cdots, n,$$

and the origin $(0, 0)$.

The cumulative sum diagram is simply the linear interpolation of the points $P_{(0)}, P_{(1)}, \ldots, P_{(m)}$, and the greatest convex minorant is the greatest convex function that is below this linear interpolation.

The proof of Theorem 3.7 is completely analogous to that of Proposition 1.4 of Groeneboom and Wellner (1992). This theorem gives an iterative procedure to compute $\Lambda_n(\cdot;\theta)$ for any fixed θ. It proceeds as follows. Suppose $\Lambda^{(k)}(\cdot;\theta)$ is obtained at the kth iteration; then $\Lambda^{(k+1)}(\cdot;\theta)$ is computed as the left derivative of the convex minorant of the cumulative sum diagram, consisting of the points

$$(G_{\Lambda^{(k)}(\cdot;\theta)}(U_{(j)}), V_{\Lambda^{(k)}(\cdot;\theta)}(U_{(j)})), \qquad j = 1, \ldots, n;$$

and the origin $(0, 0)$.

In step (ii), the Newton-Raphson method can be used. Specifically, for any fixed Λ, let

$$s_1(\theta) = (\partial/\partial\theta) l_n(\theta, \Lambda).$$

By concavity, the solution to $s_1(\theta) = 0$ is the unique maximizer of $l_n(\theta, \Lambda)$ (for fixed Λ).

The approach described above can be used for computing $(\widehat{\theta}_n, \widehat{\alpha}_n)$ in the proportional odds model. This is because $l_n(\theta, \alpha)$ is a concave function of (θ, α), so proposition 3.1 holds for the proportional odds model. To see this, verify that $-\log(1 + \exp(x))$ is a concave function. Thus the concavity of $l_n(\theta, \alpha)$ follows from Theorem 5.7 of Rockafellar (1970), page 38, or it can be verified directly.

4. Regression models with interval censoring, case 2.

Again we assume conditions (A1) to (A3) stated in the beginning of section 3. The joint density of $X = (\delta_1, \delta_2, \delta_3, U, V, Z)$ where $\delta_i \in \{0, 1\}$ for $i =$

1, 2, 3 and $\delta_1 + \delta_2 + \delta_3 = 1$ is

$$p(x) = F(u|z)^{\delta_1}[F(v|z) - F(u|z)]^{\delta_2}(1 - F(v|z))^{\delta_3} h(u, v, z); \tag{4.1}$$

here h is the joint density function of (U, V, Z) with respect to the product of Lebesgue measure on R^2 and a fixed measure μ on R^d. The log-likelihood likelihood function of an independent sample $(\delta_{1i}, \delta_{2i}, \delta_{3i}, U_i, V_i, Z_i)$, $i = 1, \ldots, n$ with the same distribution as $(\delta_1, \delta_2, \delta_3, U, V, Z)$ is, up to an additive term not depending on $F(\cdot|Z)$,

$$\begin{aligned} l_n &= \sum_{i=1}^n \{\delta_{1i} \log F(U_i|Z_i) \\ &\quad + \delta_{2i} \log[F(V_i|Z_i) - F(U_i|Z_i)] + \delta_{3i} \log(1 - F(V_i|Z_i))\}. \end{aligned}$$

For proofs of the distributional results of maximum likelihood estimators discussed below, we also need the following regularity conditions.

(B4) (a) There exists a positive number η such that $P(V - U \geq \eta) = 1$; (b) the union of the support of U and V is contained in an interval $[\tau_0, \tau_1]$, where $0 < \tau_0 < \tau_1 < \infty$.

(B5) F_0 has strictly positive and bounded continuous derivative on $[\tau_0, \tau_1]$.

(B6) The conditional density $g(u, v|z)$ of (U, V) given Z has bounded partial derivatives with respect to u and v. The bounds of these partial derivatives do not depend on z.

4.1. The proportional hazards model.

In the case of the proportional hazards model, the log-likelihood function for the regression parameter θ and the baseline cumulative hazard function Λ is

$$\begin{aligned} l_n(\theta, \Lambda) &= \sum_{i=1}^n \Big\{ \delta_{1i} \log(1 - \exp(-\Lambda(U_i) e^{\theta' Z_i})) \\ &\quad + \delta_{2i} \log[\exp(-\Lambda(U_i) e^{\theta' Z_i}) - \exp(-\Lambda(V_i) e^{\theta' Z_i})] \\ &\quad - \delta_{3i} \Lambda(V_i) e^{\theta' Z_i} \Big\}. \end{aligned} \tag{4.2}$$

The maximum likelihood estimator is then the $(\widehat{\theta}_n, \widehat{\Lambda}_n)$ that maximizes $l_n(\theta, \Lambda)$ under the constraint that $\widehat{\Lambda}_n$ is a nonnegative and nondecreasing function.

4.1.1. Information for θ.

Let $x = (\delta_1, \delta_2, \delta_3, u, v, z)$. Denote the log-likelihood function for one observation by

$$\begin{aligned} l(x; \theta, \Lambda) &= \delta_1 \log\{1 - \exp(-\Lambda(u) e^{\theta' z})\} \\ &\quad + \delta_2 \log[\exp(-\Lambda(u) e^{\theta' z}) - \exp(-\Lambda(v) e^{\theta' z})] \\ &\quad - \delta_3 \Lambda(v) e^{\theta' z}\}. \end{aligned} \tag{4.3}$$

Let f and F be the density and distribution corresponding to Λ, and let f_s be a one-dimensional smooth curve through f, where the smoothness is with respect to s. Denote $a = \frac{\partial}{\partial s} \log f_s \big|_{s=0}$. Then $a \in L_2^0(F) \equiv \{a : \int a dF = 0 \text{ and } \int a^2 dF < \infty\}$. Let

$$h = \frac{\partial}{\partial s}\Lambda_s \bigg|_{s=0} = \frac{\int_0^{\cdot} a dF}{1-F} = \frac{-\int_{\cdot}^{\infty} a dF}{1-F}; \tag{4.4}$$

it follows from Hardy's inequality that $h \in L_2(F)$; see Bickel, Klaassen, Ritov, and Wellner (1993), page 423 [hereafter referred to as BKRW (1993)].

The score function for θ is

$$\dot{l}_\theta(x) = \frac{\partial}{\partial \theta} l(x; \theta, \Lambda),$$

and the score operator for Λ is

$$\dot{l}_\Lambda a(x) = \frac{\partial}{\partial s} l(x; \theta, \Lambda_s)\bigg|_{s=0}.$$

Explicit expressions for $\dot{l}_\theta(x)$ and $\dot{l}_\Lambda a(x)$ can be obtained by carrying out the differentiation. The score operator $\dot{l}_\Lambda$ maps $L_2^0(F)$ to $L_2^0(P)$, where P is the joint probability measure of $(\delta_1, \delta_2, U, V, Z)$ and $L_2^0(P)$ is defined similarly as $L_2^0(F)$. Let $\dot{l}_\Lambda^T : L_2^0(P) \to L_2^0(F)$ be the adjoint operator of $\dot{l}_\Lambda$, i.e., for any $a \in L_2^0(F)$ and $b \in L_2^0(P)$, define h as in (4.7),

$$\langle b, \dot{l}_\Lambda a\rangle_P = \langle \dot{l}_\Lambda^T b, a\rangle_F \ ,$$

where $\langle\cdot,\cdot\rangle_P$ and $\langle\cdot,\cdot\rangle_F$ are the inner products in $L_2^0(P)$ and $L_2^0(F)$, respectively. We need to find a_* such that $\dot{l}_\theta - \dot{l}_\Lambda(a_*)$ is orthogonal to $\dot{l}_\Lambda(a)$ in $L_2^0(P)$. This amounts to solving the following normal equation:

$$\dot{l}_\Lambda^T \dot{l}_\Lambda(a_*) = \dot{l}_\Lambda^T \dot{l}_\theta. \tag{4.5}$$

The value of $\dot{l}_\Lambda^T$ at any $b \in L_2^0(P)$ can be computed by

$$\dot{l}_\Lambda^T b(t) = E[b(X)|T = t] = E_Z E[b(X)|T = t, Z];$$

see e.g. BKRW (1993), pages 271-272, or Groeneboom and Wellner (1992), pages 8 and 9.

The proof of the following theorem is deferred to the appendix.

THEOREM 4.1. *For $\theta \in R$, under conditions (A1)-(A3) and (B4)-(B6), equation (4.5) has a unique solution a_*. Moreover, the corresponding function h_* (given by (4.4) with a replaced by a_*) has a bounded derivative. In general, for $\theta \in R^d$, $h_* \in L_2(F)^d$ is a d-dimensional vector and each component has a bounded derivative. The efficient score for θ is*

$$l_\theta^*(x) = \dot{l}_\theta(x) - \dot{l}_\Lambda a_*(x),$$

and the information bound for θ is

$$I(\theta) = E[l_\theta^*(X)]^{\otimes 2},$$

where $a^{\otimes 2} = aa'$ for any column vector $a \in R^d$. Under conditions (A1)-(A5), $I(\theta)$ is a positive definite matrix with finite entries. The efficient influence function is

$$\tilde{l}_\theta(x) = I(\theta)^{-1} l_\theta^*(x).$$

4.1.2. Distributional results.

Let G be the joint distribution function of (U, V) and let G_1 and G_2 be the marginal distribution functions of U and V, respectively. The following assumption is needed in the consistency result.

THEOREM 4.2. *(Consistency) Under the same conditions as in theorem 3.2, we have*

$$\widehat{\theta}_n \to_{a.s} \theta_0, \quad \text{and} \quad \widehat{S}_n(t) \to_{a.s} S_0(t) \quad (G_1 + G_2) - \text{ almost everywhere }.$$

Two interesting special cases of this theorem are: (i) If both U and V are discrete, this implies $\widehat{S}_n(t) \to_{a.s} S_0(t)$ at all the mass points of U and V. (ii) If at least one of U and V has a continuous distribution function whose support contains the support of S_0, then

$$\sup_{0 \le t < \infty} |\widehat{S}_n(t) - S_0(t)| \to_{a.s} 0.$$

This implies for any finite $M > 0$,

$$\sup_{0 \le t \le M} |\widehat{\Lambda}_n(t) - \Lambda_0(t)| \to_{a.s} 0.$$

With regularity conditions (B4) to (B6), we have the following theorem.

THEOREM 4.3. *(Asymptotic normality and efficiency) Suppose that θ_0 is an interior point of the bounded set Θ and that conditions (A1)-(A3) and (B4)-(B6) hold. Then*

$$\sqrt{n}(\widehat{\theta}_n - \theta_0) \to_d N(0, I(\theta_0)^{-1}),$$

where $I(\theta_0)$ is the generalized Fisher information for θ taking into account that the baseline cumulative hazard function Λ_0 is unknown; hence $\widehat{\theta}_n$ is asymptotically normal and efficient. For $\widehat{\Lambda}_n$,

$$||\widehat{\Lambda}_n - \Lambda_0||_2 = O_p(n^{-1/3}).$$

Proofs of theorems 4.2 and 4.3 can be found in Huang and Wellner (1995b). Notice that in that paper, a boundedness restriction is imposed on

$\widehat{\Lambda}_n$. We recently found that such restriction can be removed with additional arguments. These will appear in the revision of that paper.

It is noteworthy that, with both "case 1" and "case 2" interval censoring, although the nonparametric component $\widehat{\Lambda}_n$ only has $n^{1/3}$ rate of convergence, the parametric component can be estimated at $n^{1/2}$ rate. With "case 1" interval censoring, Theorem 3.3 provides an explicit form of the asymptotic variance which can be used to estimate the variance of $\widehat{\theta}_n$. With "case 2" interval censoring, although theorem 4.3 does not help in estimating the variance of $\widehat{\theta}_n$ because of the implicitness of $I(\theta)$, it lends support for the use of the observed Fisher information or the curvature of the profile likelihood to estimate the variance of $\widehat{\theta}_n$.

We emphasize that imposition of the conditions for theorems 3.3 and 4.3 is for mathematical rigor of the proofs. In applications, implementation of the estimation procedure described in section 3.1 does not require that these conditions be satisfied. (Of course, the independent censorship assumption (A1) is needed for our likelihood based approach to be sensible.) Although heuristic arguments suggest that conditions (A3)-(A4) and (B4)-(B6) can be weakened, at present, we have not yet been able to prove these two theorems without these conditions. On the other hand, these conditions are not too restrictive in many practical situations. For example, assumption (B4) means that there is a positive time interval between two examination times. This is often the case since two examination times are usually separated by a positive time interval. We believe that asymptotic normality of $\widehat{\theta}_n$ in Theorems 3.3 and 4.3 continue to hold without(A3) or (B4). However, the proofs will be considerably more difficult.

4.1.3. Variance estimation.

With the "case 2" or general interval censoring, there is no close form expression for the asymptotic variance of $\widehat{\theta}_n$. Direct estimation appears to be difficult. We suggest two approaches which are straightforward extensions of two well-known methods used for parametric models. Let $l_n(\widehat{\theta}_n, \widehat{\Lambda}_n)$ be the log-likelihood function of $\widehat{\theta}_n$ and the *distinct* values of $\widehat{\Lambda}_n$. Let

$$\widehat{\Sigma}_{11} = \frac{\partial^2}{\partial \widehat{\theta}_n^2} l_n(\widehat{\theta}_n, \widehat{\Lambda}_n), \quad \widehat{\Sigma}_{12} = \frac{\partial^2}{\partial \widehat{\theta}_n \partial \widehat{\Lambda}_n} l_n(\widehat{\theta}_n, \widehat{\Lambda}_n), \quad \widehat{\Sigma}_{22} = \frac{\partial^2}{\partial \widehat{\Lambda}_n^2} l_n(\widehat{\theta}_n, \widehat{\Lambda}_n),$$

and

$$\widehat{\Sigma}_n = \begin{pmatrix} \widehat{\Sigma}_{11} & \widehat{\Sigma}_{12} \\ \widehat{\Sigma}_{12}' & \widehat{\Sigma}_{22} \end{pmatrix}.$$

Let $\widehat{\Sigma}^{11}$ be the upper left corner corresponding to $\widehat{\Sigma}_{11}$ of the inverse of $\widehat{\Sigma}_n$. Then $\widehat{\Sigma}^{11}$ can be used as an estimator of the variance-covariance matrix of $\widehat{\theta}_n$. It is well known that

$$\widehat{\Sigma}^{11} = (\widehat{\Sigma}_{11} - \widehat{\Sigma}_{12}\widehat{\Sigma}_{22}^{-1}\widehat{\Sigma}_{12}')^{-1}.$$

Computing the inverse of Σ_{22}^{-1} is not as bad as it looks even for large sample size n. $\widehat{\Lambda}_n$ is blockwise constant, and by the large sample properties of $\widehat{\Lambda}_n$ (Theorem 4.3) presented in the next section, the number of distinct values of $\widehat{\Lambda}_n$ is on the order of $n^{1/3}$. So the dimension of Σ_n is in the order of $O(n^{1/3})$.

The second approach is via profile likelihood: for low dimensional θ, it is sometimes more convenient to compute the curvature of the profile likelihood function $l_n(\theta, \widehat{\Lambda}_n(\cdot, \theta))$. The inverse of the curvature can be used as an estimator of the variance covariance matrix of $\widehat{\theta}_n$. Limited simulations and the example in section 4 suggest that $l_n(\theta, \widehat{\Lambda}_n(\cdot, \theta))$ is probably a smooth function of θ. But a rigorous proof of this smoothness is not yet available. In general, we can interpolate $l_n(\theta, \widehat{\Lambda}_n(\cdot, \theta))$ in a neighborhood of $\widehat{\theta}_n$ by a quadratic function. The second derivative of this quadratic function can be used as an approximation to the curvature of $l_n(\theta, \widehat{\Lambda}_n(\cdot, \theta))$.

It should be noted, however, that the asymptotic validity of both of the above approaches remains to be verified.

4.2. The proportional odds model.

In the case of the proportional odds model, the density of one observation is as in (4.1) with

$$F(t|z) = \exp(\alpha(t) + \theta' z)/(1 + \exp(\alpha(t) + \theta' z)). \tag{4.6}$$

Hence the log-likelihood function of the regression parameter θ and the baseline log-odds function α is

$$l_n(\theta, \alpha) = \sum_{I=1}^{n} \{\delta_{1i} \log F(U_i|Z_i) + \delta_{2i} \log[F(V_i|Z_i) - F(U_i|Z_i)] + \delta_{3i} \log[1 - F(V_i|Z_i)]\}$$

where the conditional distribution function $F(t|z)$ is given as in (4.6). The maximum likelihood estimator is the $(\widehat{\theta}_n, \widehat{\alpha}_n)$ that maximizes $l_n(\theta, \alpha)$ under the constraint that $\widehat{\alpha}_n$ is a nondecreasing function.

4.2.1. Information for θ.

The log-likelihood function for one observation is, up to an additive constant not dependent on (θ, α),

$$l(x, \theta, \alpha) = \delta_1 \log F(u|z) + \delta_2 \log[F(v|z) - F(u|z)] + \delta_3 \log[1 - F(v|z)],$$

with $F(t|z) = \exp(\alpha(t)+\theta' z)/(1+\exp(\alpha(t)+\theta' z))$. The score function $\dot{l}_\theta(x)$ for θ is simply the vector of partial derivatives of $l(x, \theta, \alpha)$ with respect to θ. In the following, we carry out the calculation for $\theta \in R$; the general case $\theta \in R^d$ only requires repeating the same calculation for each component of $\dot{l}_\theta(x)$.

Recall that $\alpha(t) = \text{logit } F(t) = \log F(t) - \log(1 - F(t))$, and let f be the density corresponding to F with f_s a one-dimensional smooth curve through f. Denote $a = \frac{\partial}{\partial s} \log f_s \big|_{s=0}$. Then $a \in L_2^0(F) \equiv \{a : \int a dF = 0 \text{ and } \int a^2 dF < \infty\}$. Let

$$h = \frac{\partial}{\partial s}\alpha_s\bigg|_{s=0} = \frac{\int_0^\cdot a dF}{F} - \frac{\int_\cdot^\infty a dF}{1-F} = \int_0^\cdot a dF \left\{\frac{1}{F} + \frac{1}{1-F}\right\} \tag{4.7}$$

since $\int_0^\infty a dF = 0$. The score operator for α, $\dot{l}_\alpha$, maps $L_2^0(F)$ to $L_2^0(P)$, where P is the joint probability measure of $(\delta_1, \delta_2, \delta_3, U, V, Z)$ and $L_2^0(P)$ is defined similarly as $L_2^0(F)$. This can be computed as

$$\dot{l}_\alpha a(x) = \frac{\partial}{\partial s} l(x, \theta, \alpha_s)\bigg|_{s=0} . \tag{4.8}$$

Let $\dot{l}_\alpha^T : L_2^0(P) \to L_2^0(F)$ be the adjoint operator of $\dot{l}_\alpha$, i.e., for any $a \in L_2^0(F)$ and $b \in L_2^0(P)$, define h as in (4.7),

$$\langle b, \dot{l}_\alpha a\rangle_P = \langle \dot{l}_\alpha^T b, a\rangle_F ,$$

where $\langle \cdot, \cdot\rangle_P$ and $\langle \cdot, \cdot\rangle_F$ are inner products in $L_2^0(P)$ and $L_2^0(F)$, respectively. We need to find a_* such that $\dot{l}_\theta - \dot{l}_\alpha(a_*)$ is orthogonal (in $L_2^0(P)$) to $\dot{l}_\alpha(a)$ for all a in $L_2^0(F)$. This amounts to solving the following normal equation:

$$\dot{l}_\alpha^T \dot{l}_\alpha(a_*)(t) = \dot{l}_\alpha^T \dot{l}_\theta(t). \tag{4.9}$$

THEOREM 4.4. *For $\theta \in R$, under conditions (A1)-(A3) and (B4)-(B6), equation (4.9) has a unique solution a_*. Moreover, the corresponding function h_* (given by (4.7) with a replaced by a_*) has bounded derivative. In general, for $\theta \in R^d$, h_* is a d-dimensional vector of functions in $L_2^0(F)$ and each component has a bounded derivative. The efficient score for θ is*

$$l_\theta^*(x) = \dot{l}_\theta(x) - \dot{l}_\alpha a_*(x).$$

The information bound for θ is

$$I(\theta) = E[l_\theta^*(X)^{\otimes 2}].$$

$I(\theta)$ is a positive definite matrix with finite entries. The efficient influence function is

$$\tilde{l}_\theta(x) = I(\theta)^{-1} l_\theta^*(x).$$

The proof of this theorem is given in Huang and Rossini (1996).

4.2.2. Distributional results.

Consistency of $(\hat{\theta}_n, \hat{\alpha}_n)$ can be shown similarly to theorem 4.2. For brevity, we only state the distributional results for $\hat{\theta}_n$ which can be proved in the same way as theorem 4.3. See Huang and Wellner (1995b).

THEOREM 4.5. *(Asymptotic normality and efficiency) Under the same conditions as in theorem 4.3,*

$$\sqrt{n}(\widehat{\theta}_n - \theta_0) \rightarrow_d N(0, I^{-1}(\theta_0)).$$

Hence $\widehat{\theta}_n$ is asymptotically normal and efficient.

4.2.3. Variance estimation.

This can be done similarly as in the proportional hazards model with "case 2" interval censoring, by using the observed Fisher information or the curvature of the profile likelihood function in a neighborhood of $\widehat{\theta}_n$.

4.3. The accelerated failure time model.

The log-likelihood function is, up to an additive constant,

$$\begin{aligned} l_n(\theta, F) &= \sum_{i=1}^{n} \{\delta_{1i} \log F(U_i - \theta' Z_i) \\ &\quad + \delta_{2i} \log[F(V_i - \theta' Z_i) - F(U_i - \theta' Z_i)] \\ &\quad + \delta_{3i} \log(1 - F(V_i - \theta' Z_i))\}. \end{aligned}$$

The maximum (profile) likelihood estimator of (θ_0, F_0) is the $(\widehat{\theta}_n, \widehat{F}_n)$ that maximizes $l_n(\theta, F)$. $(\widehat{\theta}_n, \widehat{F}_n)$ can be computed as in the "case 1" interval censoring, using the maximum profile likelihood approach of Cosslett (1983).

Asymptotic distribution theory for $\widehat{\theta}_n$ appears to be a more difficult problem than the corresponding one in the proportional hazards model, and is apparently still unknown. The main difficulty is that two parameters θ and F are "bundled" together: the result is that $\widehat{F}_n(\cdot; \theta)$ and consequently also the profile likelihood function are not smooth functions of θ. This model provides an example in which estimation of the variance-covariance matrix of $\widehat{\theta}_n$ by observed information fails since $l_n(\theta, F)$ is not a twice-differentiable function of θ.

4.3.1. Information.

For "case 2" interval censoring, as far as we know, the information calculation for θ has not been carried out and no efficient estimator of θ has been constructed. (For some work on inefficient estimators based on natural families of estimating equations, see Rabinowitz, Tsiatis, and Aragon (1995). Below we show that the information $I(\theta)$ for θ is positive. This suggests that it is possible to construct efficient estimators for θ. However, in contrast to the "case 1" setting, $I(\theta)$ does not have an explicit expression. The least favorable direction which defines the efficient score and $I(\theta)$ is determined by an integral equation with no closed form solution in general. This equation is similar to the one encountered in calculating the information for smooth functionals of the distribution function in the nonparametric setting (with no covariates). This problem is solved by Geskus and Groeneboom (1996a,b,c). Their approach and results can be used in the present setting. It should be noted that although the information calculations for this model with "case 1" interval censoring is quite straightforward, the same problem with "case 2" interval censored data is considerably more difficult. It involves an integral equation with a singular kernel (in general), and hence the Fredholm theory of integral equations cannot be applied directly.

We need the following conditions to carry out the information calculations. These conditions are similar (and slightly stronger for simplicity) to those given in Geskus and Groeneboom (1996c) in calculating the information for smooth functionals of the distribution function in the nonparametric setting. Let $U_1 = U - Z'\theta$ and $V_1 = V - Z'\theta$, and let h_1 be the joint density of (U_1, V_1). Let h_{11} and h_{12} be the marginal densities of U_1 and V_1, respectively. Finally, let

$$\begin{aligned} h_2(u_1, v_1) &= E\{Z|U_1 = u_1, V_1 = v_1\}h_1(u_1, v_1) \\ &= \int z h(u_1 + z'\theta, v_1 + z'\theta, z)d\mu(z). \end{aligned}$$

Assumption (D1): the union of the supports of U_1 and V_1 is a bounded interval $[-\tau_1, \tau_2]$, where $0 < \tau_1, \tau_2 < \infty$.

Assumption (D2): F has a continuous density f which is bounded on $[-\tau_1, \tau_2]$.

Assumption (D3): $h_1(u_1, v_1)$ is continuous and has continuous partial derivatives which are bounded, uniformly over $-\tau_1 \le u_1 \le v_1 \le \tau_2$. In addition, $h_{11}(s) + h_{12}(s) > 0$ for all $s \in [-\tau_1, \tau_2]$.

Assumption (D4): $h_2(u_1, v_1)$ is continuous and uniformly bounded over $-\tau_1 \le u_1 \le v_1 \le \tau_2$.

THEOREM 4.6. *Suppose that assumptions (D1) to (D4) hold. Then the efficient score for θ is*

$$l_\theta^*(x) = \dot{l}_\theta(x) - \dot{l}_f a_*(x), \tag{4.10}$$

where $\dot{l}_\theta$ and $\dot{l}_f a$ are defined by (6.9) and (6.10) below, and a_* is the unique solution to equation (6.12). The information bound for θ is:

$$I(\theta) = E[l_\theta^*(X)]^{\otimes 2}. \tag{4.11}$$

In section 6 we show how this theorem can be reduced to the situation considered by Geskus and Groeneboom (1996c) (or to Geskus and Groeneboom (1996a, 1996b)).

4.3.2. Consistency.

We prove a consistency result for the accelerated life model with "case 2" interval censored data; again the generalization to "case k" is straightforward.

THEOREM 4.7. *(i) The true value of the regression parameter is $\theta_0 \in \Theta$, where Θ is a bounded subset of R^d.*
(ii) The distribution of Z is not concentrated on a hyperplane.
(iii) F_0 is continuous. Then

$$\widehat{\theta}_n \to_{a.s.} \theta_0, \tag{4.12}$$

and

$$F_n(t; \widehat{\theta}_n) \to_{a.s.} F_0(t)\,. \tag{4.13}$$

for almost all t except on a set with $G_1 + G_2$-measure zero where G_1 and G_2 are probability measures corresponding to the marginal distribution functions G_1 and G_2 of U and V, respectively. In particular, if F_0 and at least one of G_1 and G_2 are continuous, then

$$\sup_{-\infty<t<\infty} |F_n(t; \widehat{\theta}_n) - F_0(t)| \to_{a.s.} 0.$$

The proof of this theorem is given in section 6.

4.4. Computation.

We focus on the proportional hazards model. Computation of the maximum likelihood estimator in the proportional odds model can be done in a similar way. For one or two-dimensional θ, the maximum profile likelihood approach can be used to compute $(\widehat{\theta}_n, \widehat{\Lambda}_n)$. For high-dimensional θ, the iterative algorithm described in section 3.4 is needed. The likelihood function with "case 2" interval censoring enjoys similar properties as the likelihood function for "case 1" data. Specifically, for any fixed θ, $l_n(\theta, \Lambda)$ is concave in Λ, and for any fixed Λ, $l_n(\theta, \Lambda)$ is concave in θ. So steps (i) and (ii) of the algorithm described in section 3.4 are again well-defined concave maximization problems, and the algorithm converges. Proposition 3.1 continues to hold in the "case 2" interval censoring setting.

To implement step (i), we use the iterative convex minorant algorithm introduced by Groeneboom (1991). To describe it in the present setting, we first introduce some notation. Define functions a_i, $i = 1, 2, 3$, by

$$a_1(x; \Lambda) = \frac{e^{\theta' z} \exp(-\Lambda(u) e^{\theta' z})}{1 - \exp(-\Lambda(u) e^{\theta' z})},$$

$$a_2(x; \Lambda) = \frac{e^{\theta' z} \exp(-\Lambda(u) e^{\theta' z})}{\exp(-\Lambda(u) e^{\theta' z}) - \exp(-\Lambda(v) e^{\theta' z})},$$

and

$$a_3(x;\Lambda) = \frac{e^{\theta' z}\exp(-\Lambda(v)e^{\theta' z})}{\exp(-\Lambda(u)e^{\theta' z}) - \exp(-\Lambda(v)e^{\theta' z})}.$$

Let

$$\begin{aligned} W_\Lambda(t) &= \sum_{i=1}^{n}\{\delta_{1i}a_1(X_i,\Lambda) - \delta_{2i}a_2(X_i,\Lambda)\}1_{[U_i\le t]} \\ &\quad + \sum_{i=1}^{n}\{\delta_{2i}a_3(X_i,\Lambda) - \delta_{3i}e^{\theta' Z_i}\}1_{V_i\le t]}, \end{aligned}$$

$$\begin{aligned} G_\Lambda(t) &= \sum_{i=1}^{n}\{\delta_{1i}a_1^2(X_i,\Lambda) + \delta_{2i}a_2^2(X_i,\Lambda)\}1_{[U_i\le t]} \\ &\quad + \sum_{i=1}^{n}\{\delta_{2i}a_3^2(X_i,\Lambda) + \delta_{3i}e^{2\theta' Z_i}\}1_{V_i\le t]}, \end{aligned}$$

and

$$V_\Lambda(t) = W_\Lambda(t) + \int_{[0,t]} \Lambda(s)\,dG_\Lambda(s).$$

Then $\widehat{\Lambda}_n$ is the left derivative of the greatest convex minorant of the self-induced cumulative sum diagram formed by the points $P_0 = (0,\ 0)$ and

$$P_j = (G_{\widehat{\Lambda}_n}(Y_{(j)}),\ V_{\widehat{\Lambda}_n}(Y_{(j)})), \qquad j = 1,\ldots,m.$$

With this characterization, the iterative convex minorant algorithm can again be used in step (i). Finally, the Newton-Raphson method can be used in step (ii).

5. Further problems.

1. **Regularity conditions.** In the regression setting, it would be very desirable to relax conditions (A4) and (A5) for case 1 data, and conditions (B4) to (B6) for case 2 data. It appears that removing the positive separation hypothesis (B4a) is the most difficult task. Recently, Groeneboom and Geskus (1996) succeeded in proving asymptotic normality and efficiency of the NPMLE of a smooth functional of the distribution function in the nonparametric setting (with no covariates) in the case when the two examination times U and V are arbitrarily close and the joint density $h(u,v)$ is strictly positive along the diagonal $u = v$. The techniques developed by Groeneboom and Geskus (1996) can probably be extended to regression problems.

2. **The accelerated failure time model with interval censoring.** Although with case 1 data, efficient estimators have been constructed by Klein and Spady (1993), it seems that no efficient estimators with case 2 or general interval-censored data have been constructed in the literature. It is shown in theorem 4.6 that the Fisher information for θ with case 2 data is positive under appropriate conditions. This suggests that it is possible to construct efficient estimators of θ. We are not able to prove that the maximum profile likelihood estimator is asymptotically normal and efficient. The main difficulty is that the profile likelihood function is not smooth in θ. We are investigating regularized (or penalized) profile maximum likelihood estimation approaches with case 2 data. It seems likely that these approaches will yield efficient estimators.
3. **Estimation of $I(\theta)$ and Confidence Sets for θ.** In the case 2 or the general interval censoring, we lack an explicit formula for the information $I(\theta)$ and do not know how to estimate $I(\theta)$ directly. In section 4.1.3 we suggested the use of observed information or the curvature of the profile likelihood function as estimators of the information and hence of the asymptotic variance of the estimators. It is reasonable to conjecture is that these two approaches provide consistent estimators of $I(\theta)$, but this remains to be proved. For some progress in this direction, see Murphy and van der Vaart (1996a, 1996b).
4. **Testing Hypotheses with interval-censored data.** It would be desirable to be able to test hypotheses about θ; for example, consider testing $H : c'(\theta - \theta_0) = 0$ versus $K : c'(\theta - \theta_0) \neq 0$ for fixed vectors θ_0 and c. Furthermore, inversion of the family of likelihood ratio tests of the hypotheses $H : \theta = \theta_0$ versus $K : \theta \neq \theta_0$ provides confidence sets for θ in the usual way. [Note that this method apparently circumvents direct estimation of the asymptotic variances of estimators.] Likelihood ratio type tests for semiparametric models have recently been studied by Murphy and Van der Vaart (1996). Roughly their result asserts that the natural likelihood ratio test for the finite-dimensional parameter θ in a semiparametric model "works as would be expected from finite-dimensional parametric" theory if the corresponding maximum likelihood estimator $\widehat{\theta}_n$ is asymptotically normal and efficient. Murphy and Van der Vaart (1995) consider several examples, including the proportional hazards model with case 1 interval censored data (see our section 3.1 and Huang (1996a)). It remains to implement their proposal in a range of interval censoring models and to compare the resulting confidence intervals with those obtained by other methods, including those suggested in sections 3.1.3 and 4.1.3.

Score tests seem feasible and useful in all of the interval censoring regression models considered here. Tests of this type have been previously suggested in the literature (see e.g. Hoel and Walburg (1972) and Peto et al. (1980)), but apparently have not yet been studied thoroughly and still lack justification from the current viewpoint. Fay (1996) considered a class of rank invariant tests for interval-censored data which appears to be closely related to score tests. The methods of Huang (1994) and (1996) seem to apply directly to these natural score tests, and we intend to pursue this elsewhere.

Another interesting problem concerns nonparametric tests for two or k populations with interval-censored data. We are currently investigating test statistics generated by score functions based on certain regression models; see e.g. Huang, Wang, and Wellner (1996).

5. **Regression models with time-dependent covariates.** In this review we have restricted attention to regression models with covariates *not* depending on time. Time-dependent covariates also often arise in practice. Extension of the maximum likelihood approach considered in this article seems feasible and potentially useful. It may also be possible to extend the methods for proving distributional properties with time-independent covariates to the case of time-dependent covariates.

6. **Small sample comparisons and refinements.** We have emphasized maximum likelihod procedures in this review. It remains to carry out thorough small-sample comparisons of these methods with alternative procedures involving sieved maximum likelihood or smoothing methods. For a bit of work in this direction in the case of the proportional odds model, see Rossini and Tsiatis (1996). [Rossini and Tsiatis (1996) concluded that "Nonetheless, we did find that the automatic choice of intervals resulting by the use of the NPMLE behaved well, for the most part, over a variety of models and sample sizes."]

6. Appendix.

Proof of Theorem 3.1. Without loss of generality, we only prove the theorem for $Z \in R$. The general case can be proved similarly. We first compute the score function for θ and F. The score function for θ is simply the derivative of the log-likelihood with respect to θ, that is,

$$\dot{l}_\theta(x) = ze^{\theta z}\Lambda(u)\left[\delta\frac{\overline{F}(u|z)}{1-\overline{F}(u|z)} - (1-\delta)\right].$$

Now suppose $\mathcal{F}_0 = \{F_\eta, |\eta| < 1\}$ is a regular parametric sub-family of $\mathcal{F} = \{F : F \ll \mu, \mu = \text{Lebesgue measure}\}$. Set $\frac{\partial}{\partial\eta}\log f_\eta(t)|_{\eta=0} = a(t)$, then $a \in L_2^0(F)$ and $\frac{\partial}{\partial\eta}\overline{F}_\eta(t)|_{\eta=0} = \int_t^\infty a dF$. The score operator for f is

$$\dot{l}_f(a)(x) \quad = \quad e^{\theta z}\frac{\int_u^\infty a dF}{\overline{F}(u)}\left[-\delta\frac{\overline{F}(u \mid z)}{1-\overline{F}(u \mid z)} + (1-\delta)\right].$$

Let $Q(u,\delta,z)$ be defined by (3.5), then

$$\dot{l}_\theta(x) = ze^{\theta z}\Lambda(u)Q(u,\delta,z) \qquad \text{and} \qquad \dot{l}_f(a)(x) = -e^{\theta z}\frac{\int_u^\infty a dF}{\overline{F}(u)}Q(u,\delta,z).$$

With conditions (A2) and (A3), $\dot{l}_\theta$ is square integrable, and for any $a \in L_2(F)$, $\dot{l}_f(a)$ is square integrable. To calculate the efficient score l_θ^* for θ, we need to find a function a_* so that

$$\dot{l}_\theta - \dot{l}_f a_* \perp \dot{l}_f a \text{ for all } a \in L_2^0(F), \tag{6.1}$$

that is, $E(\dot{l}_\theta - \dot{l}_f a_*)(\dot{l}_f a) = 0.$ for all $a \in L_2^0(F)$. Let $r(z) = e^z$. Some calculation yields

$$\begin{aligned} -E(\dot{l}_\theta - \dot{l}_f a_*)(\dot{l}_f a) &= E_U\left\{\frac{\int_U^\infty a dF}{\overline{F}(U)} E[r(2\theta Z)O(U \mid Z)[Z\Lambda(U) \right. \\ &\qquad \left. + \frac{\int_U^\infty a_* dF}{\overline{F}(U)}] \mid U]\right\}, \end{aligned}$$

where $O(U|Z)$ is defined by (3.6). Let

$$E\left\{r(2\theta Z)O(U \mid Z)[Z\Lambda(U) + \frac{\int_U^\infty a_* dF}{\overline{F}(U)}] \mid U\right\} = 0,$$

then

$$\Lambda(U)E[r(2\theta Z)O(U \mid Z)Z \mid U] = -\frac{\int_U^\infty a_* dF}{\overline{F}(U)} E[r(2\theta Z)O(U \mid Z) \mid U],$$

thus with a_* determined by

$$\int_u^\infty a_* dF = -\frac{\Lambda(u)\overline{F}(u)E[Zr(2\theta Z)O(U \mid Z)|U = y]}{E[r(2\theta Z)O(U \mid Z) \mid U = y]},$$

(6.1) holds. Notice that a_* is only determined on the support of U. However, $\dot{l}_f a_*$ is a square integrable function with expectation zero. So the efficient score function for θ is

$$\begin{aligned} \dot{l}_\theta^*(x) &= \dot{l}_\theta(x) - (\dot{l}_f a_*)(x) \\ &= r(\theta z)Q(y, \delta, z)\Lambda(u)\left\{z - \frac{E[(Zr(2\theta Z)O(U \mid Z) \mid U = y]}{E[(r(2\theta Z)O(U \mid Z) \mid U = y]}\right\}. \end{aligned}$$

The information for θ is

$$I(\theta) = E[\dot{l}_\theta^*(X)]^2 = E\left\{R(U, Z)\left[Z - \frac{E(ZR(U, Z) \mid U)}{E(R(U, Z) \mid U)}\right]^2\right\}, \tag{6.2}$$

where $R(U, Z) = \Lambda^2(U \mid Z)O(U \mid Z)$. Λ

Proof of Theorem 3.6. For simplicity, we will prove (3.14) and (3.15) for $\theta \in R$. The generalization to $\theta \in R^d$ is straightforward.

We first compute the score function for θ and F. The score function for θ is simply the derivative of the log-likelihood with respect to θ. That is,

$$\dot{l}_\theta(x) = -\delta \frac{f(u-z'\theta)z}{F(u-z'\theta)} + (1-\delta)\frac{f(u-z'\theta)z}{1-F(u-z'\theta)}.$$

Now suppose $\mathcal{F}_0 = \{F_\eta, |\eta| < 1\}$ is a regular parametric sub-family of $\mathcal{F} = \{F : F \ll \mu, \mu = \text{Lebesgue measure}\}$. Set

$$\frac{\partial}{\partial\eta} \log f_\eta(t) \mid_{\eta=0} = a(t).$$

Then $a \in L_2^0(F)$, and

$$\frac{\partial}{\partial\eta} F_\eta(t)|_{\eta=0} = \int_{-\infty}^t a dF, \quad \frac{\partial}{\partial\eta}\overline{F}_\eta(t)|_{\eta=0} = -\int_{-\infty}^t a dF.$$

The score operator for f is:

$$\dot{l}_f(a)(x) = \delta \frac{\int_{-\infty}^{u-z'\theta} a(t)dF(t)}{F(u-z'\theta)} - (1-\delta)\frac{\int_{-\infty}^{u-z'\theta} a(t)dF(t)}{1-F(u-z'\theta)}.$$

To calculate the information for θ in this semiparametric model, we follow the general theory of Bickel, Klaassen, Ritov, and Wellner (1993). We first need to compute the efficient score function $\dot{l}_\theta^*$ for θ. Geometrically, $\dot{l}_\theta^*$ can be interpreted as the residual of $\dot{l}_\theta$ projected in the space spanned by $\dot{l}_f a$, where $a \in L_2^0(F) = \{a : \int a dF = 0 \text{ and } \int a^2 dF < \infty\}$. Thus we need to find a function a_* with $\int a_* dF = 0$ so that $\dot{l}_\theta^* = \dot{l}_\theta - \dot{l}_f a_* \perp \dot{l}_f a$ for all $a \in L_2^0(F)$. That is

$$E(\dot{l}_\theta - \dot{l}_f a_*)(\dot{l}_f a) = 0 \tag{6.3}$$

for all $a \in L_2^0(F)$. Now we proceed to find a_* such that (6.3) is true. We have

$$\begin{aligned}
&\dot{l}_\theta(x) - (\dot{l}_f a_*)(x) \\
&= \left[f(u-z'\theta)z + \int_{-\infty}^{u-z'\theta} a_*(t)dF(t)\right]\left[\frac{1-\delta}{1-F(u-z'\theta)} - \frac{\delta}{F(u-z'\theta)}\right].
\end{aligned}$$

Thus

$$\begin{aligned}
&-E((\dot{l}_\theta - \dot{l}_f a_*)(\dot{l}_f a)) \\
&= E\Bigg\{\left[\frac{1-\delta}{1-F(U-Z'\theta)} - \frac{\delta}{F(U-Z'\theta)}\right]^2 \\
&\quad \times \left[f(U-Z'\theta)Z + \int_{-\infty}^{U-Z'\theta} a_*(t)dF(t)\right]\int_{-\infty}^{U-Z'\theta} a(t)dF(t)\Bigg\}
\end{aligned}$$

$$= E\left\{\left[f(U-Z'\theta)Z + \int_{-\infty}^{U-Z'\theta} a_*(t)dF(t)\right] \times \left[\frac{1}{1-F(U-Z'\theta)} + \frac{1}{F(U-Z'\theta)}\right] \int_{-\infty}^{U-Z'\theta} a(t)dF(t)\right\}.$$

So, with $U_1 \equiv U - Z'\theta$, we can set

$$E\left[(f(U_1)Z + \int_{-\infty}^{U_1} a_*(t)dF(t))\,|U_1\right] = 0$$

to ensure that (6.3) is true. We obtain a_* by solving the following equation:

$$f(s)E(Z|U - Z'\theta = s) + \int_{-\infty}^{s} a_*(t)dF(t) = 0.$$

In other words, we can choose any a_* that satisfies the above equation. In particular, if $f(s)$ and $k(s) = E(Z|U - Z'\theta = s)$ are differentiable, and $f(s) > 0$ for any $s \in R$, then we have an explicit expression for a_*:

$$a_*(s) = -k'(s) - k(s)\frac{f'(s)}{f(s)}.$$

By the assumptions, we have

$$\int a_*(t)dF(t) = \lim_{s\to\infty}\int_{-\infty}^{s} a_*(t)dF(t) = \lim_{s\to\infty} f(s)E(Z|U - Z'\theta = s) = 0.$$

It follows that the efficient score function for θ is

$$\begin{aligned} l_\theta^*(x) &= \dot{l}_\theta(x) - \dot{l}_f a_*(x) \\ &= f(u - z'\theta)[z - E(Z|U - Z'\theta = u - z'\theta)] \\ &\qquad \times \left[\frac{1-\delta}{1 - F(u - z'\theta)} - \frac{\delta}{F(u - z'\theta)}\right]. \end{aligned}$$

The information for θ is

$$\begin{aligned} I(\theta) &= E[\dot{l}_\theta^*(X)]^2 \\ &= E\left\{\left[\frac{f(U - Z'\theta)^2}{F(U - Z'\theta)(1 - F(U - Z'\theta))}\right][Z - E(Z|U - Z'\theta)]^2\right\}. \end{aligned}$$

Hence the information for θ is positive unless $Z = E(Z|U - Z'\theta)$ with probability one. Λ

Proof of Theorem 4.1. Recall that a_1, a_2 and a_3 are defined in section 4.4. By conditions (A3) and (A4), a_1, a_2 and a_3 are positive functions of

(u, v, z). To make formulas shorter, we will drop the arguments of a's in the following. Let $x = (\delta, \gamma, u, v, z)$, and let $l(x; \theta, \Lambda)$ be the logarithm of the density function of X. The score function for θ is

$$\dot{l}_\theta(x) = \frac{\partial}{\partial \theta} l(x; \theta, \Lambda) = z e^{\theta' z}\{\delta\Lambda(u)a_1 - \gamma[\Lambda(u)a_2 - \Lambda(v)a_3] - (1-\delta-\gamma)\Lambda(v)\}.$$

The score operator for Λ is

$$\begin{aligned} \dot{l}_\Lambda a(x) &= \left.\frac{\partial}{\partial s} l(x; \theta, \Lambda_s)\right|_{s=0} \\ &= e^{\theta' z}\{\delta h(u)a_1 - \gamma[h(u)a_2 - h(v)a_3] - (1-\delta-\gamma)h(v)\} \end{aligned}$$

where $h \equiv h[a]$ is defined by (4.4). In the following, we carry out the calculation for $\theta \in R$. For the general case $\theta \in R^d$, we only need to repeat the same calculation for each component of $\dot{l}_\theta(x)$.

The score operator $\dot{l}_\Lambda$ maps $L_2^0(F)$ to $L_2^0(P)$, where P is the joint probability measure of $(\delta, \gamma, U, V, Z)$ and $L_2^0(P)$ is defined similarly as $L_2^0(F)$. Let $\dot{l}_\Lambda^T : L_2^0(P) \to L_2^0(F)$ be the adjoint operator of $\dot{l}_\Lambda$, i.e., for any $a \in L_2^0(F)$ and $b \in L_2^0(P)$, define h as in (4.7),

$$\langle b, \dot{l}_\Lambda a\rangle_P = \langle \dot{l}_\Lambda^T b, a\rangle_F \ ,$$

where $\langle\cdot,\cdot\rangle_P$ and $\langle\cdot,\cdot\rangle_F$ are the inner products in $L_2^0(P)$ and $L_2^0(F)$, respectively. We need to find a_* such that $\dot{l}_\theta - \dot{l}_\alpha(a_*)$ is orthogonal to $\dot{l}_\alpha(a)$ in $L_2^0(P)$. This amounts to solving the following normal equation:

$$\dot{l}_\Lambda^T \dot{l}_\Lambda(a_*) = \dot{l}_\Lambda^T \dot{l}_\theta. \tag{6.4}$$

We know that

$$\dot{l}_\Lambda^T \dot{l}_\Lambda(a)(t) = E[\dot{l}_\Lambda(a)(X)|T = t] = E_Z E[\dot{l}_\Lambda(a)(X)|T = t, Z];$$

see e.g. BKRW (1993), pages 271-272, or Groeneboom and Wellner (1992), pages 8 and 9. By conditions (A1) and (A4),

$$\begin{aligned} &E[\dot{l}_\Lambda(a)(X)|T = t, Z = z] \\ &= \int_{u=t}^{\tau_1}\int_{v=u+\eta}^{\tau_1} e^{\theta' z} h(u)a_1 g(u, v|z)dvdu \\ &\quad - \int_{u=\tau_0}^{t}\int_{v=t}^{\tau_1} e^{\theta' z}(h(u)a_2 - h(v)a_3)g(u, v|z)1_{[v-u\geq\eta]}dvdu \\ &\quad - \int_{u=\tau_0}^{t}\int_{v=u+\eta}^{t} e^{\theta' z} h(v)g(u, v|z)dudv, \end{aligned}$$

where $g(u, v|z)$ is the conditional density of (U, V) given Z. Let

$$b_1(u, v) = E_Z[e^{\theta' Z} a_1(u, v, Z)g(u, v|Z)],$$

$$b_2(u,v) = E_Z[e^{\theta' Z} a_2(u,v,Z) g(u,v|Z)],$$
$$b_3(u,v) = E_Z[e^{\theta' Z} a_3(u,v,Z) g(u,v|Z)],$$

and

$$b_4(u,v) = E_Z[e^{\theta' Z} g(u,v|Z)].$$

By the their definition, all the functions A_j and B_j are positive functions, and

$$b_2(u,v) = b_3(u,v) + b_4(u,v). \tag{6.5}$$

We calculate

$$\begin{aligned} L(t) &\equiv \dot{l}_\Lambda^T \dot{l}_\Lambda(a)(t) \\ &= \int_{u=t}^{\tau_1} \int_{v=u+\eta}^{\tau_1} h(u) b_1(u,v)\,dv\,du \\ &\quad - \int_{u=\tau_0}^{t} \int_{v=t}^{\tau_1} [h(u) b_2(u,v) - h(v) b_3(u,v)] 1_{[v-u\geq\eta]}\,dv\,du \\ &\quad - \int_{u=\tau_0}^{t} \int_{v=u+\eta}^{t} h(v) b_4(u,v)\,dv\,du; \end{aligned}$$

recall that h is defined in terms of a by (4.4). Let

$$c_1(u,v) = E_Z[Z e^{\theta' Z} a_1(u,v,Z) g(u,v|Z)],$$
$$c_2(u,v) = E_Z[Z e^{\theta' Z} a_2(u,v,Z) g(u,v|Z)],$$
$$c_3(u,v) = E_Z[Z e^{\theta' Z} a_3(u,v,Z) g(u,v|Z)],$$

and

$$c_4(u,v) = E_Z[Z e^{\theta' Z} g(u,v|Z)].$$

Then further calculation yields

$$\begin{aligned} R(t) \equiv \dot{l}_\Lambda^T \dot{l}_\theta(t) &= \int_{u=t}^{\tau_1} \int_{v=u+\eta}^{\tau_1} \Lambda(u) c_1(u,v)\,dv\,du \\ &\quad - \int_{u=\tau_0}^{t} \int_{v=t}^{\tau_1} [\Lambda(u) c_2(u,v) - \Lambda(v) c_3(u,v)] 1_{[v-u\geq\eta]}\,dv\,du \\ &\quad - \int_{u=\tau_0}^{t} \int_{v=u+\eta}^{t} \Lambda(v) c_4(u,v)\,dv\,du. \end{aligned}$$

Let

$$b(t) = \int_{t+\eta}^{\tau_1} b_1(t,x)\,dx + \int_{t+\eta}^{\tau_1} b_2(t,x)\,dx + \int_{\tau_0}^{t-\eta} b_3(x,t)\,dx + \int_{\tau_0}^{t-\eta} b_4(x,t)\,dx.$$

After some straightforward calculations, the derivative of $L(t)$ is

$$L'(t) = -b(t)h(t) + \int_{\tau_0}^{t-\eta} h(x) b_2(x,t)\,dx + \int_{t+\eta}^{\tau_1} h(x) b_3(t,x)\,dx.$$

Similarly, let

$$c(t) = \int_{t+\eta}^{\tau_1} c_1(t,x)dx + \int_{t+\eta}^{\tau_1} c_2(t,x)dx + \int_{\tau_0}^{t-\eta} c_3(x,t)dx + \int_{\tau_0}^{t-\eta} c_4(x,t)dx.$$

The derivative of $R(t)$

$$r(t) \equiv R'(t) = -c(t)\Lambda(t) + \int_{\tau_0}^{t-\eta} \Lambda(x)c_2(x,t)dx + \int_{t+\eta}^{\tau_1} \Lambda(x)c_3(t,x)dx.$$

By conditions (A3)–(A5), r has a bounded derivative r' on $[\tau_0, \tau_1]$. So equation (6.4) reduces to

$$-b(t)h(t) + \int_{\tau_0}^{t-\eta} h(x)b_2(x,t)dx + \int_{t+\eta}^{\tau_1} h(x)b_3(t,x)dx = r(t). \qquad (6.6)$$

By conditions (A3) and (A4), $\inf_{\tau_0 \le t \le \tau_1} b(t) > 0$. Let $d(t) = -r(t)/b(t)$ and

$$K(t,x) = [b_2(x,t)1_{[\tau_0 \le x \le t-\eta]} + b_3(t,x)1_{[t+\eta \le x \le \tau_1]}]/b(t).$$

We obtain a Fredholm integral equation of the second kind,

$$h_*(t) - \int K(t,x)h_*(x)dx = d(t).$$

Since $K(t,x)$ is a bounded kernel (K being a L_2 kernel suffices), by lemma 6.1 below and the classical results on Fredholm integral equations (Kanwal 1971), Sections 4.2 and 4.3; or Kress (1989), Chapter 4), there exists a resolvent $\Gamma(t,x)$ (completely determined by K) such that

$$h_*(t) = d(t) + \int \Gamma(t,x)d(x)dx. \qquad (6.7)$$

From equations (6.6) and (6.7), we can derive properties of h_*. By (6.7), h_* is bounded on $[\tau_0, \tau_1]$. By (6.6), this implies h_* is continuous. This in turn implies h_* is differentiable. Since b is bounded away from zero, the partial derivative of $b_2(x,t)$ with respect to t is bounded on $[\tau_0, t-\eta]$, the partial derivative of $b_3(t,x)$ with respect to t is bounded on $[t+\eta, \tau_1]$, and the derivative of r is bounded, it follows that the derivative of h_* is bounded. Λ

The following lemma is used in the proof of theorem 4.1.

LEMMA 6.1. *If*

$$h(t) - \int K(t,x)h(x)dx = 0,$$

then $h(t) \equiv 0$ *on* $[\tau_0, \tau_1]$.

Proof of Lemma 6.1. According to the definition of K, the equation can be written as

$$\begin{aligned}
&\int_{t+\eta}^{\tau_1} B_3(t,x)h(x)dx + \int_{\tau_0}^{t-\eta} B_2(x,t)h(x)dx \\
&= h(t)\left[\int_{t+\eta}^{\tau_1} B_1(t,x)dx + \int_{t+\eta}^{\tau_1} B_2(t,x)dx \right.\\
&\qquad \left. + \int_{\tau_0}^{t-\eta} B_3(x,t)dx + \int_{\tau_0}^{t-\eta} B_4(x,t)dx\right]. \qquad (6.8)
\end{aligned}$$

If there exists a point x_0 such that $h(x_0) > 0$, let y be the point in $[\tau_0, \tau_1]$ such that $h(u) = \sup_{\tau_0 \le x \le \tau_1} h(x)$, then by equation (6.5),

$$\begin{aligned}
&\int_{u+\eta}^{\tau_1} h(x)B_3(u,x)dx + \int_{\tau_0}^{u-\eta} h(x)B_2(x,y)dx \\
&\le h(u)\left[\int_{u+\eta}^{\tau_1} B_3(u,x)dx + \int_{\tau_0}^{u-\eta} B_2(x,y)dx\right] \\
&< h(u)\left[\int_{u+\eta}^{\tau_1} B_2(u,x)dx + \int_{\tau_0}^{u-\eta} B_2(x,y)dx\right] \\
&\le h(u)\left[\int_{u+\eta}^{\tau_1} B_1(u,x)dx + \int_{u+\eta}^{\tau_1} B_2(u,x)dx + \int_{\tau_0}^{u-\eta} B_2(x,y)dx\right] \\
&\le h(u)\left[\int_{u+\eta}^{\tau_1} B_1(u,x)dx + \int_{u+\eta}^{\tau_1} B_2(u,x)dx \right.\\
&\qquad \left. + \int_{\tau_0}^{u-\eta} B_3(x,y)dx + \int_{\tau_0}^{u-\eta} B_4(x,y)dx\right].
\end{aligned}$$

This is a contradiction to equation (6.8). Since if h satisfies (6.8), $-h$ also satisfies (6.8), so there can not be a point x_0 such that $h(x_0) < 0$. Λ

Proof of Theorem 4.6. For simplicity, we will carry out the computation for one-dimensional $\theta \in R$. For $\theta \in R^d$, we only need to repeat the same calculation for each component of θ.

The score function for θ is simply the derivative of the log-likelihood with respect to θ. That is,

$$\begin{aligned}
\dot{l}_\theta(x) &= \left(-\delta_1 \frac{f(u-z'\theta)}{F(u-z'\theta)} - \delta_2 \frac{f(v-z'\theta)-f(u-z'\theta)}{F(v-z'\theta)-F(u-z'\theta)}\right.\\
&\qquad \left. + \delta_3 \frac{f(v-z'\theta)}{1-F(v-z'\theta)}\right) z. \qquad (6.9)
\end{aligned}$$

Now suppose $\mathcal{F}_0 = \{F_\eta, |\eta| < 1\}$ is a regular parametric sub-family of $\mathcal{F} = \{F : F \ll \mu, \mu = \text{Lebesgue measure}\}$. Set

$$\frac{\partial}{\partial \eta} \log f_\eta(t) \mid_{\eta=0} = a(t).$$

Then $a \in L_2^0(F) \equiv \{a : \int a dF = 0,$ and $\int a^2 dF < \infty$, and, with $\overline{F} \equiv 1 - F$,

$$\frac{\partial}{\partial \eta} F_\eta(t)|_{\eta=0} = \int_{-\infty}^{t} a dF, \qquad \frac{\partial}{\partial \eta} \overline{F}_\eta(t)|_{\eta=0} = -\int_{-\infty}^{t} a dF.$$

Let $\phi(t) = \int_{-\infty}^{t} a(s) dF(s)$. The score operator for f is:

$$\dot{l}_f(a)(x) = \delta_1 \frac{\phi(u - z'\theta)}{F(u - z'\theta)} + \delta_2 \frac{\phi(v - z'\theta) - \phi(u - z'\theta)}{F(v - z'\theta) - F(u - z'\theta)} - \delta_3 \frac{\phi(v - z'\theta)}{1 - F(v - z'\theta)}. \tag{6.10}$$

The score operator $\dot{l}_f$ maps $L_2^0(F)$ to $L_2^0(P)$, where P is the joint probability measure of $(\delta_1, \delta_2, U, V, Z)$ and $L_2^0(P)$ is defined similarly as $L_2^0(F)$. Let $\dot{l}_f^T : L_2^0(P) \to L_2^0(F)$ be the adjoint operator of $\dot{l}_f$, i.e., for any $a \in L_2^0(F)$ and $b \in L_2^0(P)$,

$$\langle b, \dot{l}_f a \rangle_P = \langle \dot{l}_f^T b, a \rangle_F \ ,$$

where $\langle \cdot, \cdot \rangle_P$ and $\langle \cdot, \cdot \rangle_F$ are the inner products in $L_2^0(P)$ and $L_2^0(F)$, respectively. We need to find a_* such that $\dot{l}_\theta - \dot{l}_f(a_*)$ is orthogonal to $\dot{l}_\alpha(h)$ in $L_2^0(P)$. This amounts to solving the following normal equation:

$$\dot{l}_f^T \dot{l}_f(a_*) = \dot{l}_f^T \dot{l}_\theta. \tag{6.11}$$

The value of $\dot{l}_f^T$ at any $b \in L_2^0(P)$ can be computed by

$$\begin{aligned} \dot{l}_f^T b(t) &= E[b(X)|T = t] \\ &= E_Z E[b(X)|T = t, Z] = E_Z E[b(X)|T - Z'\theta = t - Z'\theta, Z]; \end{aligned}$$

see e.g. BKRW (1993), pages 271-272, or Groeneboom and Wellner (1992), pages 8 and 9.

Let h_1 be the joint density of (U_1, V_1) where $U_1 = U - Z'\theta, V_1 = V - Z'\theta$. Let $s = t - z'\theta$. Then, with $\phi(t) \equiv \int_{-\infty}^{t} a dF$ as before,

$$\begin{aligned} \dot{l}_f^T \dot{l}_f a(t) = & \int_{u_1=t}^{\infty} \int_{v_1=u_1}^{\infty} \frac{\phi(u_1)}{F(u_1)} h_1(u_1, v_1) dv_1 du_1 \\ & + \int_{u_1=-\infty}^{t} \int_{v_1=t}^{\infty} \frac{\phi(v_1) - \phi(u_1)}{F(v_1) - F(u_1)} h_1(u_1, v_1) dv_1 du_1 \\ & - \int_{v_1=-\infty}^{t} \int_{u_1=-\infty}^{v_1} \frac{\phi(v_1)}{1 - F(v_1)} h_1(u_1, v_1) du_1 dv_1. \end{aligned}$$

Similarly, let

$$\begin{aligned} h_2(u_1, v_1) &= \int z h(u_1 + z'\theta, v_1 + z'\theta, z) d\mu(z) \\ &= E\{Z | U_1 \equiv U - Z'\theta = u_1, V_1 \equiv V - Z'\theta = v_1\} h_1(u_1, v_1). \end{aligned}$$

Then

$$\dot{l}_f^T \dot{l}_\theta(t) = -\int_{u_1=t}^{\infty}\int_{v_1=u_1}^{\infty}\frac{f(u_1)}{F(u_1)}h_2(u_1,v_1)dv_1du_1$$
$$-\int_{u_1=-\infty}^{t}\int_{v_1=t}^{\infty}\frac{f(v_1)-f(u_1)}{F(v_1)-F(u_1)}h_2(u_1,v_1)dv_1du_1$$
$$+\int_{v_1=-\infty}^{t}\int_{u_1=-\infty}^{v_1}\frac{f(v_1)}{1-F(v_1)}h_2(u_1,v_1)du_1dv_1\,.$$

Let h_{11} and h_{12} be the marginal densities of h_1. Using these two expressions and differentiating both sides of (6.11), we find, after multiplying across by $-d$, that

$$\begin{aligned}\phi(t) \quad &+ \quad d(t)\left[\int_{-\infty}^{t}\frac{\phi(t)-\phi(u)}{F(t)-F(u)}h_1(u,t)du - \int_t^{\infty}\frac{\phi(u)-\phi(t)}{F(u)-F(t)}h_1(t,u)du\right]\\ &= \quad -\, d(t)g(t)\end{aligned} \tag{6.12}$$

where

$$d(t) = \frac{F(t)(1-F(t))}{h_{11}(t)[1-F(t)]+h_{12}(t)F(t)},$$

$g(t)$ is the derivative of $\dot{l}_f^T \dot{l}_\theta(t)$, and $g(t)F(t)(1-F(t))$ is given by

$$\begin{aligned}g(t)F(t)\overline{F}(t) &= \{h_{21}(t)[1-F(t)] + h_{22}(t)F(t)\}f(t)\\ &+F(t)\overline{F}(t)\left[\int_{u=-\infty}^{t}\frac{\phi(t)-\phi(u)}{F(t)-F(u)}h_2(u,t)du\right.\\ &\left.-\int_{u=t}^{\infty}\frac{\phi(u)-\phi(t)}{F(u)-F(t)}h_2(t,u)du\right]\end{aligned}$$

with $h_{21}(u_1) = \int_{u_1}^{\infty} h_2(u_1,v_1)dv_1$ and $h_{22}(v_1) = \int_{-\infty}^{v_1} h_2(u_1,v_1)du_1$.

Equation (6.12) has exactly the same structure as the integral equation involved in computing the information for a smooth functional of the distribution function studied in Geskus and Groeneboom (1996a,b,c).

In particular, Geskus and Groeneboom (1996c) showed that (6.12) has a unique solution $\phi_*(s)$ and that the Radon-Nikodyn derivative $d\phi_*/dF$ is a.e.$-[F]$ bounded under hypotheses which are implied by our assumptions (D1) - (D4); see their Theorem 2.2 and Corollary 2.1. This implies (6.11) has a unique solution. Λ

Proof of Theorem 4.7. We apply Theorem 3.1 of Van de Geer (1993). Let

$$p(x;\theta,F) \quad = \quad F(u-\theta'z)^{\delta_1}[F(v-\theta'z)-F(u-\theta'z)]^{\delta_2}(1-F(v-\theta'z))^{\delta_3}.$$

Denote the class of functions

$$\mathcal{H} = \{p(x;\theta, F) : (\theta, F) \in \Theta \times \mathcal{F}\}.$$

Clearly $\mathcal{H}$ is uniformly bounded by 1. Here we change notation from section 4 slightly by letting μ be the product of counting measure on $\{(0,0),(0,1),(1,0)\}$ and the joint measure induced by the joint distribution H of (U, V, Z). Then

$$\int p(x;\theta_0, F_0) d\mu(x) \leq 1.$$

Furthermore, the class $\mathcal{H}$ is a VC-hull class; see Dudley (1987), or Van der Vaart and Wellner (1996), section 2.6.3 and lemma 2.6.19, page 145. Thus the entropy condition of Theorem 3.1 of Van de Geer (1993) are satisfied. This implies

$$\int \left(\sqrt{p(x;\widehat{\theta}_n, \widehat{F}_n)} - \sqrt{p(x;\theta_0, F_0)}\right)^2 d\mu(x) \rightarrow_{a.s.} 0.$$

This in turn implies,

$$\int \left(\sqrt{\widehat{F}_n(u - \widehat{\theta}_n{}' z;\widehat{\theta}_n)} - \sqrt{F_0(u - \theta_0' z)}\right)^2 dG(u, v, z) \rightarrow_{a.s.} 0.$$

Changing variables in the integration and integration over v yields

$$\int \left(\sqrt{\widehat{F}_n(s;\widehat{\theta}_n)} - \sqrt{F_0(u + \widehat{\theta}_n - \theta_0' z)}\right)^2 dG_{13}(s + \widehat{\theta}_n{}' z, z) \rightarrow_{a.s.} 0 \quad (6.13)$$

where G_{13} is the marginal distribution of (U, Z). Since Θ is bounded, for any subsequence of $\widehat{\theta}_n$, we can find a further subsequence converging to $\theta_* \in \overline{\Theta}$, the closure of Θ. On the other hand, by Helly's selection theorem, for any subsequence of $F_n(\cdot;\widehat{\theta}_n)$, we can find a further subsequence converging in distribution to some subdistribution function F_*; i.e. pointwise convergence at continuity points of F_*. Apparently, we can choose the convergent subsequence of $\widehat{\theta}_n$ and the convergent subsequence of $F_n(\cdot;\widehat{\theta}_n)$ so that they have the same indices. Without causing confusion, we assume that $\widehat{\theta}_n$ converges to θ_* and that $F_n(\cdot;\widehat{\theta}_n)$ converges to F_*. To prove the theorem, it suffices to prove that $\theta_* = \theta_0$ and $F_* = F_0$. By continuity of F_0 and (6.13), F_* and θ_* satisfy

$$F_*(u) = F_0(u + \theta_*' z - \theta_0' z) \quad \text{for } G - \text{almost all } (u, z).$$

Thus

$$\theta_*' z = \theta_0' z \quad \text{and} \quad F_*(u) = F_0(u) \quad G - \text{almost all } (u, z).$$

Condition (ii) implies $\theta_* = \theta_0$. (4.12) follows. Similarly, we have

$$F_*(v) = F_0(v) \quad G_2 - \text{almost all } v,$$

and (4.13) follows. Λ

ACKNOWLEDGEMENTS: The authors owe thanks to Piet Groeneboom and Tony Rossini for many helpful discussions about interval censoring. We also owe thanks to John Crowley for pointing out several references in connection with testing with interval censored data. The research of Jon A. Wellner was supported in part by National Science Foundation grant DMS-9306809 and by NIAID grant 2R01 AI291968-04.

REFERENCES

Aragón, J. and Eberly D. (1992). On convergence of convex minorant algorithms for distribution estimation with interval-censored data. *J. of Computational and Graphical Stat.* **1**, 129-140.

Bennett, S. (1983). Analysis of survival data by the proportional odds model. *Stat. in Medicine* **2**, 273 - 277.

Bickel, P. J., Klaassen, C. A. J., Ritov, Y., and Wellner, J. A. (1993). *Efficient and Adaptive Estimation for Semiparametric Models.* Johns Hopkins University Press, Baltimore.

Chamberlain, G. (1986). Asymptotic efficiency in semi-parametric models with censoring. *J. Econometrics* **32**, 189-218.

Chernoff, H. (1954). On the distribution of the likelihood ratio. *Ann. Math. Statist.* **25**, 573 - 578.

Cosslett, S. R. (1983) Distribution-free maximum likelihood estimator of the binary choice model. *Econometrica* **51**, 765-782.

Cosslett, S. R. (1987). Efficiency bounds for distribution-free estimators of the binary choice and censored regression models. *Econometrica* **55**, 559-585.

Cox, D. R. (1972). Regression models and life-tables (with discussion). *Journal of the Royal Statistical Society. B* **34**, 187-220.

De Gruttola, V. and Lagakos, S. W. (1989). Analysis of doubly-censored survival data, with application to AIDS. *Biometrics* **45**, 1-11.

Diamond, I. D. and McDonald, J. W. (1991). Analysis of current status data. In *Demographic Applications of Event History Analysis*, eds. J. Trussell, R. Hankinson, and J. Tilton, Oxford, U.K., Oxford University Press, 231-252.

Diamond, I. D. and McDonald, J. W. and Shah, I. H. (1986). Proportional hazards models for current status data: application to the study of differentials in age at weaning in Pakistan. *Demography* **23**, 607-620.

Dinse, G. E. and Lagakos, S. W. (1983). Regression analysis of tumour prevalence data. *Applied Statistics* **32**, 236 - 248. (Corr. **33**, 79-80.)

Fay, M. (1996). Rank invariant tests for interval censored data under the grouped continuous model. *Biometrics* **52**, to appear.

Finkelstein, D. M. and Wolfe R. A. (1985). A semiparametric model for regression analysis of interval-censored failure time data. *Biometrics* **41**, 933-945.

Finkelstein, D. M. (1986). A proportional hazards model for interval-censored failure time data. *Biometrics* **42**, 845-854.

Geskus, R. and Groeneboom, P. (1996a). Asymptotically optimal estimation of smooth functionals for interval censoring, part 1. *Statistica Neerl.* **50**, to appear. Technical Report, Delft University of Technology.

Geskus, R. and Groeneboom, P. (1996b). Asymptotically optimal estimation of smooth functionals for interval censoring, part 2. *Statistica Neerl.* **50**, to appear. Technical Report, Delft University of Technology.

Geskus, R. and Groeneboom, P. (1996c). Asymptotically optimal estimation of smooth functionals for interval censoring, case 2. Technical Report, Delft University of Technology. Submitted to *Ann. Statist.*

Gill, R. D. and Levit, B. Y. (1995). Applications of the van Trees inequality: a Bayesian Cramér-Rao bound. *Bernoulli* **1**, 59 - 79.

Groeneboom, P. (1985). Estimating a monotone density. *Proceedings of the Berkeley Conference in Honor of Jerzy Neyman and Jack Kiefer*, Vol. II, Lucien M. LeCam and Richard A. Olshen eds.

Groeneboom, P. (1987). Asymptotics for interval censored observations. Technical Report 87-18, Department of Mathematics, University of Amsterdam.

Groeneboom, P. (1991). Nonparametric maximum likelihood estimators for interval censoring and deconvolution. *Technical Report 378*, Department of Statistics, Stanford University.

Groeneboom, P. (1996). Inverse problems in statistics.. Proceedings of the St. Flour Summer School in Probability, 1994. *Lecture Notes in Math.*, to appear. Springer Verlag, Berlin.

Groeneboom, P. and Wellner, J. A. (1992). *Information Bounds and Nonparametric Maximum Likelihood Estimation.* DMV Seminar Band 19, Birkhäuser, Basel.

Han, A. K. (1987). Non-parametric analysis of a generalized regression model. *J. Econometrics*, **35**, 303-316.

Hoel, D. G. and Walburg, H. E. (1972). Statistical analysis of survival experiments. *J. of the National Cancer Institute* **49**, 361-372.

Huang, J. (1993). Maximum scored likelihood estimation of a linear regression with interval censored data. *Technical Report* **256**, Department of Statistics, University of Washington.

Huang, J. (1995). Maximum Likelihood Estimation for Proportional Odds Regression Model with Current Status Data. *Analysis of Censored Data*, IMS Lecture Notes - Monograph Series **27**, 129 - 146. H.L. Koul and J. V. Deshpande, editors. IMS, Hayward.

Huang, J. (1996). Efficient estimation for the Cox model with interval censoring. *Ann. Statist.* **24**, 540 - 568.

Huang, J. and Rossini, A. J. (1995). Sieve estimation for the proportional odds failure-time regression model with interval censoring. Technical Report, Department of Statistics, University of Iowa. Submitted to *J. Amer. Statist. Assoc.*.

Huang, J., Wang, K., and Wellner, J. A. (1996). Nonparametric tests with interval-censored data. Submitted to *J. Amer. Statist. Assoc.*.

Huang, J. and Wellner, J. A. (1995a). Asymptotic Normality of the NPMLE of Linear Functionals for Interval Censored Data, Case 1. *Statistica Neerlandica* **49**, 153 - 163.

Huang, J. and Wellner, J. A. (1995b). Efficient estimation for the proportional hazards model with "case 2" interval censoring. *Technical Report No.* **290**, Department of Statistics, University of Washington, Seattle. Submitted to *Biometrika*.

Huang, J. and Wellner, J. A. (1996). Regression models with interval censoring. To appear, *Proceedings of the Kolmogorov Seminar, Euler Mathematics Institute,* St. Petersburg, Russia. *Technical Report* **261**, Department of Statistics, University of Washngton.

Jewell, N. P. , Malani, H. M. and Vittinghoff, E. (1994) Nonparametric estimation for a form of doubly censored data, with application to two problems in AIDS. *Journal of the American Satistical Association* **89**, 7-18.

Jongbloed, G. (1995a). *Three Statistical Inverse Problems.* Unpublished Ph.D. dissertation, Delft University of Technology, Delft.

Jongbloed, G. (1995b). The iterative convex minorant algorithm for nonparametric estimation. Submitted to *J. of Computational and Graphical Stat.*

Kim, J. and Pollard, D. (1990). Cube root asymptotics. *Ann. Statist.* **18**, 191-219.

Klein, R. W. and Spady, R. H. (1993). An efficient semiparametric estimator for binary response models. *Econometrica* **61**, 387-421.

Kongerud, J. and Samuelsen, S.O. (1991). A longitudinal study of respiratory symptoms in aluminum potroom workers. *American Review of Respiratory Diseases,* **144**, 10-16.

Kress, R. (1989). *Linear Integral Equations.* Springer-Verlag, Berlin.

Manski, C. F. (1975). Maximum score estimation of the stochastic utility model of choice. *J. Econometrics* **3**, 205-228.

Manski, C. F. (1985). Semiparametric analysis of discrete response: Asymptotic properties of the maximum score estimator. *J. Econometrics* **27**, 313-333.

Murphy, S. A. and Van der Vaart, A. W. (1996a). Semiparametric likelihood ratio inference. Technical Report, Pennsylvania State University.

Murphy, S. A. and Van der Vaart, A. W. (1996b). Observed information in semiparametric models. Preprint.

Peto, R., Pike, M., Day, N., Gray, R., Lee, P., Parish, S., Peto, J., Richards, S., and Wahrendorf, J. (1980). Guidelines for simple, sensitive significance tests for carcinogenic effects in long term animal experiments. In *IARC Monographs on the Evaluation of the Carcinogenic Risk of Chemicals to Humans, Supplement 2, Long Term and Short-Term Screening Assays for Carcinogens: A Critical Appraisal*, pages 311 - 346. International Agency for Research in Cancer, Lyon.

Parzen, M. (1993). *Computational Issues in Robust Regression.* Ph. D. dissertation, Harvard University.

Pettitt, A. N. (1984). Proportional odds models for survival data and estimates using ranks. *Applied Statistics* **33**, 169 - 175.

Politis, D. N. and Romano, J. P. (1994). Large sample confidence regions based on subsamples under minimal assumptions. *Ann. Statist.* **22**, 2031 - 2050.

Rabinowitz, D., Tsiatis, A., and Aragon, J. (1995). Regression with interval censored data. *Biometrika* **82**, 501 - 514.

Robertson, T., Wright, F. T., Dykstra, R. L. (1988). *Order Restricted Statistical Inference.* Wiley, New York.

Rockafellar, R. T. (1970). *Convex Analysis.* Princeton University Press.

Rossini, A. J. and Tsiatis, A. A. (1996). A semiparametric proportional odds regression model for the analysis of current status data. *J. Amer. Statist. Assoc.* **91**, 713 - 721.

Samuelsen, S. O. and Kongerud, J. (1993). Evaluation of applying interval censoring on longitudinal data on Asthmatic symptoms. *Statistical Research Report* **2**, Institute of Mathematics, University of Oslo.

Self, S. G. and Grossman, E. A. (1986). Linear rank tests for interval-censored data with application to PCB levels in adipose tissue of transformer repair workers. *Biometrics* **42**, 521-530.

Shiboski, S. C. and Jewell, N. P. (1992). Statistical analysis of the time dependence of HIV infectivity based on partner study data. *J. Amer. Statist. Assoc.* **87**, 360 - 372.

Silverman, B. M. (1986). *Density Estimation for Statistics and Data Analysis.* Chapman and Hall, New York.

Stone, C. J. (1977). Consistent nonparametric regression (with discussion). *Ann. Statist.* **5**, 595-645.

Turnbull, B. W. (1976). The empirical distribution function with arbitrarily grouped, censored and truncated data. *J. Roy. Statist. Soc. B* **38**, 290 - 295.

Van de Geer, S. (1993). Hellinger-consistency of certain nonparametric maximum likelihood estimators. *Ann. Statist.* **21**, 14-44.

Van der Vaart, A. W. and Wellner, J. A. (1996). *Weak Convergence and Empirical Processes.* Springer Verlag, New York.

Van Eeden, C. (1956). Maximum likelihood estimation of ordered probabilities. *Proc. Kon. Ned. Akad. Wet.* **A59**, 444 - 455.

Van Eeden, C. (1957). Maximum likelihood estimation of partially ordered or completely ordered probabilities, I. *Proc. Kon. Ned. Akad. Wet.* **A60**, 128 - 136.

Wellner, J. A. (1995). Interval censoring case 2: alternative hypotheses. *Analysis of Censored Data*, IMS Lecture Notes - Monograph Series **27**, 271 - 291. H.L. Koul and J. V. Deshpande, editors. IMS, Hayward.

Wellner, J. A. and Zhan, Y. (1997). A hybrid algorithm for computation of the NPMLE from censored data. *J. Amer. Statist. Assoc.* **92**, to appear.

Zhan, Y. and Wellner, J. A. (1995). Double censoring: characterization and computation of the nonparametric maximum likelihood estimator. *Technical Report No.* **292**, Department of Statistics, University of Washington.

Department of Statistics
and Acturial Science
University of Iowa
Iowa City, Iowa 52242
U.S.A.
e-mail: jian@stat.uiowa.edu

University of Washington
Statistics
Box 354322
Seattle, Washington 98195-4322
U.S.A.
e-mail: jaw@stat.washington.edu

Singly and Doubly Censored Current Status Data with Extensions to Multi-State Counting Processes

Nicholas P. Jewell
Mark van der Laan

ABSTRACT In estimation of a survival function, current status data arises when the only information available on sampled individuals is their survival status at a single monitoring time. Here, we briefly review recent developments in nonparametric estimation techniques for two extensions of current status data, namely (i) doubly censored current status data, where there is incomplete information on the origin of the failure time random variable, and (ii) current status information on more complicated stochastic processes.

1 Introduction

In survival analysis, interest focuses on estimation of the distribution function of the random variable, T, which measures the length of time between two successive events, the so-called initiating event and subsequent event. If the initiating and subsequent events occur in chronological time at times I and J, respectively, then, of course, $T = J - I$. Throughout our discussion we assume that the random variables I and T are independent. Recently, the analysis of current status data on the random variable T has attracted considerable attention. In this case, information on T for the ith sampled individual is limited to the status of T at a single monitoring time B. That is, we observe whether either or both of the events of interest have occurred at a *single point in time*, B. Then, when the time of the initiating event I is known, the data consists of observations of the random variable $Y = (C, \Delta)$, where $C = B - I$ (the induced monitoring time on T), and $\Delta = 1$ if $T \leq C$ and is zero otherwise. Note that $\Delta = 1$ if and only if $J \leq B$. Conditioning on the monitoring times, the conditional likelihood for a sample of n independent observations of this kind is given by

$$L = \prod_{i=1}^{n} G(C_i)^{\Delta_i}(1 - G(C_i))^{1-\Delta_i} \tag{1}$$

where G is the distribution function of T. Interest then focuses on estimation of G based on the likelihood function (1). In the literature, this case, where the time of the initiating event, that is, the origin of the random variable T, is known, is referred to as *current status data*.

In Section 2, we briefly review what is known about nonparametric estimation of the failure time distribution G based on the likelihood function (1). Subsequently, we discuss several extensions of this data structure and the associated nonparametric estimation problem. In Section 3, we consider the situation where there is incomplete information on the time of the initiating event, I. In Section 4, we introduce problems when the data structure consists of single cross-sectional observations on simple stochastic counting processes whose sample paths move through more than the standard two states ("alive" and "dead"). Throughout, we will pay particular attention to the attributes of the nonparametric maximum likelihood estimator (NPMLE). Our style is discursive—full details of the results can be found in several papers which we reference at appropriate points.

2 Singly Censored Current Status Data

The structure of current status data was described briefly in the Introduction. Estimation of G based on (1) represents a standard likelihood analysis problem if a parametric structure is assumed for G. When nothing is assumed regarding G, the corresponding nonparametric problem is non-standard—the nonparametric maximum likelihood estimator of G, and algorithms for its computation are well known. See, for example, Ayer et al. (1955), Barlow et al. (1972), and Groeneboom and Wellner (1992). Previous work and examples of current status data can be found in Diamond, McDonald, and Shah (1986), Diamond and McDonald (1991), Jewell and Shiboski (1990, 1993), Keiding (1991), and Shiboski and Jewell (1992).

2.1 Theoretical Results

The density of the random variable Y, as defined in the Introduction, is easily seen to be

$$f(y) \equiv f(c,\delta) = [G(c)\delta + (1 - G(c))(1 - \delta)]h(c) \tag{2}$$

where h is the density of the monitoring time random variable C and whose support we assume to be the entire positive half-line. Note that $G(C) = E(\Delta|C)$. It follows that nonparametric estimation of G is akin to a nonparametric monotone regression problem. An immediate consequence is that a continuous distribution function G can be consistently estimated—in a nonparametric fashion—at an optimal rate of $n^{1/3}$; from a practical point of view this represents a considerably slower rate of convergence than the familiar $n^{1/2}$ asymptotics. On the other hand, it should be recognized that, in most applied situations, we are most interested in functionals of G rather than the shape of the entire distribution function. With current status data, smooth functionals of G can still be nonparametrically estimated at the $n^{1/2}$ asymptotic rate (van der Laan, Bickel and Jewell, 1996)—in particular, this is true for regression coefficients in standard semiparametric regression models for survival data (Klein and Spadey, 1993; Rabinowitz, Tsiatis, and Aragon, 1995; Huang, 1994).

3 Doubly Censored Current Status Data

In the description of current status data arising from observation of the times of an initiating and subsequent event, briefly described in Sections 1 and 2, it is assumed that the initiating time I is observed. We now generalize this data structure by allowing these times to remain unknown—instead we assume that only the distribution of I is known. The latter piece of information is necessary, in general, if the distribution function G is to remain identifiable. (Jewell, Malani, and Vittinghoff, 1994).

This kind of data arises when, at the single monitoring time B, we only observe whether either or both of the initiating and subsequent events have occurred, but not the times of either event. We assume that only cases for which $I \leq B$, that is, the initiating event has occurred by the monitoring time, are included in the sample.

For simplicity we assume that, for the ith case, the known distribution function of I_i is supported on the interval (A_i, B_i) and is denoted by F_i, the index allowing the distribution of initiating event times to vary across sampled units. Now writing $C_i = B_i - A_i$, and with $\Delta_i = 1$ if $J_i \leq B_i$ and zero otherwise, the probability that $\Delta_i = 1$, conditional on A_i and B_i, denoted by P_i, is given by

$$P_i = \int_0^{C_i} F_i(B_i - T)dG(T). \tag{3}$$

Thus, the (conditional) likelihood of n independent observations of this kind is

$$L = \prod_{i=1}^{n} P_i^{\Delta_i}(1 - P_i)^{1-\Delta_i} \tag{4}$$

with P_i given in (3).

Rabinowitz and Jewell (1996) refer to this data structure as *doubly censored current status data* , a special case of doubly censored data as described by DeGruttola and Lagakos (1989). An informative case of (3) arises when F_i is given by the Uniform distribution on $[A_i, B_i]$. Then,

$$P_i \equiv P_G(C_i) = \int_0^{C_i} \frac{(C_i - T)}{C_i} dG(T) = \frac{1}{C_i} \int_0^{C_i} G(T)dT. \tag{5}$$

Here P_i only depends on $C_i = B_i - A_i$ and is a distribution function with argument C_i. Thus, this case of doubly censored current status data is a *submodel* of current status data since the likelihood (4) can be obtained from the likelihood (1) by substitution of $P_G(C_i)$ for $G(C_i)$ for all i. The doubly censored current staus data model is not equivalent to the current status data model since not every distribution function P can be expressed as in (5) for some distribution function G.

Returning to the general case, we now consider estimation of G based on the likelihood function (4) with P_i given by (3). Parametric estimation of G is again straightforward, in principal. Nonparametric maximum likelihood estimation of G can be approached by viewing the model as a nonparametric mixture estimation problem (Jewell, Malani, and Vittinghoff, 1994). Some remarks regarding the properties of the NPMLE will be given below. Doubly censored current status data arises in similar applications to those for current status data when there is the additional complication of incomplete observation of the time of the initiating event I. An example of this kind, arising from partner studies of HIV transmission, is analyzed in Jewell, Malani, and Vittinghoff (1994). The latter paper also describes a retrospective version of doubly censored current status data.

3.1 Theoretical Results

Consider doubly censored current status data with the Uniformity assumption on F_i, leading to (5) invoked. It follows from (5) that

$$G(C_i) = P_G(C_i) + C_i P_G{}'(C_i) \tag{6}$$

as shown in Jewell, Malani, and Vittinghoff (1994). The analogous development to (2) for doubly censored current status data shows that the density of the random variable $Y = (C, \Delta)$, where $C = B - A$, is

$$f(y) \equiv f(c, \delta) = [P_G(c)\delta + (1 - P_G(c))(1 - \delta)]h(c) \tag{7}$$

where h is now the density of $C = B - A$ and $P_G(c)$ is given by (5). Again, we assume that the distribution of C has full support. Then, as before, nonparametric estimation of the distribution function P_G corresponds to nonparametric monotone regression estimation. But, as seen in (6), P_G is 'one degree' smoother than G. Hence, nonparametric estimation of G involves estimation of the derivative of P_G and hence is akin, asymptotically, to nonparametric estimation of the derivative of a density function. It follows that, with G continuous, the optimal rate of convergence of a nonparametric estimator of G is $n^{1/5}$ (van der Laan, Bickel, and Jewell, 1996).

The relationship (6) suggests that the NPMLE of G will be inconsistent, in general, since the likelihood function (4) identifies $P_G(C_i)$ but not $P_G{}'(C_i)$. Hence, this doubly censored current status data model provides an example where the NPMLE fails. In van der Laan, Bickel, and Jewell (1996) a regularized NPMLE—see Bickel, Klaassen, Ritov, and Wellner (1993)—is considered, based on smoothing the data prior to estimation; this estimator is shown to be consistent.

For doubly censored current status data, the properties of the NPMLE, and the results on rates of convergence essentially force the use of parametric models for G with all but extremely large data sets. On the other hand, when interest focuses on certain functionals of G, rather than G itself—in particular, the estimation of regression effects on G—it may still be viable to allow the distribution function G to remain unspecified with regard to any of its properties. As for current status data, certain smooth functionals of G and parametric regression effects on G continue to be estimable at $n^{1/2}$ rates (Rabinowitz and Jewell, 1996; van der Laan, Bickel, and Jewell, 1996) under the Uniformity assumption on F_i. Results regarding a limited class of functionals can be obtained by writing $\int \phi dG = \int \phi_1 dP_G$ where ϕ is the functional of interest and $\phi_1(t) = \phi(t) - t\phi'(t)$—this is possible under integrability conditions on phi_1 and assuming that $\lim_{t\to\infty} \phi(t)/t = 0$. In such cases, estimation of $\int \phi dG$ with doubly censored data is equivalent to estimation of $\int \phi_1 dP_G$ with current status data on P_G. This equivalency of functionals is, in fact, necessary for the functional $\int \phi dG$ to be estimable at rate $n^{1/2}$ (van der Laan, Bickel, and Jewell, 1996). An unpleasant consequence of this is that only truncated moments—and not the (untruncated) mean, for example—are estimable at the usual $n^{1/2}$ rate when G has infinite support. If the support of G is compact, however, then it is possible to construct estimators of the mean that converge at rate $n^{1/2}$. For further details, see van der Laan, Bickel, and Jewell (1996).

4 Current Status Data for Other Stochastic Processes

Another way of viewing current status data is to picture the survival time random variable T in terms of a simple counting process $X(t)$ whose sample path starts at 0 at the time origin $t = 0$ (chronological time I), remains at zero until T time units thereafter where it jumps to the value 1 (at chronological time J). The properties of the stochastic process X generated in this way are completely determined by the distribution function G and vice-versa. We can thus cast the current status data problem in terms of inference regarding the stochastic process X based on incomplete observation of n independent sample paths of X, the incomplete observation of the ith path being represented by measurement of the value of $X(t)$ at a single point in time, $t = C_i$.

Jewell and van der Laan (1995) describe generalizations of this problem for more complex stochastic processes X. For example, suppose that X

is a counting process with its realization at an arbitrary time t being the number of jumps that have occurred by time t. In particular suppose that X begins with value 0 at the origin, jumps to value 1 after T_1 time units have elapsed, jumps to value 2 after T_2 total time units have passed— with T_2 necessarily greater than T_1—and so on. Now, the stochastic properties of the process X are determined by the joint distribution of $T_1, T_2, T_3, \ldots$. Again, we wish to estimate these properties based on a set of independent cross-sectional observations of n sample paths of X, the ith sample path being observed at the single point in time C_i. In the next subsections, we describe progress on several variants of this general structure.

Practical examples of this kind of data structure arise in a number of applications. For example, in cross-sectional surveys of sexual practices, the number of lifetime sexual partners to date and current age of respondents may be ascertained. With age as the time scale, the random variable T_1 represents age at sexual contact with first partner, and so on. In employment or mobility studies the relevant question may focus on the cumulative number of jobs held or 'migrations' at the time of the survey. In each of these cases, the investigator may wish to investigate the longitudinal properties of these variables based on the available cross-sectional information.

4.1 Joint Estimation of Lifetime and Disease Onset Distributions

Dinse and Lagakos (1982) consider joint estimation of the lifetime and disease onset distribution for a simple three state counting process described, in its simplest form, as follows. Suppose, measuring from an appropriate time origin, the stochastic process $X(t)$ moves from state 0, through state 1, to an absorbing state 2. Here, state 0 might be described as "disease-free", state 1 as "diseased", and state 2 refers to "death". Thus, the time until X moves to state 1, given by the random variable T_1, is the time until disease onset; similarly, the time from the origin until X moves to state 2, given by T_2, is total lifetime. Now suppose that one observes the value of X at a random monitoring time C, and that, if $X(C) = 2$, one also knows the exact value of T_2 which is less than C, of course. In other words, the monitoring time C is a censoring time—all times to death that occur prior to censoring are observed exactly, and, at censoring, it is known whether the disease has commenced or not. The problem is to estimate the distributions of T_1 and T_2, respectively, the disease onset distribution, G_1, and the lifetime distribution, G_2, based on a sample of n independent observations denoted by $\{C_i, X(C_i) : i = 1, ..., n\}$. Note that, in some examples, when the ith lifetime, T_{2i} is known exactly for observations with $X(C_i) = 2$, the actual censoring time C_i might not be known; this will not be relevant to our subsequent discussion.

By ignoring the disease status information of a censored observation,

$X(C_i) < 2$, the data reduce to standard right censored observations of the random variable T_2—the distribution function G_2 can thus be estimated at rate $n^{1/2}$ using the Kaplan-Meier estimator. On the other hand, with appropriate assumptions regarding the censoring, as we discuss below, the censored observations provide data in the form of current status information on the conditional probability function $Pr(T_1 > t|T_2 > t)$, yielding an estimator of G_1 converging at rate $n^{1/3}$. An immediate question arises as to whether these estimators can be improved through some form of joint estimation.

We should point out that the likelihood function is easily seen to depend only on the two marginal distributions, G_1 and G_2, of the random variable (T_1, T_2) (van der Laan, Jewell, and Peterson, 1996). Thus, the data do not identify any additional information on the joint distribution of (T_1, T_2), in particular, the dependence structure of T_1 and T_2.

First, we consider estimation of the onset distribution, G_1. Let $\Delta = 1$ if $T_1 \leq C$ and be zero otherwise. Then,

$$E(1 - \Delta|C, T_2 > C) = \frac{S_1(C)}{S_2(C)} \tag{8}$$

where S_j is the survival function corresponding to G_j for $j = 1, 2$. From this it immediately follows that

$$S_1(C) = E[S_2(C)(1 - \Delta)|C, T_2 > C]. \tag{9}$$

If S_2 were known, this suggests the strategy of estimating S_1 by minimizing

$$\sum_{i=1}^{n} \{S_2(C_i)(1 - \Delta_i) - S_1(C_i)\}^2 I(T_{2i} > C_i) \tag{10}$$

under the constraint that S_1 is a survival function and hence is nonincreasing. If S_2 is not known, then an estimate—such as the Kaplan-Meier estimator discussed above— can be substituted in (10). This minimization therefore corresponds to an isotonic regression problem for which algorithms (for example, the pool adjacent violators algorithm—see Ayer et al., 1955; Barlow et al., 1972) are well-known. However, it is apparent that, in this regression context, the variance of the "outcome variable" $S_2(C)(1-\Delta)$ is not constant. In fact,

$$\text{variance}[S_2(C)(1 - \Delta)|C, T_2 > C] = S_2(C)^2 R(C)(1 - R(C)) \tag{11}$$

with $R = S_1/S_2$. Thus, an improved estimator of S_1 should result from minimization of a weighted sum of squares

$$\sum_{i=1}^{n} \{S_2(C_i)(1 - \Delta_i) - S_1(C_i)\}^2 I(T_{2i} > C) w_i. \tag{12}$$

The variance formula (11) suggests the use of the weights $w_i = [S_2(C_i)^2 R(C_i)(1-R(C_i))]^{-1}$. Since, R depends on S_1 this requires an iterative scheme. In fact, it can be shown that such an iterative scheme for estimating S_1, that is, minimizing (12) where S_2 is replaced by its NPMLE (rather than the Kaplan-Meier estimator—more on that below) is equivalent to the (joint) NPMLE of S_1 (van der Laan, Jewell, and Peterson, 1996).

In addition, a non-iterative weighted least squares estimator is obtained by avoiding the terms involving R in (11). Two alternative choices of weights have been suggested:

$$w_i = [S_2(C_i)]^{-1} \tag{13}$$

by Dinse and Lagakos (1982), and

$$w_i = [S_2(C_i)]^{-2} \tag{14}$$

by van der Laan, Jewell and Peterson (1996). Dinse and Lagakos (1982) believed mistakenly that the weights (13) corresponded to those used by nonparametric maximum likelihood. As we note below, simulations in van der Laan, Jewell and Peterson (1996) indicate improved performance with the use of (14) and so we do not consider the weights (13) further.

In comparison to these latter two estimators, based on the weights given in (13) and (14), the NPMLE attempts to update the weights in (11) by estimating R at each iterative step. Were R to be described accurately by a parametric model, this approach would be an improvement in terms of precision—albeit slight—over the simpler weight functions. However, with the shape of R unspecified, the NPMLE strategy does not acknowledge that local variations in R are not, in fact, identifiable. Thus, the estimator based on (12) using weights (14) uses weights that are, on average, *more* precise than estimates of the weights (11) used by the NPMLE. This argument is most easily explained by examining the form of the NPMLE estimator, $\hat{S}_1$, of S_1. It is straightforward to show that $\hat{S}_1$ has a finite set of support points, $t_1 < t_2 < \cdots < t_k$, which are, in fact, a subset of the observed monitoring times, $C_1, \ldots, C_n$. Since $\hat{S}_1$ is constant between its support points, it follows that

$$\hat{S}_1(t_j) = \frac{\sum_{i=1}^n S_2(C_i)(1-\Delta_i)w_i I(C_i \in (t_j, t_{j+1}])}{n^{-1}\sum_{i=1}^n w_i I(C_i \in (t_j, t_{j+1}])}, \tag{15}$$

where S_2 is replaced by its NPMLE if it is unknown. It is clear that $R = S_1/S_2$ cannot be "more identifiable" than S_1. Hence we cannot improve on an estimator of R which is piecewise constant on the support points of $\hat{S}_1$, that is, the points where S_1 is identified. For such a piecewise constant estimator $\hat{R}$, the factor $\hat{R}(1-\hat{R})$ is constant within $(t_j, t_{j+1}]$ and hence factorizes and cancels in the numerator and denominator of (15), reducing the weights w_i to those given in (14).

The latter paragraph suggests that the simpler non-iterative weighted least squares estimator, with weights (14), may outperform the NPMLE in finite samples with regard to efficiency. This is borne out in a simple set of simulations described in van der Laan, Jewell, and Peterson (1996) which also show the improvement of (14) over (13). Detailed discussion of the estimators described above is provided in van der Laan, Jewell, and Peterson (1996). This paper also gives some attention to efficient estimation of smooth functionals of G_1 despite the fact that the distribution function G_1 can not be estimated at rate $n^{1/2}$.

Now, we return to discussion of estimation of the lifetime distribution function, G_2, or equivalently, the survival function, S_2. van der Laan, Jewell and Peterson (1996) show that the Kaplan-Meier estimator is asymptotically equivalent to the NPMLE of S_2. However, simulations, even in moderately sized samples, illustrate a substantial advantage in relative efficiency, for the Kaplan-Meier estimator over the NPMLE.

In many infinite dimensional settings it has been observed that (nonparametric) maximum likelihood procedures either work as effectively as they do in parametric settings—at least in terms of asymptotic consistency and efficiency—or fail entirely due to bias that persists even asymptotically. In the situation we have described, the nonparametric maximum likelihood estimators are asymptotically efficient, but are outperformed in finite samples, sometimes substantially, by the considerably simpler estimators described in van der Laan, Jewell and Peterson (1996); that is the weighted least squares estimator for G_1 based on (12) with weights (14), and the Kaplan-Meier estimator for G_2. Here, the attempt by maximum likelihood to mimic what the technique would do in a parametric setting produces a procedure which introduces additional "noise" in the finite sampling setting. Hence, nonparametric maximum likelihood cannot be recommended for this example.

4.2 General Counting Processes with Finite Number of States

We now return to the case of a counting process $X(t)$ with k jump times given by $T_1 < T_2 < ... < T_k$, where we are only able to observe the process at a single monitoring time C, assumed independent of $(T_1, ..., T_k)$. More formally, we observe C and the j for which $T_j < C < T_{j+1}$. Let G_j be the marginal distribution of T_j, for $j = 1, ..., k$. We wish to estimate G_j for each j, without further specification, based on an independent sample of n observations, $(C_i, X(C_i))$ of the kind descriibed. Note that the likelihood again depends only on the joint distribution of $(T_1, ..., T_k)$ through the marginals $G_1, ..., G_k$, so that only the latter structure is identifiable.

Let $\Delta_j = I(T_j < C)$; then (C, Δ_j) provides current status data on the distribution function G_j. We can then obtain NPMLE estimators of G_j separately for each j using just the set of (C_i, Δ_i) for $i = 1, ..., n$, as in Section 2. An immediate question arises as to whether these estimators can be improved by using the remaining information in the data. That is, for example, in estimating G_1 based on whether $T_1 < C$ or not, does it help to know whether $T_2 < C$, or $T_3 < C$, etc.? Further, what is the full joint NPMLE of $G_1, ..., G_k$?

These questions are addressed in van der Laan and Jewell (1996). In particular, it is shown that the set of separate current status NPMLEs of $G_1, ..., G_k$ differ from the estimators provided by the full joint NPMLE of $G_1, ..., G_k$. However, the former estimators are asymptotically efficient estimators of $G_1, ..., G_k$. Thus, asymptotically, nothing is lost by only using the current status information on T_j, ignoring the information on the other random variables $\{T_m : m \neq j\}$. Apparently, the full NPMLE uses information about G_j in the likelihood (beyond that available from the current status information on T_j) which is asymptotically negligible for estimation of smooth functionals of G_j. It remains to compare the finite sample performance of the full NPMLE and the marginal current status estimators.

4.3 Current Status Data on a Renewal Process

In some cases, additional structure may be placed on the counting process X discussed in Section 4.2—for example, suppose the random variables $T_1, T_2 - T_1, T_3 - T_2 - T_1, \ldots$ are independent and identically distributed, according to the distribution function G, that is, $X(t)$ is a renewal process. In other words, $T_j = T_{j-1} + T$ for $j = 1, 2, \ldots$ where $T_0 = 0$ and T follows the distribution function G, and the successive realizations of T are independent as each T_j is generated.

In the case of a renewal process, it is clear that the marginal current status NPMLE based on a single set of current status data, say (C, Δ_1), although consistent, will be inefficient as an estimator of G. Here, the likelihood involves the distribution of $T_j = T + T + \cdots + T$ (j terms in the sum) whose distribution is the convolution of j Gs, which we will denote by G^{*j}. Hence, computation of the full NPMLE of G will be extremely computer intensive since it entails computation of estimation of various convolutions of G at each step of any iterative scheme. Thus, it would be helpful to develop simple ad hoc estimators of G that are nevertheless reasonably efficient in practical situations.

One approach, considered in van der Laan and Jewell (1996), is to estimate G^{*j}, for each j, with the marginal current status data estimator as described in Section 4.2, and then combine these to obtain an omnibus estimator of G. In particular, first consider estimation of μ, the mean of G. Let

$$\hat{\mu}^{*j} = \int t d\hat{G}^{*j}(t) \tag{16}$$

where, for each j, $\hat{G}^{*j}$ is the marginal estimator of G^{*j}, based on current status information on T_j at the monitoring time C for an independent sample of size n. Since $\mu = \mu^{*j}/j$, we thus have a set of separate estimators of μ, given by $\hat{\mu}_j = \hat{\mu}^{*j}/j$ for $j = 1, \ldots$. Letting $\hat{\sigma}_j^2$ be an estimator of ${\sigma_j}^2 = \mathrm{var}(\hat{\mu}_j)$, we can then estimate μ with

$$\hat{\mu} = \frac{\sum_j \frac{\hat{\mu}_j}{\hat{\sigma}_j^2}}{\sum_j \frac{1}{\hat{\sigma}_j^2}}. \tag{17}$$

Since $\hat{\mu}_j$ is a simple functional based on a standard current status data estimator, it can be shown to be asymptotically linear with a known influence curve. This immediately yields a formula for ${\sigma_j}^2$ which subsequently provides for a straightforward estimate, $\hat{\sigma}_j^2$. Asymtptotic properties, in principal, can also be derived based on this information; for further details and discussion, see Groeneboom and Wellner (1992), van der Laan, Bickel and Jewell (1996), and van der Laan and Jewell (1996). Note that, although it might be valuable to take advantage of the dependence structure between the estimators $\hat{G}^{*j}$, we speculate that including this information through weighting in (16) will not improve the performance of the estimator $\hat{\mu}$ substantially.

In a similar fashion, it is possible to derive combined estimators of the real and imaginary parts of the characteristic function of G, and hence an estimator of the characteristic function itself. Such an estimator can be used directly to obtain estimators of smooth functionals of G , or the distribution function G itself; see van der Laan and Jewell (1996) for details. The finite sample performance of this method remains to be examined and will be considered elsewhere.

5 Final Remarks

Additional questions of considerable interest can also be raised by allowing the stochastic process X to have a different structure than the simple counting processes considered here or by permitting more complex monitoring schemes. Consider,for example, the common situation where there is a set of m times, $C_1 < C_2 < \cdots < C_m$ at which the process X is observed (independent of the T_js, of course). Letting $C_0 = 0$ and $C_{m+1} = \infty$, we then observe that $T_j \in (C_l, C_{l+1}]$ for some $l \in \{0, 1, \ldots, m+1\}$. For a renewal process, and again assuming that the monitoring times are independent of the process, we can estimate G^{*j} using nonparametric maximum likelihood based on the marginal interval censored data available on each T_j. These estimators can then be combined to obtain omnibus estimators of the mean or characteristic function of G as in Section 4.3 so long as estimators of the variance of the appropriate smooth functionals of $\hat{G}^{*j}$ are available. Unfortunately, explicit expressions for the limiting variance of the NPMLE of smooth functionals of G^{*j} are not available for interval censored data. However, van der Laan and Hubbard (1996) provide an ad hoc estimator of a distribution function based on interval censored data which essentially has the same performance as the NPMLE and whose influence curve is explicitly known so that the relevant asymptotic variance formulae can be derived. This estimation procedure can then be used as the basis of our proposed omnibus estimators rather then the NPMLE estimators.

We can improve on these marginal estimators of G^{*j} by noting that T_2 is very informative for T_1. Similarly, data on T_1 and T_3 are informative for T_2, and, in general, T_{j-1} and T_{j+1} for T_j. It seems reasonable that information on the "neighboring" T_js captures most of the additional information on T_j in the entire sample, beyond the direct data on T_j itself. This suggests the use of an estimator for G^{*j}, and subsequently G, which uses the data on either T_{j-1}, or T_{j+1}, (or both), as a covariate with the interval censored data on T_j. Suppose, for argument, that one knows T_{j-1}, and one observes current status information on T_j at monitoring time C—then, the distribution function G can be estimated through current status data on T at the monitoring time $C - T_{j-1}$ since

$$P(T_j \leq t | T_{j-1}) = P(T \leq t - T_{j-1}). \tag{18}$$

If this is done for each j in turn, then the resulting estimators of G can be combined, as before, to provide an overall estimator. Returning to the general case of several monitoring times on a renewal process, if A_j is the midpoint of the interval in which T_j is known to fall (with a suitable ad hoc value if this interval has infinite length), we then propose estimation, for each j, of G using the interval censored data on T_j with covariate information A_{j-1} and A_{j+1}. Subsequently, these separate estimators can be combined as above. We note that van der Laan and Hubbard (1996) show how to adapt their interval censored distribution function estimation procedure to allow for covariate information.

A final problem of interest arises when we observe the status of a renewal process, $X(t)$, at several monitoring times but do not know the origin of the process. A more complex example of this kind can be found in Vittinghoff, Malani, and Jewell (1994).

Acknowledgements

This research was supported by a grant from the National Institute of Allergy and Infectious Diseases (grant number AI33831).

References

Ayer, M, Brunk, H. D., Ewing, G. M., Reid, W. T., and Silverman, E. (1955). An empirical distribution function for sampling with incomplete information. *The Annals of Statistics*, **26**, 641–647.

Barlow, R. E., Bartholomew, D. J., Bremner, J. M., and Brunk, H. D. (1972). *Statistical Inference Under Order Restrictions.* New York: Wiley.

Bickel, P. J., Klaassen, A. J., Ritov, Y., and Wellner, J. A. (1993). *Efficient and Adaptive Inference in Semi-Parametric Models.* Baltimore: Johns Hopkins University Press.

DeGruttola, V., and Lagakos, S. W. (1989). Analysis of doubly-censored survival data, with application to AIDS. *Biometrics*, **45**, 1–11.

Diamond, I. D., and McDonald, J. W. (1991). The analysis of current status data. In *Demographic Applications of Event History Analysis*, Eds. J. Trussell, R. Hankinson and J. Tilton. Oxford: Oxford University Press.

Diamond, I. D., McDonald, J. W., and Shah, I. H. (1986). Proportional hazards models for current status data: application to the study of differentials in age at weaning in Pakistan. *Demography*, **23**, 607–620.

Dinse, G. E., and Lagakos, S. W. (1982). Nonparametric estimation of lifetime and disease onset distributions from incomplete observations. *Biometrics*, **38**, 921–932.

Groeneboom, P., and Wellner, J. A. (1992). *Nonparametric Maximum Likelihood Estimators for Interval Censoring and Deconvolution.* Boston: Birkhaüser–Boston.

Huang, J. (1994). *Estimation in Regression Models with Interval Censoring.* Ph.D. Thesis, University of Washington.

Jewell, N. P., Malani, H.M., and Vittinghoff, E. (1994). Nonparametric estimation for a form of doubly censored data with application to two problems in AIDS. *Journal of the American Statistical Association*, **89**, 7–18.

Jewell, N. P., and Shiboski, S. C. (1990). Statistical analysis of HIV infectivity based on partner studies. *Biometrics*, **46**, 1133–1150.

Jewell, N. P., and Shiboski, S. C. (1993). The design and analysis of partner studies of HIV transmission. In *Methodological Issues in AIDS Behavioral Research*, Eds. D. G. Ostrow and R. Kessler. New York: Plenum Publishing Company.

Jewell, N. P., and van der Laan, M. (1995). Generalizations of current status data with applications. *Lifetime Data Analysis*, **1**, 101–109.

Keiding, N. (1991). Age-specific incidence and prevalence (with discussion). *Journal of the Royal Statistical Society, Series A*, **154**, 371–412.

Klein, R. W., and Spadey, R. H. (1993). An efficient semiparametric estimator for binary response models. *Econometrica*, **61**, 387–421.

van der Laan, M., Bickel, P., and Jewell, N. P. (1996). Singly and doubly censored current status data: estimation, asymptotics and regression. Submitted for publication.

van der Laan, M., and Hubbard, A. (1996). Estimation with interval censored data and covariates. Submitted for publication.

van der Laan, M., Jewell, N. P., and Peterson, D. (1996). Efficient estimation of the lifetime and disease onset distributions. Submitted for publication.

van der Laan, M., and Jewell, N. P. (1996). Current status data for renewal processes and general multi-State counting processes. Submitted for publication.

Rabinowitz, D., and Jewell, N. P. (1996). Regression with doubly censored current status data. *Journal of the Royal Statistical Society, Series B*, to appear.

Rabinowitz, D., Tsiatis, A and Aragon, J. (1995). Regression with interval censored data. *Biometrika*, **82**, 501–513.

Shiboski, S. C., and Jewell, N. P. (1992). Statistical analysis of the time dependence of HIV infectivity based on partner study data. *Journal of the American Statistical Association*, **87**, 360–372.

Vittinghoff, E., Malani, H.M., and Jewell, N. P. (1994). Estimating patterns of CD4 lymphocyte decline using data from a prevalent cohort of HIV infected individuals. *Statistics in Medicine*, **13**, 1101–1118.

Additive Hazards Regression Models for Survival Data

D. Y. Lin
Zhiliang Ying

ABSTRACT The additive hazards regression model relates the conditional hazard function of the failure time linearly to the covariates. This formulation complements the familiar proportional hazards model in that it describes the association between the covariates and failure time in terms of the risk difference rather than the risk ratio. In this paper, we provide a closed-form semiparametric estimator for the (vector-valued) regression parameter of the additive hazards model with right-censored data, which is consistent and asymptotically normal with a simple variance estimator. We also demonstrate how the additive hazards framework can be used effectively to incorporate frailty and to handle interval-censored data, the resulting semiparametric inference procedures being much simpler than their counterparts under the proportional hazards framework.

1 Introduction

Since the fundamental paper of Cox (1972), the proportional hazards model has been the popular choice for the regression analysis of right-censored survival data. This model formulates the relationship between the failure time T and the possibly time-dependent covariate vector Z through the conditional hazard function of T,

$$\lambda(t|Z(s), s \leq t) = \lambda_0(t)e^{\beta' Z(t)}, \tag{1.1}$$

where λ_0 is a completely unspecified baseline hazard function and β is a (vector-valued) regression parameter. The exponential link function may be replaced by a more general smooth function (Prentice and Self, 1983).

The proportional hazards model belongs to a family of hazard-based generalized linear regression models in which the conditional hazard function of T takes the form

$$\lambda(t|Z(s), s \leq t) = L(\lambda_0(t), \beta' Z(t)),$$

where λ_0 is again an unspecified function while L is a known function. Clearly, the choice of $L(x, y) = xe^y$ yields model (1.1). Another important example is the additive hazards model, which is obtained by taking

$L(x,y) = x + y$, i.e.,

$$\lambda(t|Z(s), s \leq t) = \lambda_0(t) + \beta' Z(t). \tag{1.2}$$

Model (1.2) has been dealt with by Aranda-Ordaz (1983), Cox and Oakes (1984), Thomas (1986), Breslow and Day (1987) and Lin and Ying (1994). A related nonparametric additive hazards model has been studied by Aalen (1980; 1989) and Huffer and McKeague (1991).

The hazard-based regression models are natural in dealing with right-censored survival data, especially in the presence of time-dependent covariates. The counting-process martingale theory can be used to construct parameter estimators and to study their properties. In particular, a large-sample theory was developed elegantly by Andersen and Gill (1982) for model (1.1), while a similar theory was recently established by Lin and Ying (1994) for model (1.2).

Despite the enormous success of the proportional hazards model, it is desirable to study and utilize the additive hazards model for several reasons. First, the risk difference is complementary to and, from the public health point of view, more important than the risk ratio in describing the association between the risk factor and disease occurrence. Secondly, biological and empirical evidence suggests that the additive hazards model fits certain types of data better than the proportional hazards model. These two justifications were elaborated by Breslow and Day (1980, pp. 53-59; 1987, pp. 122-131). A further justification, as will become evident in this paper, is that the additive hazards model provides a simple structure for studying frailty models and interval-censored data, which are very difficult to deal with under the proportional hazards framework.

The rest of this paper is organized as follows. In the next section, we summarize the main results of Lin and Ying (1994) for model (1.2) and provide some extensions. In Section 3, we introduce additive hazards frailty models and demonstrate their advantages in analyzing multivariate failure time data. In Section 4, we develop simple inference procedures for interval-censored data under the additive hazards model. The material presented in Sections 3 and 4 consists of preliminary findings from ongoing investigations; the final results will be communicated in separate reports. Section 5 discusses more general hazard-based regression models and contains some concluding remarks.

2 Additive hazards models for right-censored data

For $i = 1, \ldots, n$, let (T_i, C_i, Z_i) be the failure time, censoring time and covariate vector for the ith subject. Under the additive hazards model, the regression relationship between T_i and Z_i is given by model (1.2). The observations consist of $\{\tilde{T}_i, \delta_i, Z_i(t), t \leq \tilde{T}_i\}$ $(i = 1, \ldots, n)$, where $\tilde{T}_i = \min(T_i, C_i)$ and $\delta_i = I(T_i \leq C_i)$.

It is convenient to introduce the notation

$$N_i(t) = \delta_i I(\tilde{T}_i \leq t), \; Y_i(t) = I(\tilde{T}_i \geq t), \; \bar{Z}(t) = \sum_{i=1}^{n} Z_i(t)Y_i(t) \Big/ \sum_{i=1}^{n} Y_i(t), \tag{2.1}$$

where $0/0 = 0$. Lin and Ying (1994) proposed the following estimating function

$$U(\beta) = \sum_{i=1}^{n} \int \{Z_i(t) - \bar{Z}(t)\}\{dN_i(t) - \beta' Z_i(t) Y_i(t) dt\}, \tag{2.2}$$

which is based on the simple fact that, when β takes the true parameter value, $U(\beta)$ is a martingale integral and therefore has mean 0. In (2.2) and throughout the rest of this paper, the integration is taken from 0 to ∞ unless otherwise indicated. Since U is linear in β, the resulting estimator may be expressed explicitly, i.e.,

$$\hat{\beta} = \left[\sum_{i=1}^{n} \int \{Z_i(t) - \bar{Z}(t)\}^{\otimes 2} Y_i(t) dt \right]^{-1} \sum_{i=1}^{n} \int \{Z_i(t) - \bar{Z}(t)\} dN_i(t), \tag{2.3}$$

where $a^{\otimes 2} = aa'$.

Applying the standard counting-process martingale techniques, Lin and Ying (1994) showed that $\hat{\beta}$ is asymptotically normal with mean β and with a variance-covariance matrix consistently estimated by $\hat{A}^{-1}\hat{V}\hat{A}^{-1}$, where

$$\begin{aligned} \hat{A} &= \sum_{i=1}^{n} \int \{Z_i(t) - \bar{Z}(t)\}^{\otimes 2} Y_i(t) dt, \\ \hat{V} &= \sum_{i=1}^{n} \int \{Z_i(t) - \bar{Z}(t)\}^{\otimes 2} dN_i(t). \end{aligned}$$

More precisely, $(\hat{A}^{-1}\hat{V}\hat{A}^{-1})^{-1/2}(\hat{\beta} - \beta)$ converges in distribution to $\mathcal{N}(0, I)$. Interestingly, neither $\hat{A}$ nor $\hat{V}$ involves the regression parameter.

By redefining the covariate process in model (1.2) as $Z(t)q(t)$, where q is a completely specified function, we obtain the following model

$$\lambda(t|Z) = \lambda_0(t) + \beta' Z(t) q(t). \tag{2.4}$$

With appropriate choices of q, formulation (2.4) may be used to adjust for non-additive hazards and to constrain the right-hand side of (1.2) to be positive.

It is desirable to consider parameterization of q in (2.4). Let $q(\gamma, t)$ be a known function of γ and t. For simplicity, assume both β and γ are scalars. Model (2.4) then becomes

$$\lambda(t|Z) = \lambda_0(t) + \beta Z(t) q(\gamma, t).$$

Provided that certain identifiability conditions are satisfied, it is possible to estimate both β and γ reasonably well by properly choosing the corresponding estimating functions. For example, let $q(\gamma, t) = t^\gamma$. Then a plausible pair of estimating functions is

$$\begin{cases} \sum_{i=1}^n \int \{Z_i(t) - \bar{Z}(t)\}\{dN_i(t) - \beta Z_i(t) t^\gamma Y_i(t) dt\}, \\ \sum_{i=1}^n \int t \{Z_i(t) - \bar{Z}(t)\}\{dN_i(t) - \beta Z_i(t) t^\gamma Y_i(t) dt\}. \end{cases}$$

In general, one can choose appropriate weight functions $W_i(t)$, which are $(\dim(\beta) + \dim(\gamma)) \times 1$ vectors, and consider the following class of estimating functions

$$U(\beta, \gamma) = \sum_{i=1}^n \int \{W_i(t) - \bar{W}(t)\}\{dN_i(t) - \beta' Z_i(t) q(\gamma, t) Y_i(t) dt\},$$

where $\bar{W}(t) = \sum W_i(t) Y_i(t) / \sum Y_i(t)$. Let $(\hat{\beta}', \hat{\gamma}')'$ be the solution to $U(\beta, \gamma) = 0$, and $A(\beta, \gamma)$ the derivative of $U(\beta, \gamma)$ with respect to $(\beta', \gamma')'$. Then, under certain regularity conditions, $n^{1/2}[(\hat{\beta} - \beta)', (\hat{\gamma} - \gamma)']'$ is asymptotically zero-mean normal whose variance-covariance matrix is the limit of $nA^{-1}(\beta, \gamma)\left[\sum_{i=1}^n \int \{W_i(t) - \bar{W}(t)\}^{\otimes 2} dN_i(t)\right]^{-1} A^{-1}(\beta, \gamma)$.

3 Additive frailty models for right-censored data

Frailty or random effect models are frequently used to incorporate intraclass association, serial correlation and other forms of dependence. These models have received tremendous attention in the last decade (Clayton and Cuzick, 1985; Hougaard, 1987; Oakes, 1989; Nielsen, Gill, Andersen and Sorensen, 1992; Murphy, 1995). Frailty can also be used to induce multivariate survival distributions, whose parameters have natural interpretations, and to model heterogeneity in the context of regression (Heckman and Singer, 1982).

The traditional approach to introducing frailty into a survival model is by assuming the existence of a latent random variable (frailty) which multiplicatively affects a "baseline" hazard function. By varying the distributional form of the frailty, different families of frailty models may be generated. Inference procedures for (multiplicative) frailty models are complicated so far as non- or semi-parametric survival models are concerned. The main reasons for the complications are that, when marginalized over the frailty, the covariate effect on the hazard function is no longer of the multiplicative form and that the inference procedures typically rely on the E-M algorithm, which is not easy to deal with under non- and semi-parametric models. Consequently, the analytic properties are also difficult to establish. The numerical and theoretical difficulties, together with awkward interpretation of the regression parameter, have hindered the use of (multiplicative) frailty models in applications.

An alternative way to construct frailty models is through what we call additive frailty. This formulation assumes that a latent random variable (frailty) is being added (instead of multiplied) to the hazard function. The resulting models are different from and complementary to the multiplicative frailty models. To introduce this new class of frailty models, we consider first the general hazard-based family of frailty models. Let ξ be a possibly time-dependent frailty process which is independent of the covariate process Z and which affects the hazard function of failure time T in the following way

$$\lambda(t|Z(s), \xi(s), s \leq t) = L(\lambda_0(t), \beta' Z(t), \xi(t)). \tag{3.1}$$

The specification of $L(u, v, w) = e^v uw$ in (3.1) yields the familiar Cox regression model with a multiplicative frailty, i.e.,

$$\lambda(t|Z(s), \xi(s), s \leq t) = e^{\beta' Z(t)} \lambda_0(t) \xi(t). \tag{3.2}$$

Suppose that both Z and ξ are time-independent. The special case of gamma frailty has received the most attention. If ξ has a gamma distribution with scale parameter α and shape parameter k, then model (3.2) implies that

$$P(T \geq t|Z) = \{1 + \alpha e^{\beta' Z} \Lambda_0(t)\}^{-k}, \tag{3.3}$$

where $\Lambda_0(t) = \int_0^t \lambda_0(s) ds$. The model specified by (3.3) is no longer of the proportional hazards form and is difficult to deal with when Λ_0 is completely unspecified (Andersen et al., 1993; Murphy, 1995).

The additive frailty model, on the other hand, is obtained by letting $L(u, v, w) = L_1(u, v) + w$, where L_1 specifies a hazard-based regression model without frailty. For definiteness, let L_1 also be additive, i.e., $L(u, v, w) = u + v + w$, or

$$\lambda(t|Z(s), \xi(s), s \leq t) = \lambda_0(t) + \xi(t) + \beta' Z(t). \tag{3.4}$$

An interesting feature of the additive frailty model (3.4) is that, when marginalized over ξ, it is still an additive hazards model. To see this, we compute

$$\begin{aligned} P(T \geq t|Z(s), s \leq t) &= E\left[P\{T \geq t|Z(s), \xi(s), s \leq t\} \Big| Z(s), s \leq t\right] \\ &= E\left\{e^{-\Lambda_0(t) - \beta' \int_0^t Z(s) ds} e^{-\int_0^t \xi(s) ds} \Big| Z(s), s \leq t\right\} \\ &= e^{-\Lambda_0^\dagger(t) - \beta' \int_0^t Z(s) ds}, \end{aligned}$$

where $\Lambda_0^\dagger(t) = \Lambda_0(t) - \log E\{e^{-\int_0^t \xi(s) ds}\}$. Denoting the derivative of $\Lambda_0^\dagger(t)$ by $\lambda_0^\dagger(t)$, we then have

$$\lambda(t|Z(s), s \leq t) = \lambda_0^\dagger(t) + \beta' Z(t), \tag{3.5}$$

which is the same as model (1.2) since λ_0 (and therefore $\lambda_0^\dagger$) is completely unspecified.

Because of (3.5), the regression parameter β has a more straightforward interpretation under model (3.4) than under model (3.2). Furthermore, simple inference procedures can be obtained by combining the ideas of Lin and Ying (1994) with those of Wei, Lin and Weissfeld (1989) and Lee, Wei and Amato (1993). It is instructive to consider the case of paired data. For $i = 1, \ldots, n$, and $j = 1, 2$, let T_{ij} and Z_{ij} be the failure time and covariate vector of the jth member of the ith pair. Also, let ξ_i be the frailty of the ith pair, representing the dependence within the pair. The additive frailty model takes the form

$$\lambda(t|Z_{ij}(s), \xi_i(s), s \leq t) = \lambda_0(t) + \xi_i(t) + \beta' Z_{ij}(t),$$

the corresponding marginal model being

$$\lambda(t|Z_{ij}(s), s \leq t) = \lambda_0^\dagger(t) + \beta' Z_{ij}(t), \tag{3.6}$$

As usual, let C_{ij}, δ_{ij}, N_{ij} and Y_{ij} denote the censoring time, failure indicator, counting process and at-risk indicator. In view of (2.2), a natural estimating function for β under model (3.6) is

$$U^\dagger(\beta) = \sum_{i=1}^{n} \sum_{j=1}^{2} \int \{Z_{ij}(t) - \bar{Z}^\dagger(t)\}\{dN_{ij}(t) - \beta' Z_{ij}(t) Y_{ij}(t) dt\},$$

where $\bar{Z}^\dagger(t) = \sum_{i=1}^n \sum_{j=1}^2 Z_{ij}(t) Y_{ij}(t) / \sum_{i=1}^n \sum_{j=1}^2 Y_{ij}(t)$. The difference between $U^\dagger$ and U is that the N_{ij} are correlated within each pair. The solution to $U^\dagger(\beta) = 0$ is

$$\hat{\beta}^\dagger = \left[\sum_{i=1}^{n} \sum_{j=1}^{2} \int \{Z_{ij}(t) - \bar{Z}^\dagger(t)\}^{\otimes 2} Y_{ij}(t) dt\right]^{-1} \sum_{i=1}^{n} \sum_{j=1}^{2} \int \{Z_{ij}(t) - \bar{Z}^\dagger(t)\} dN_{ij}(t).$$

The variance-covariance matrix of $\hat{\beta}^\dagger$ may be estimated by $\hat{A}^{\dagger -1} \hat{V}^\dagger \hat{A}^{\dagger -1}$, where

$$\hat{A}^\dagger = \sum_{i=1}^{n} \sum_{j=1}^{2} \int \{Z_{ij}(t) - \bar{Z}^\dagger(t)\}^{\otimes 2} Y_{ij}(t) dt,$$

and

$$\hat{V}^\dagger = \sum_{i=1}^{n} \left[\sum_{j=1}^{2} \int \{Z_{ij}(t) - \bar{Z}^\dagger(t)\}\{dN_{ij}(t) - Y_{ij}(t) d\hat{\Lambda}_0^\dagger(t) - \hat{\beta}^{\dagger\prime} Z_{ij}(t) Y_{ij}(t) dt\}\right]^{\otimes 2}, \tag{3.7}$$

where

$$\hat{\Lambda}_0^\dagger(t) = \sum_{i=1}^{n} \sum_{j=1}^{2} \int_0^t \frac{dN_{ij}(s) - \hat{\beta}^{\dagger\prime} Z_{ij}(s) Y_{ij}(s) ds}{\sum_{l=1}^n \sum_{k=1}^2 Y_{lk}(s)}$$

estimates the "baseline" cumulative hazard function $\Lambda_0^\dagger(t)$. Notice that $\hat{V}^\dagger$ incorporates the within-pair dependence.

In some applications, it is of interest to assess the within-group association. For the special case of paired observations, we show next that a simple test for such association may be constructed under the additive frailty model. Let us define the following statistic

$$W = n^{-1/2} \sum_{i=1}^{n} \hat{M}_{i1} \hat{M}_{i2},$$

where $\hat{M}_{ij} = \int [dN_{ij}(t) - Y_{ij}(t)\{d\hat{\Lambda}_0^\dagger(t) + \hat{\beta}^{\dagger\prime} Z_{ij}(t)dt\}]$. It is not difficult to show that, under the null hypothesis that there is no frailty, i.e., $\xi_i \equiv 0$ for all i,

$$W = n^{-1/2} \sum_{i=1}^{n} M_{i1} M_{i2} + o_p(1), \tag{3.8}$$

where the M_{ij} are the same as the $\hat{M}_{ij}$ except that $\hat{\Lambda}_0^\dagger$ and $\hat{\beta}^\dagger$ are replaced by the true parameter values $\Lambda_0^\dagger$ and β. From (3.8) it follows that W converges in distribution to $\mathcal{N}(0, \sigma_w^2)$, where a consistent estimator of σ_w^2 is

$$\hat{\sigma}_w^2 = n^{-1} \sum_{i=1}^{n} (\hat{M}_{i1} \hat{M}_{i2})^2.$$

Thus, an asymptotically valid test can be performed by referring $W/\hat{\sigma}_w$ to the standard normal distribution. It can be shown that, with proper scaling of the $\xi_i(t)$, $n^{-1/2}W$ converges to $E\left\{\int Y_{i1}(t)\xi_i(t)dt \int Y_{i2}(t)\xi_i(t)dt\right\}$, which is typically non-zero in the presence of frailty.

4 Additive hazards models for interval-censored data

In some biomedical studies, the failure time is only known to be below or above a random monitoring time, or to lie in a random interval, which yields the current-status, or interval-censored data. In this section, we demonstrate that the additive hazards regression provides a useful and convenient framework for developing simple inference procedures for otherwise a rather difficult problem.

Non- and semi-parametric models and related methods for current-status and interval-censored data have been proposed and studied by a number of authors. For the one-sample case, Turnbull (1976) gave an iterative algorithm for computing a nonparametric maximum likelihood estimator of the underlying distribution function, while Groeneboom and Wellner (1992) provided a comprehensive asymptotic theory for this estimator. Klein and

Spady (1993) and Rabinowitz, Tsiatis and Aragon (1995) proposed methods to deal with linear regression problems. Huang (1996) developed an asymptotic theory for the maximum likelihood estimation of the Cox regression model with current status data.

For the regression problem with current status data, one again has the random triplets (T_i, C_i, Z_i) $(i = 1, \ldots, n)$. But the observations on the ith subject only consist of C_i, $d_i = I(C_i \leq T_i)$ and $Z_i(t)$, $t \leq C_i$. We assume that T_i and Z_i are related through model (1.2). In addition, we assume for the moment that the C_i are independent of the (T_i, Z_i) with a common cumulative hazard function Λ_c.

To introduce our estimating function, we define the counting processes for the monitoring times C_i by

$$N_i^c(t) = d_i I(C_i \leq t) = I\{C_i \leq \min(t, T_i)\}, \quad i = 1, \ldots, n.$$

By the counting-process martingale theory (Fleming and Harrington, 1991; Andersen et al., 1993),

$$N_i^c(t) - \int_0^t I(T_i \geq s) I(C_i \geq s) d\Lambda_c(s), \quad i = 1, \ldots, n, \tag{4.1}$$

are martingales with respect to an appropriate σ-filtration. Furthermore, the additive model assumption entails

$$P(T_i \geq t | Z_i(s),\ s \leq t) = e^{-\Lambda_0(t) - \beta' Z_i^*(t)}, \quad i = 1, \ldots, n, \tag{4.2}$$

where $Z_i^*(t) = \int_0^t Z_i(s) ds$. Since the T_i are unobservable, we modify (4.1) to be

$$M_i^c(t) = N_i^c(t) - \int_0^t I(C_i \geq s) e^{-\beta' Z_i^*(s)} e^{-\Lambda_0(s)} d\Lambda_c(s), \quad i = 1, \ldots, n. \tag{4.3}$$

Because of (4.2), $M_i^c(t)$ has zero mean for every t and every i. Note that all the random quantities in (4.3) are constructed from the observable data.

The structure of (4.3) is very similar to that of the basic martingales in the Cox regression model with right-censored failure time data. Mimicking the Cox partial likelihood score function, we obtain the following estimating function for β

$$S(\beta) = \sum_{i=1}^n \int \{Z_i^*(t) - \bar{Z}^*(t, \beta)\} dN_i^c(t), \tag{4.4}$$

where $\bar{Z}^*(t, \beta) = \sum_j Z_j^*(t) e^{-\beta' Z_j^*(t)} I(C_j \geq t) / \sum_j e^{-\beta' Z_j^*(t)} I(C_j \geq t)$. Note that the construction of (4.4) is free of the knowledge of Λ_0 and Λ_c. The estimating function $S(\beta)$ is the first derivative of a concave function and therefore must have a unique root. In fact, it is straightforward to show that $-\partial S(\beta)/\partial \beta$ is a positive definite matrix.

Let $\hat{\beta}^*$ be the unique solution to $S(\beta) = 0$. It can be shown that, under certain regularity conditions, $\hat{\beta}^*$ is consistent and asymptotically normal. More specifically, let $\hat{A}^*(\beta) = -n^{-1}\partial S(\beta)/\partial\beta$, and let

$$\hat{V}^*(\beta) = n^{-1}\sum_{i=1}^{n}\left[\int \{Z_i^*(t) - \bar{Z}^*(t,\beta)\}d\hat{M}_i^c(t)\right]^{\otimes 2},$$

where $\hat{M}_i^c(t)$ is obtained from $M_i^c(t)$ by replacing $e^{-\Lambda_0(s)}d\Lambda_c(s)$ with

$$\frac{\sum_{i=1}^{n} dN_i^c(s)}{\sum_{j=1}^{n} e^{-\hat{\beta}' Z_j^*(s)} I(C_j \geq s)}.$$

Let $A^*(\beta)$ and $V^*(\beta)$ be the limits of $\hat{A}^*(\beta)$ and $\hat{V}^*(\beta)$ as $n \to \infty$. By the counting process-martingale techniques (Andersen et al, 1993) and some other probability arguments, $n^{1/2}(\hat{\beta}^* - \beta)$ converges in distribution to $\mathcal{N}(0, A^{*-1}(\beta)V^*(\beta)A^{*-1}(\beta))$, and the limiting variance-covariance matrix may be estimated consistently by $\hat{A}^{*-1}(\hat{\beta}^*)\hat{V}^*(\hat{\beta}^*)\hat{A}^{*-1}(\hat{\beta}^*)$.

The validity of the preceding results relies on the assumption that the monitoring time C is independent of the covariate vector Z. When this assumption fails, we may model the relationship between C and Z. A convenient and technically advantageous way is to relate C to Z through the Cox model

$$\Lambda_c(t|Z_i) = e^{\gamma' Z_i(t)}\Lambda_{c,0}(t), \tag{4.5}$$

where $\Lambda_{c,0}$ denotes the baseline cumulative hazard function of C.

Similar to (4.3), a crucial observation is that

$$N_i^c(t) - \int_0^t I(C_i \geq s)e^{-\beta' Z_i^*(s)}e^{\gamma' Z_i(s)}e^{-\Lambda_0(s)}d\Lambda_{c,0}(s)$$

has zero mean for every i and every t. Thus, we extend (4.4) to be

$$S(\beta;\hat{\gamma}) = \sum_{i=1}^{n}\int \{Z_i^*(t) - \bar{Z}^*(t,\beta,\hat{\gamma})\}dN_i^c(t), \tag{4.6}$$

where $\hat{\gamma}$ is the maximum partial likelihood estimator of γ based on (C_i, Z_i) $(i = 1, \ldots, n)$, which satisfy model (4.5), and

$$\bar{Z}^*(t,\beta,\gamma) = \frac{\sum_j Z_j^*(t)e^{-\beta' Z_j^*(t)+\gamma' Z_j(t)}I(C_j \geq t)}{\sum_j e^{-\beta' Z_j^*(t)+\gamma' Z_j(t)}I(C_j \geq t)}.$$

We estimate β by solving $S(\beta;\hat{\gamma}) = 0$. It is not difficult to show that the resulting estimator is consistent and asymptotically normal, and the limiting variance-covariance matrix can be estimated easily.

An alternative approach to accommodating the dependence of C on Z is through stratification. For simplicity, we assume that C only takes finitely many values, say $t_1, \ldots, t_L$. One can then use the following estimating function

$$S_{st}(\beta) = \sum_{l=1}^{L} \sum_{i:C_i=t_l} d_i \left\{ Z_i^*(C_i) - \bar{Z}_l^*(C_i, \beta) \right\}, \tag{4.7}$$

where $\bar{Z}_l^*(t,\beta) = \sum_{j:C_j=t_l} e^{-\beta' Z_j^*(t)} Z_j^*(t) / \sum_{j:C_j=t_l} e^{-\beta' Z_j^*(t)}$. The resulting estimator can be shown to be consistent and asymptotically normal. Furthermore, the extension to the case of continuous C is possible. The details, however, are omitted.

Under general interval-censorship, each subject may be monitored more than once. Suppose that there are K monitoring times. For each of the K monitoring times, we may construct an estimating function similar to (4.4), (4.6) or (4.7). (Slight modifications will be required if the subjects are not monitored after the detection of the event of interest.) These K estimating functions may then be combined to yield a single estimating function for β_0. The resulting estimator of β_0 can be shown to be consistent and asymptotically normal. The details are again omitted.

5 Further results and concluding remarks

Model (1.2) has the limitation that the linear predictor $\beta' Z(t)$ needs to be constrained to ensure positivity for the right side of (1.2). One may avoid this constraint by replacing $\beta' Z(t)$ in (1.2) by $e^{\beta' Z(t)}$, in which case $\lambda_0(t)$ pertains to the hazard function under $\beta' Z(t) = -\infty$ rather than under $Z(t) = 0$. The ideas presented in this paper can be applied to any regression function $g\{\beta' Z(t)\}$, where $g(\cdot)$ is a smooth function.

The proportional and additive hazards regression models may be viewed as the two extremes of a family of hazard-based generalized linear regression models. Aranda-Ordaz (1983) and Breslow (1985) argued that a compromise between the proportional and additive models is often desirable. In particular, they proposed a Box-Cox-type transformation family in the form of

$$\frac{\lambda^{\rho}(t|Z) - 1}{\rho} = \lambda_0(t) + \beta' Z(t), \tag{5.1}$$

in which $\rho = 0$ corresponds to the Cox model and $\rho = 1$ the additive model. The modeling capacity of (5.1) may be illustrated by the British doctors study, which was concerned with the effects of cigarette smoking on mortality (Breslow and Day, 1987, pp. 336–338). The data from this study suggest that the difference of the two hazard functions increases in t whereas their ratio decreases (Breslow and Day, 1987, p. 112, Table 3.15). Thus, neither the additive model nor the Cox model adequately describes the data. Yet, it can be shown easily that (5.1) with $\rho \approx 1/2$ provides such a feature when β and λ_0 are properly chosen. Although a discretized version may be dealt with as a special case of generalized linear models, (5.1) itself appears to be intractable.

An alternative extension of the multiplicative and additive models was proposed recently by Lin and Ying (1995). A basic feature of their extension is that the conditional hazard function is linear with respect to an unspecified baseline hazard. With the special exponential-linear link, the model takes the following form

$$\lambda(t|Z(s), s \le t) = e^{\beta' Z_m(t)}\{\lambda_0(t) + \gamma' Z_a(t)\}. \tag{5.2}$$

The covariate process in the additive part, Z_a, can be the same as, related to or different from that in the multiplicative part, Z_m. For example, one may take $Z_a(t) = Z_m(t)q(t)$ with an appropriately chosen q. For the two-sample case with $Z_m = Z_a = 1$ or 0, it is not difficult to show that for certain β, γ and λ_0, (5.2) also provides the feature that the risk difference increases while the risk ratio decreases.

Lin and Ying (1995) proposed to estimate the regression parameters β and γ under model (5.2) by combining the partial likelihood score function with estimating function (2.2). Specifically, they introduced the following estimating function

$$\sum_{i=1}^{n} \int \{X_i(t;\beta,\gamma) - \bar{X}(t;\beta,\gamma)\} \left\{dN_i(t) - Y_i(t)e^{\beta' Z_{mi}(t)}\gamma' Z_{ai}(t)dt\right\},$$

where the N_i and Y_i are the same as those defined in (2.1) and the X_i are suitably chosen $\{\dim(\beta) + \dim(\gamma)\}$-dimensional weight processes and $\bar{X}(t;\beta,\gamma) = \sum Y_i(t)e^{\beta' Z_{mi}(t)}X_i(t;\beta,\gamma)/\sum Y_i(t)e^{\beta' Z_{mi}(t)}$. They established the large-sample properties of the resulting estimator via the counting-process martingale approach and provided simple inference procedures.

In this article, we demonstrate the potential usefulness of the additive hazards regression model and its variations. The simplicity of the inference procedures is evident not only from (2.3), which gives an explicit form for the estimator of the regression parameter, but also from the availability of the appealing model-driven methods for interval-censored data, which are known to be difficult to deal with. Furthermore, its ability to incorporate additive frailty with resulting model invariance (when marginalized over the frailty) shows that the additive model provides an effective tool to handle dependent data. We hope that further developments of the results outlined in this paper will lead to simplification of statistical inference for some nonstandard problems in survival analysis such as multivariate failure times and interval censored data.

Acknowledgements

This research was supported by the NIH Grant R29 GM-47845 (for Lin) and by the NSF Grants DMS-92-07730 and DMS-96-26750 and the NSA Grant MDA-90496-10034 (for Ying).

References

Aalen, O. O. (1980). A model for nonparametric regression analysis of counting processes. In *Lecture Notes in Statistics*, Vol. 2, pp. 1–25. New York: Springer.

Aalen, O. O. (1989). A linear regression model for the analysis of life times. *Statist. Med.*, **8**, 907–925.

Andersen, P. K., Borgan, Ø., Gill, R. D. and Keiding, N. (1993). *Statistical Models Based on Counting Processes.* New York: Springer-Verlag.

Andersen, P. K. and Gill, R. D. (1982). Cox's regression model for counting processes: a large sample study. *Ann. Statist.*, **10**, 1100–1120.

Aranda-Ordaz, F. J. (1983). An extension of the proportional-hazards model for grouped data. *Biometrics*, **39**, 109–117.

Breslow, N. E. (1985). Cohort analysis in epidemiology. In *A Celebration of Statistics*, Eds. A. C. Atkinson and S. E. Fienberg. pp. 109–143. New York: Springer.

Breslow, N. E. and Day, N. E. (1980). *Statistical Methods in Cancer Research, Vol. I, The Design and Analysis of Case-Control Studies.* Lyon: IARC.

Breslow, N. E. and Day, N. E. (1987). *Statistical Methods in Cancer Research, Vol. II, The Design and Analysis of Cohort Studies.* Lyon: IARC.

Clayton, D. and Cuzick J. (1985). Multivariate generalizations of the proportional hazards model (with discussion). *J. Roy. Statist. Soc.* **A**, **148**, 82–117.

Cox, D. R. (1972). Regression models and life-tables (with discussion). *J. R. Statist. Soc.* **B**, **34**, 187–220.

Cox, D. R. and Oakes, D. (1984). *Analysis of Survival Data.* London: Chapman and Hall.

Fleming, T. R. and Harrington, D. P. (1991). *Counting Processes and Survival Analysis.* New York: Wiley.

Groeneboom, P. and Wellner, J.A. (1992). *Information Bounds and Nonparametric Maximum Likelihood estimation.* Basel: Birkhäuser Verlag.

Heckman, J.J. and Singer, B. (1982). Population heterogeneity in demographic models. In *Multidimensional Mathematical Demography*, Ch. 12, pp. 567–599, Eds. K. C. Land and A. Rogers. New York: Academic Press.

Hougaard, P. (1987). Modelling multivariate survival. *Scand. J. Statist.*, **14**, 291–304.

Huang, J. (1996). Efficient estimation for the Cox model with interval censoring. *Ann. Statist.*, **24**, 540–568.

Huffer, F. W. and McKeague, I. W. (1991). Weighted least squares estimation for Aalen's additive risk model. *J. Amer. Statist. Assoc.*, **86**, 114–129.

Klein, R.W. and Spady, R.H. (1993). An efficient semiparametric estimator for binary response models. *Econometrica*, **61**, 387–421.

Lee, E.W., Wei, L.J. and Amato, D. (1992). Cox-type regression analysis for large number of small groups of correlated failure time observations. In *Survival Analysis, State of the Art*, Eds. J. P. Klein and P. K. Goel, pp. 237–247. Netherlands: Kluwer Academic Publishers.

Lin, D. Y. and Ying, Z. (1994). Semiparametric analysis of the additive risk model. *Biometrika*, **81**, 61–71.

Lin, D.Y. and Ying, Z. (1995). Semiparametric analysis of general additive-multiplicative intensity models for counting processes. *Ann. Statist.*, **23**, 1712–1734.

Murphy, S. (1995). Asymptotic theory for the frailty model. *Ann. Statist.*, **23**, 182–198.

Nielsen, G.G., Gill, R.D., Andersen, P.K. and Sorensen, T.I.A. (1992). A counting process approach to maximum likelihood estimation in frailty models. *Scand. J. Statist.*, **19**, 25–43.

Oakes, D. (1989). Bivariate survival models induced by frailties. *J. Amer. Statist. Assoc.*, **84**, 487–493.

Prentice, R. L. and Self, S. G. (1983). Asymptotic distribution theory for Cox-type regression models with general risk form. *Ann. Statist.*, **11**, 804–813.

Rabinowitz, D., Tsiatis, A. and Aragon, J. (1995). Regression with interval censored data. *Biometrika*, **82**, 501–513.

Thomas, D. C. (1986). Use of auxiliary information in fitting nonproportional hazards models. In *Modern Statistical Methods in Chronic Disease Epidemiology*, Ed. S. H. Moolgavkar and R. L. Prentice. pp. 197–210. New York: Wiley.

Turnbull, B.W. (1976). The empirical distribution function with arbitrarily grouped, censored and truncated data. *J. Roy. Statist. Soc. Ser.* **B**, **38**, 290–295.

Wei, L. J., Lin, D. Y. and Weissfeld, L. (1989). Regression analysis of multivariate incomplete failure time data by modeling marginal distributions. *J. Amer. Statist. Assoc.*, **84**, 1065–1073.

Some Exploratory Tools for Survival Analysis

John Crowley
Michael LeBlanc
Joth Jacobson
Sidney E. Salmon

ABSTRACT Graphical and computer-intensive methods for exploring survival data are becoming more widely available, though perhaps not yet widely used. We review some recent developments in this area, including running quantile plots, local estimation of the relative risk function, and recursive partitioning. The use of Martingale residuals in some of these methods will be contrasted with other approaches. The methods are applied to data on patients with multiple myeloma treated on trials conducted by the Southwest Oncology Group.

1 Introduction

Exploratory methods for survival data are becoming more widely available but are still underutilized. The frequent presence of censoring has hampered development to some extent, but another difficulty is that with the regression model most used in survival analysis, the Cox (1972) proportional hazards model, the covariates act on survival through the hazard function and not as a simple shift in location or scale. In this article we present several graphical approaches for exploring the relationship of a single covariate to survival, as well as a variety of techniques for forming prognostic groups or staging systems for cancer patients. We will focus on the proportional hazards model and will contrast methods based on what are now called Martingale residuals (Fleming and Harrington, 1991) with various other suggestions.

We will illustrate the methods using data on patients with multiple myeloma, a malignancy of the plasma cells of the blood and bone marrow. The disease is characteristized by various complications including bone breakage, kidney failure, and infections. Myeloma patients have an extremely variable course but with the use of combination chemotherapy and steroid hormones the median survival in most series of patients is from 30-36 months. The data for our examples come from two successive trials done by the Southwest Oncology Group (SWOG), a national cancer clinical

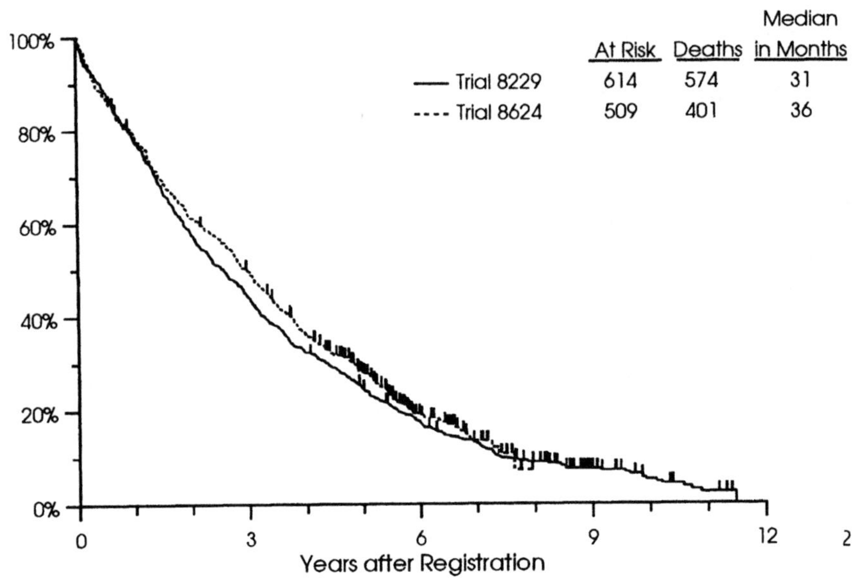

FIGURE 1. Survival Curves From Two Successive SWOG Myeloma Trials

trials consortium in the United States. SWOG study 8229 (Salmon *et al*, 1990) compared two schedules of administration of the same set of agents, finding no difference in survival. SWOG study 8624 (Salmon *et al*, 1994) compared one of the arms of 8229 with two other regimens, finding some advantage to higher doses of steroids. Product-limit (Kaplan and Meier, 1958) survival curves from these two studies, collapsed over treatment arms, are presented in Figure 1 (the tic marks represent right-censored observations). We will use the 8229 data as a training set and the 8624 data for external validation.

The covariates we will use are age (in years) at entry into the trial, performance status (ambulatory vs. non-ambulatory), calcium (reflecting bone loss), albumin (nutritional status), creatinine (kidney function) and serum β_2 microglubulin (which reflects tumor volume and kidney failure in a single measurement). A generally used staging system was developed by Durie and Salmon (1975), and is based on an indirect quantification of the number of tumor cells (I, II, or III) and a classification of kidney function (A,B), which is shown in Figure 2 collapsed into 3 groups (I-II, IIIA, IIIB). Many of the exploratory analyses we will present have as a goal the improvement of this staging system.

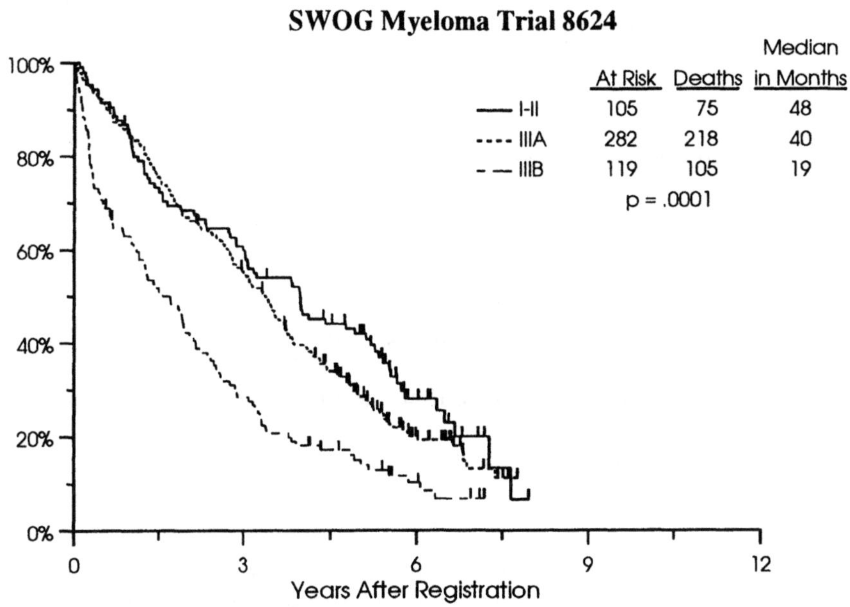

FIGURE 2. Survival by Durie-Salmon Stage

2 Methods for a Single Quantitative Covariate

We begin this section by presenting some notation which will be used throughout the remainder of this article. Let X be a random variable with survivor function

$$S(t) = P(X > t), 0 \leq t < \infty,$$

and distribution function $F(t) = 1 - S(t)$. The data are typically subject to right-censoring, which for simplicity we will model using a random censoring time C. The observables $T = X \wedge C$, $\Delta = I(T = X)$, and Z, where $\wedge$ stands for minimum, $I(\cdot)$ is the indicator function, and Z is a possibly vector-valued covariate. Given Z, X and C are independent.

The hazard function is given by

$$\lambda(t) = \lim_{h \downarrow 0} \frac{P(t \leq X < t + h \mid X \geq t)}{h}$$

and the cumulative hazard function is

$$\Lambda(t) = \int_0^t \lambda(u)du = -\ln S(t).$$

Denote the observed data by

$$Y = [(t_i, \delta_i, z_i)]_{i=1}^n ,$$

where for convenience throughout we assume there are no ties among the observed failure times.

The proportional hazards model states that

$$\lambda(t, z) = \lambda_0(t) f(z), \tag{1.1}$$

where $\lambda_0(t)$ is an unspecified baseline hazard function, and $f(z)$ is often assumed to be of the form $\exp(z\beta)$. This can be contrasted with what is sometimes called the accelarated failure time model (*cf* Kalbfleisch and Prentice, 1980), which posits the familiar linear regression form for the log of survival time:

$$\log(T) = z\beta + \varepsilon. \tag{1.2}$$

Elsewhere in this volume, Professor Cox argues for fully parametric models instead of the semiparametric Cox proportional hazards model, and Professor Wei advocates the use of equation 1.2, so we will hereafter refer to equation 1.1 as the Seattle model.

2.1 Scatterplots

Residuals can be defined for the Seattle model based on the work of Cox and Snell (1968), as elucidated by Crowley and Hu (1977) and Kay (1977).

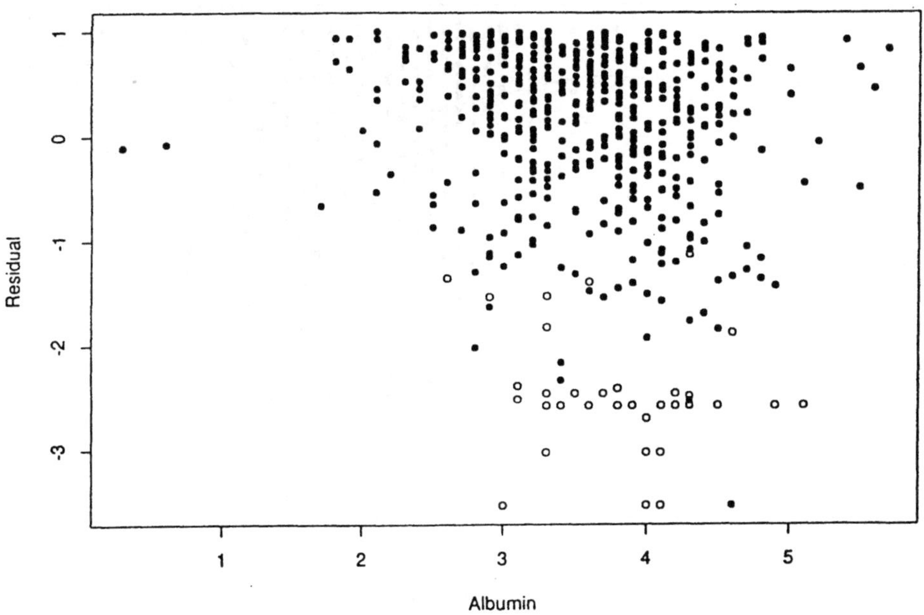

FIGURE 3. Martingale residuals vs albumin

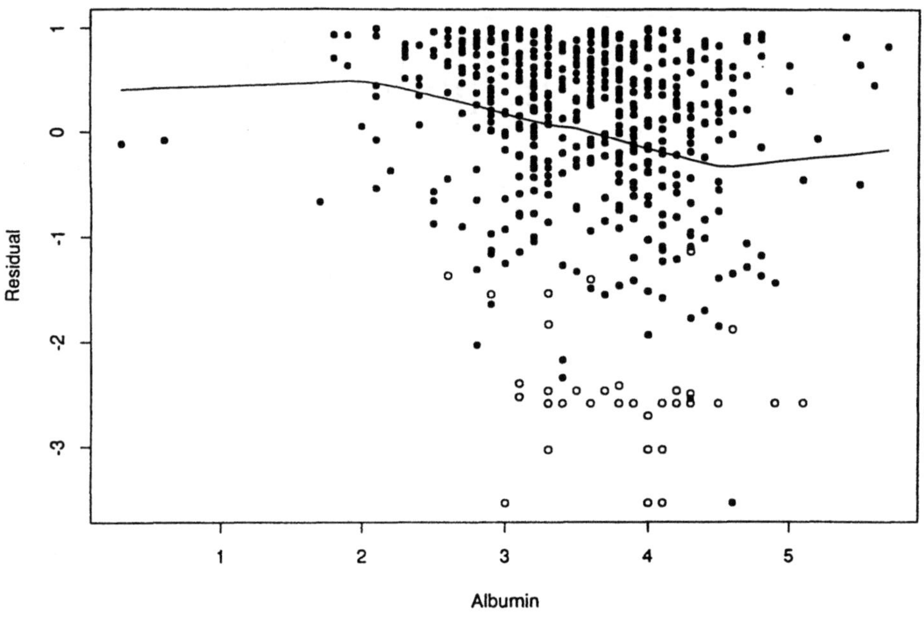

FIGURE 4. Smooth of martingale residuals vs albumin

If the Seattle model represents the truth, then

$$\Lambda_0(t_i)\exp(z_i\beta)$$

are unit exponentials, and

$$\widehat{\Lambda}_0(t_i)\exp(z_i\widehat{\beta})$$

are (approximate) censored unit exponentials. If $\widehat{\Lambda}_0$ is the Breslow (1974) estimator and $\widehat{\beta}$ is the maximum partial likelihood estimator of Cox (1972, 1975) and one adds 1 (the conditional expected additional survival for unit exponentials) to the censored observations, the result is a linear transformation of what are now called Martingale residuals (*cf* Fleming and Harrington, 1991):

$$\delta_i - \widehat{\Lambda}_0(t_i)\exp(z_i\widehat{\beta}).$$

The replacement of Λ by $\widehat{\Lambda}$ adds considerably more structure to these residuals (with no covariates, their placement, though not their order, is completely predictable!) than does the estimation of β, as pointed out by Lagakos (1981), but Arjas (1988) has shown that subsets of residuals defined by stratification behave sensibly in large samples. Crowley and Storer (1983) show by simulation that these residuals, when averaged over many samples, can identify omitted covariates in scatterplots, but that in single samples such plots tend not to be very informative. Figure 3 is a scatterplot of the Martingale residuals from a null model (no covariates) fit to our SWOG 8229 data set against the variable albumin, with censored observation shown with open circles. There is no apparent trend in this plot, though it is clear that the residuals need to be transformed and/or smoothed before one could conclude there is no information beneath the noise. Figure 4 shows a smooth (using lowess as implemented in S-Plus^{TM}) superimposed on the scatterplot of null residuals vs albumin, and again the impression is one of no obvious trend.

A more direct technique is a simple scatterplot of survial times vs albumin, as shown in Figure 5, where again censored observations are indicatated by open circles (no attempt is made to put the censored observations where they "should" be, as was done for Martingale residuals). Again, there is no obvious trend in these data. Imposing some sort of trend line on this scatterplot might be helpful, however. Since the mean can not be reliably estimated with right-censored data, running quantile plots have been suggested by Gentleman and Crowley (1991a), building on earlier work by Beran (1981) and Doksum and Yandell (1983). The basic tool is a local application of the product-limit estimator (Kaplan and Meier, 1958), which for the complete data set is given by

$$\widehat{S}(t) = \prod_{i:t_i \le t} \left[1 - \frac{\delta_i}{r_i}\right]^{\delta_i},$$

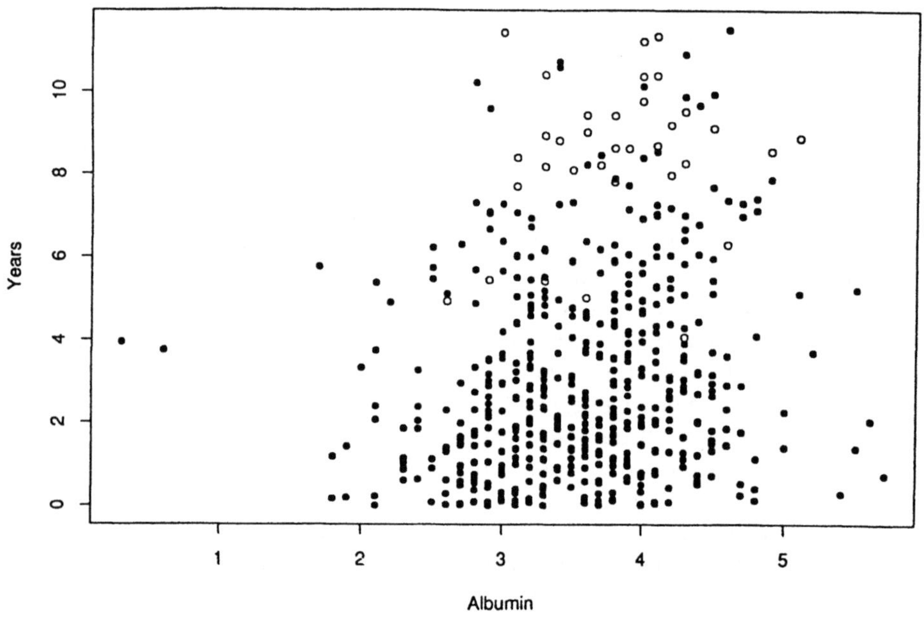

FIGURE 5. Survival times vs albumin

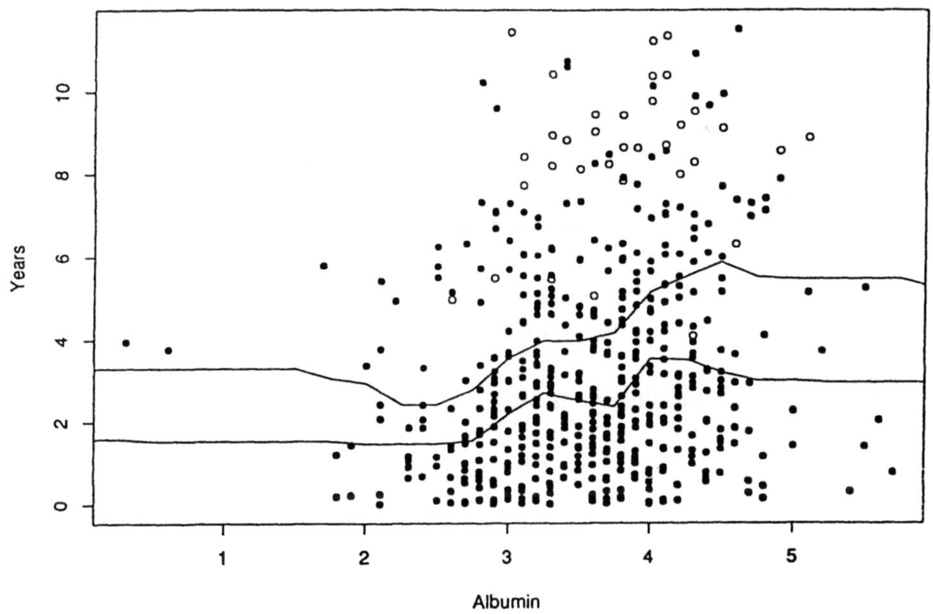

FIGURE 6. 50% and 30% running quantiles of survival vs albumin

where r_i is the number of individuals at risk at t_i (the number with observed survival times, censored or not, which are at least t_i). For a given covariate Z, their approach is to use the k nearest neighbors (in the covariate space) of a point z to estimate the distribution function, and to plot selected quantiles of this distribution superimposed on the scatterplot of survival vs covariate. Figure 6 shows the scatterplot of survival and albumin with the 30% and 50% quantiles of the distribution superimposed, using a symmetric neighborhood of about one-third of the data, and now the impression is one of increasing survival with increasing albumin. This impression can be confirmed by formal application of the Seattle model, and by relative risk plots as described in the next section.

2.2 Relative Risk Plots

In the Seattle model the usual method for estimation of the regression parameters is the partial likelihood of Cox (1975). Given the data $(t_1, \delta_1, z_1), \ldots, (t_n, \delta_n, z_n)$, the risk set at time x_i is denoted by R_i and consists of the indices j such that $x_i \leq x_j$. The partial likelihood is then given by

$$L(\beta) = \prod_{i=1}^{n} \left\{ \frac{\exp \beta z_i}{\sum_{j \in R_i} \exp \beta z_j} \right\}^{\delta_i},$$

and the log partial likelihood by

$$l(\beta) = \sum_{i=1}^{n} \delta_i \left\{ \beta z_i - \log \left\{ \sum_{j \in R_i} \exp \beta z_j \right\} \right\}.$$

A basic aspect of exploratory survival analysis is the assessment of the form of the regression function (or relative risk) $\exp \beta z$. One approach is to replace this parametric form with $\exp s(z)$, where $s(z)$ is some smooth function, and to estimate $s(z)$ using local techniques. Tibshirani and Hastie (1987) proposed a local partial likelihood approach, while Gentleman and Crowley (1991b) suggested an alternating scheme to estimate both the baseline cumulative hazard function and the regression function. In the latter approach the baseline hazard function $\Lambda_0(t)$ is first estimated, using all the data and a preliminary estimate of $\widehat{s}_0(z)$ of $s(z)$, employing the Breslow (1974) estimator:

$$\widehat{\Lambda}_{01}(t) = \sum_{i: t_i \leq t} \frac{\delta_i}{\sum_{l \in R_j} \exp \widehat{s}_0(z_l)}. \tag{1.3}$$

Then with $\widehat{\Lambda}_{01}(t)$ fixed, the full likelihood is used, locally, to give an updated estimate $\widehat{s}_1(z)$, assuming s is given by $s(z) = \alpha + \beta z$ in a neighborhood of

z. The local log likelihood for this step is given by

$$l(\alpha_i, \beta_i) = \sum_{j \in N_i} \left[\delta_j(\alpha_i + \beta_i z_j) - \widehat{\Lambda}_{01}(t_j) \exp(\alpha_i + \beta_i z_j) \right],$$

where N_i denotes the indices of the nearest neighbors of z_i. Thus for each value of the covariate a linear approximation to s is made and the appropriate parameters are estimated. This estimate $\widehat{s}_1(z)$ is then used in 1.3 to give an updated estimate $\widehat{\Lambda}_{02}(t)$, and the cycle is repeated until convergence. For this presentation we have chosen to implement a slight modification of the above method, where the local likelihood esimation step is approximated by applying a smoothing spline to the residuals $\delta_j - \widehat{\Lambda}_{01}(t_j) \exp(s(z))$ and interating until convergence.

Figure 7 shows the result of estimating $s(z)$ for the myeloma data from SWOG 8229 using the covariate serum β_2 microglobulin, and a neighborhood of about one-third of the data. A histogram of serum β_2 microglobulin is superimposed on the plot, as well as the straight line fit (linear on the log scale) from assuming that $s(z) = \alpha + \beta z$ (α is absorbed into $\lambda_0(t)$). It can be seen that the regression function is decidedly nonlinear, suggesting a transformation such as taking logs (advisable on other grounds as well). Figure 8 show the result of fitting a Seattle model using the log of serum β_2 microglobulin, showing a more nearly linear relationship.

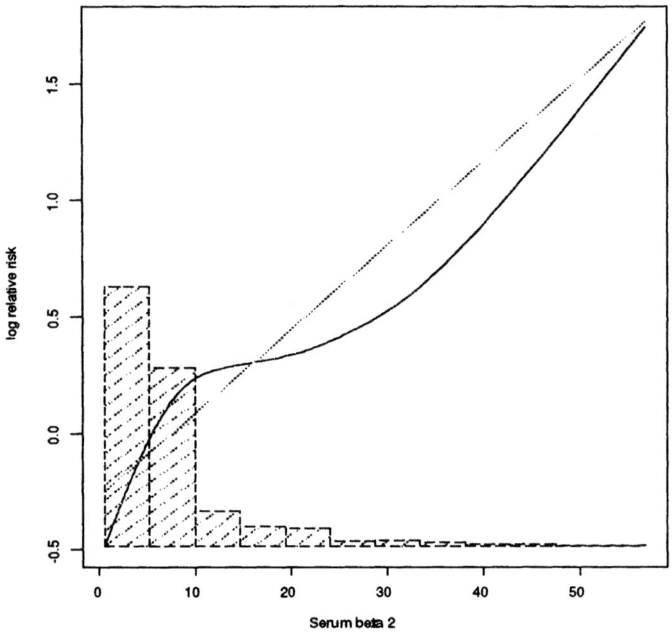

FIGURE 7. Smooth Relative Risk vs Serum β_2 Microglobulin

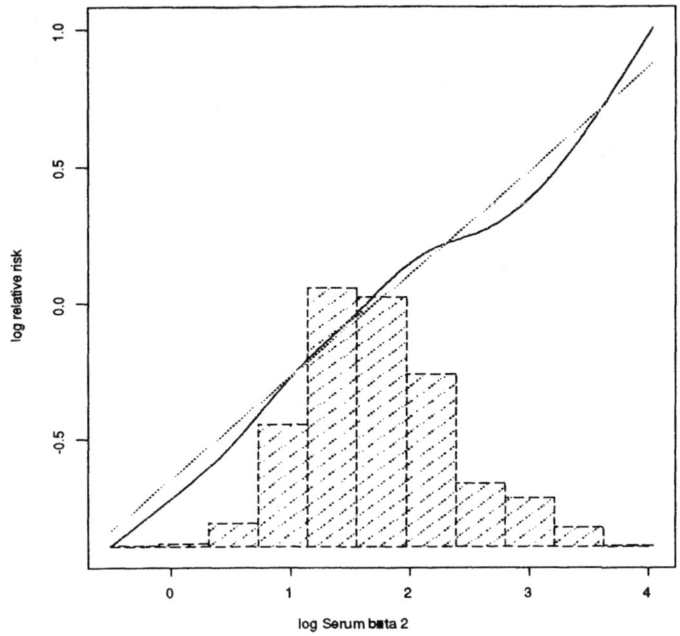

FIGURE 8. Smooth Relative Risk vs Log Serum β_2 Microglobulin

3 Forming Prognostic Groups

3.1 Regression Methods

Prognostic groupings, or staging systems, are valuable not just for making prognostic statements but also for use in clinical trial design. If patients can be divided into a few groups which have similar survival between members of the group but different survival between groups, then different treatment strategies might be derived for each group, or the groups might be used in a stratified randomization scheme. A typical method for forming prognostic groups based on survival is to fit a Seattle model to the data, then use the resulting regression equation to divide the predictor space into regions. To illustrate, we fit such a model to the SWOG 8229 data, using a stepwise procedure as implemented in the SAS^{TM} "proc" PHREG (after first tranforming serum β_2 by taking logs; relative risk plots of the other 4 continuous variables indicated no need for transformation). The selected

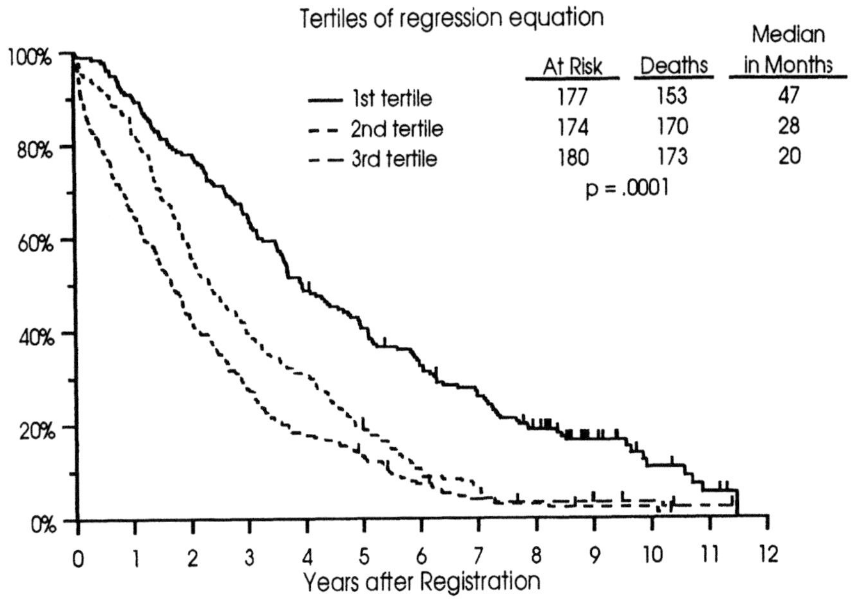

FIGURE 9. Prognosis Groups: SWOG 8229

model omitted only calcium, and produced the regression equation

$$\begin{aligned}\beta z &= 1.175(\log \text{serum } \beta_2) + 1.018(\text{age}) + 1.161(\text{performance status}) \\ &\quad +1.086(\text{creatinine}) + 0.833(\text{albumin})\end{aligned}$$

(where the variables are listed in the order of their stepwise selection).

We used the regression equation to divide the SWOG 8229 data into three groups with roughly equal numbers of observations; the result is given in Figure 9. The same equation was then applied to the data on SWOG 8624 as a validation step, as shown in Figure 10.

As can be seen, the procedure worked quite well in terms of prediction, as might have been expected since relatively little adaptive fitting went into the model selection (a fair amount of experience went into the choice of the 6 variables!). However, it is difficult to characterize the 3 groups in terms of covariate values other than being high, intermediate and low on a prognostic index. A variation which carries a little more intuition is to dichotomize the continuous covariates before fitting a regression model, then group the patients by the "number of bad things." The dichotomization can be done using the median, using historicial cutpoints if they exist, or by adaptive methods. One such adaptive method is illustrated in Figure

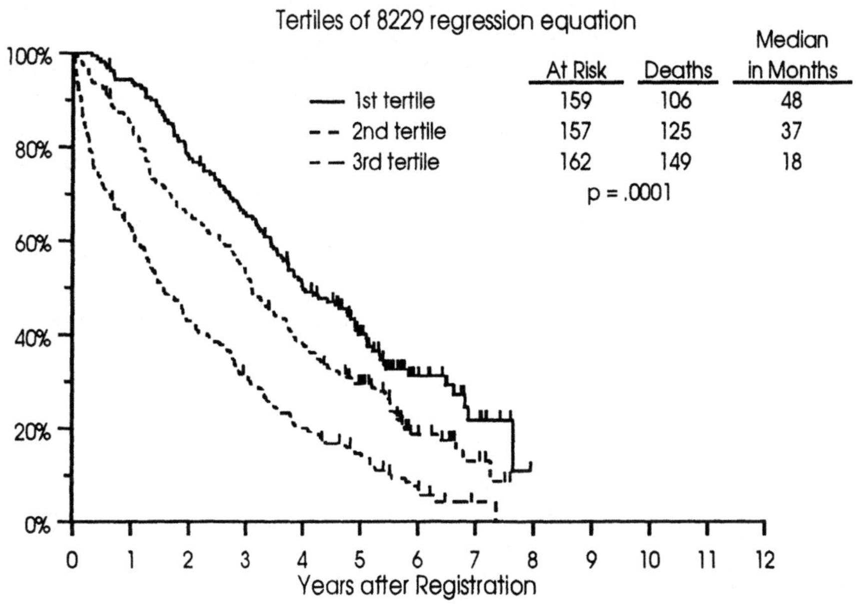

FIGURE 10. Prognosis Groups: SWOG 8624

11, which gives the value of the logrank statistic (Mantel, 1966) in χ^2 form as a function of covariate cutpoints (used to split the sample into 2 groups) for each of the 6 covariates (performance status is already dichotomous). Superimposed on the plots for the continuous covariates are the 5% and 1% points of the distribution of the logrank statistic as derived by sampling from the permutation distribution of the maximum.

The cutpoints chosen ($<$ vs $\geq$) were age (67), serum β_2 (5.4), albumin (3.3), calcium (10.7), and creatinine (1.8, because a minimum group size of 100 was imposed). High values of albumin are a good prognostic sign, while high values for the other continuous variables are bad. A Seattle model was again fit to the SWOG 8229 data in a stepwise fashion using these dichotomized covariates, and this time only creatinine was eliminated. (A procedure which adapts at each stage of a stepwise regression fit, instead of fixing the cutpoints adaptively in advance, as done here, is given in LeBlanc and Crowley, 1995.) Rather than using the actual regression equation, one typically forms groups by counting up the number of bad prognostic signs a patient has. This is illustrated in Figure 12 for the SWOG 8229 data, and validated in Figure 13 for the SWOG 8624 data. Again, the prediction works well, and the procedure is easy to explain, but the groups are still not easily characterized; those with 0-1 bad signs might have low performance status as their only poor prognsotic factor, or high serum β_2, etc. A staging procedure which results in easily characterized groups is described in Section 3.2.

3.2 Recursive Partitioning

Recursive partitioning proceeds by first splitting the predictor space into two regions or nodes, based on a specified rule. The rule is then applied recursively to the resulting nodes until the space has been split into a large number of nodes with a few observations each. The resulting structure is depicted graphically in the form of a tree. As this large tree represents an over fit to the data, an algorithm for pruning back the branches of the tree to choose the "best" subtree is then applied. The statistical underpinnings of this procedure were detailed in the monograph by Breiman, Friedman, Olshen and Stone (1984), who coined the phrase classification and regression trees, or $CART^{TM}$.

More formally, a cost function R is defined for each node h of a tree $\mathcal{T}$, for example the sum of squares around the mean. The splitting of a node into two is then based examining each covariate and each possible cutpoint, and finding that covariate and cutpoint which gives the least cost. Define the cost of tree $\mathcal{T}$ as

$$R(\mathcal{T}) = \sum_{h \in \widetilde{\mathcal{T}}} R(h),$$

where $\widetilde{\mathcal{T}}$ represents the terminal nodes of the tree. This is minimized for a

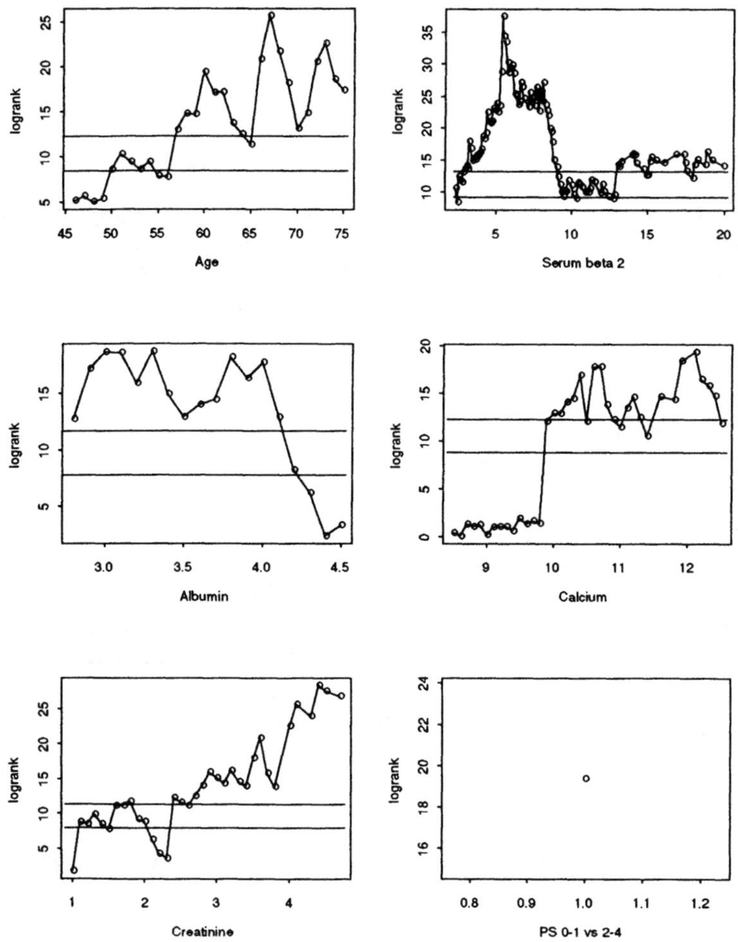

FIGURE 11. Running Logrank Test Statistics for Six Covariates

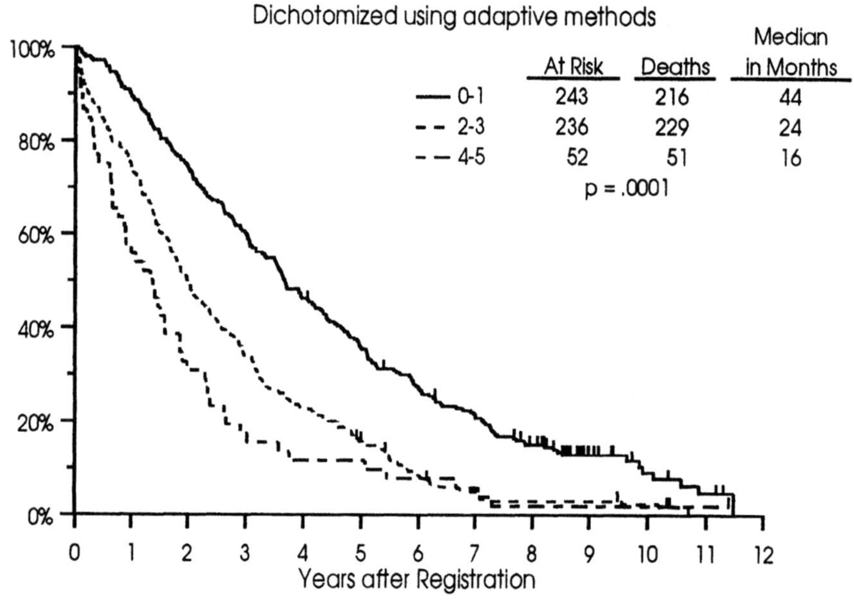

FIGURE 12. Number of Poor Prognosis Groups: SWOG 8229

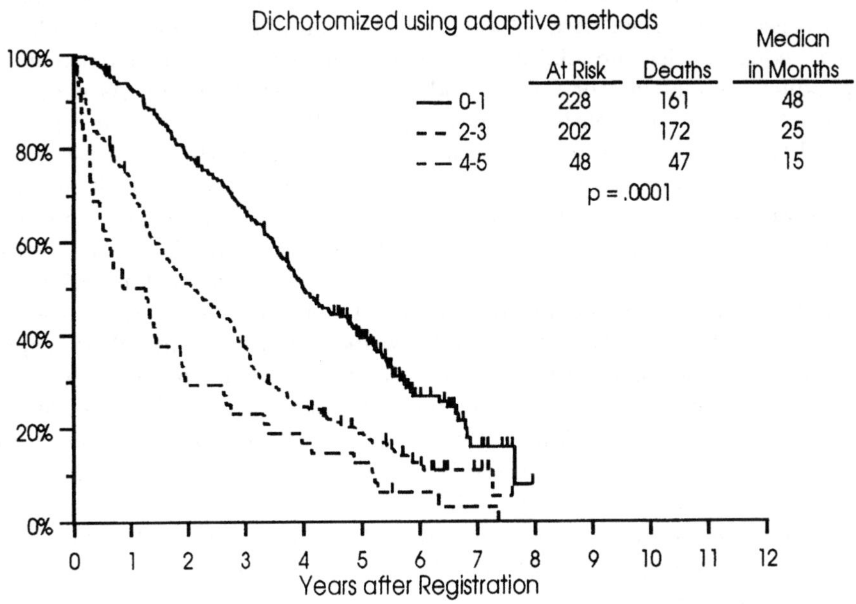

FIGURE 13. Number of Poor Prognosis Groups: SWOG 8624

function such as the sum of squares by having each observation be its own node; if a minimum node size is imposed, the cost is minimized as the terminal nodes approach that size. Thus a penalized cost, or cost complexity, is introduced as

$$R\alpha(\mathcal{T}) = R(\mathcal{T}) + \alpha\,|\mathcal{T}|,$$

where $|\mathcal{T}|$ is the number of terminal nodes. Breiman, Friedman, Olshen and Stone show that varying α defines a nested sequence of subtrees, and that there is an optimally pruned subtree for each α. They propose choosing α by cross-validation using prediction error.

Several methods have been proposed for recursive partitioning based on censored survival data. Gordon and Olshen (1985) suggested replacing least squares with the minimum horizontal or vertical distance between the product-limit estimator for the node and any step function. As shown by Crowley, LeBlanc, Gentleman and Salmon (1995) however, these distances tend to find structure in the censoring variable when there is no structure in the underlying covariate space. Davis and Anderson (1989) fit an exponential model and use likelihood deviance as the cost function. Therneau, Grambsch and Fleming (1990) use the Martingale residuals $\delta_i - \widehat{\Lambda}_0(t_i)$ from the null Seattle model, and the sum of squares for the cost function (the residuals can thus be plugged into standard recursive partitioning software).

Segal (1988) and Ciampi, Thiffault, Nakache and Asselain (1986) propose splitting on between node test statistics rather than on reducing within node variability. This allows one to construct trees using for example the familiar logrank test (Mantel, 1966), but leaves open the question of how to prune such trees (since there is no obvious analog to prediction error with between node splitting). LeBlanc and Crowley (1993) define $G(h)$ as the value of the splitting statistic at node h, and $G(\mathcal{T})$ as the sum of all the two sample splits in the tree $\mathcal{T}$. They then define "goodness of split complexity" as

$$G\alpha(\mathcal{T}) = \sum_{h \in \mathcal{T} - \widetilde{\mathcal{T}}} G(h) - \alpha\,|\,\mathcal{T} - \widetilde{\mathcal{T}}\,|,$$

where $\mathcal{T} - \widetilde{\mathcal{T}}$ represents the interior nodes of the tree. The second term is a penalty for the complexity (number of splits) of the tree. LeBlanc and Crowley (1993) show that an optimally pruned subtree for any α can be found by a weakest link pruning algorithm which is analogous to the cost complexity pruning algorithm in CART^{TM}. However, the selection of a reasonable complexity parameter α is made difficult because splits have been adaptively chosen based on the response. If the tree were grown on one data set and the test statistics evaluated on a separate data set, then choosing a value for α of 4 (with the statistic G in χ^2 form) would correspond roughly to a 5% p-value, and a value of 2 corresponds to the AIC method of model selection.

Since the same data set is used both to grow the tree and to evaluate the splitting statistics, we suggest a bias-corrected version of the split complexity described above to correct for the adaptive split selection. A bias corrected version of the split-complexity is

$$G\alpha^*(\mathcal{T}) = G^*(\mathcal{T}) - \alpha\,|\mathcal{T} - \tilde{\mathcal{T}}|,$$

where $G^*(\mathcal{T})$ is the bias corrected sum of the test statistics in the tree $\mathcal{T}$. This is the bias-corrected split complexity used in LeBlanc and Crowley (1993), except that the method for estimating $G^*(\mathcal{T})$ we use here involves estimating the bias correction at each node conditional on the previous splits in the tree, then summing over nodes to form the bias corrected split complexity, rather than bias correcting pruned subtrees as is done in the CART^{TM} algorithm.

For each node h define

$$G^*(h) = G(h) - \Delta(G(h)), \tag{1.4}$$

where $\Delta(G(h)) = E_{Y^*}G(h, Y^*, Y^*) - E_{Y^*}G(h, Y^*, Y)$. Note that we have augmented the notation so that $G(h, X^*, X)$ denotes that standardized test statistic for which Y was the sample used to choose the split point and Y^* is used to calculate the test statistic.

Typically, however, resampling methods are needed to estimate $\Delta(G(h))$. We have explored both permutation and bootstrap methods for bias correction. Permutation methods only apply when the effect sizes at a node are small, since one does the bias correction under the assumption that the survival times and censoring are not related to any of the predictor variables. Bootstrap bias correction is more generally applicable when survival times are associated with covariates. We implement a bootstrap method (bootstrap .632, Efron and Tibshirani, 1993) to obtain an estimate of $\Delta(G(h))$ for every internal node in the tree. Let $\tilde{G}(h)$ denote the estimate of $G^*(h)$ using $\Delta(\widehat{G(h)})$ and $\tilde{G}(\mathcal{T}) = \sum \tilde{G}(h)$.

Then weakest link pruning is applied to using the estimate of bias corrected split complexity pruning measure

$$\tilde{G}\alpha(\mathcal{T}) = \tilde{G}(\mathcal{T}) - \alpha\,|\,\mathcal{T} - \tilde{\mathcal{T}}\,|\,.$$

We find that setting $\alpha = 4$ performs well in selecting tree size in our simulation experiments.

Simuated Data

Use of Martingale residuals in the CART^{TM} algorithm with 10-fold cross-validation was compared to bias-corrected split complexity. Both bootstrap bias correction using the .632 method and permutation based bias correction were considered for the split complexity method.

We simulated data from two survival models with uncensored data and with approximately 50% censoring. The failure time distributions were members of the $G\rho$ family (Harrington and Fleming, 1982),

$$S(t;\psi,\rho) \equiv P(X > t;\psi,\rho) = \begin{cases} (1+\rho t\psi)^{-\frac{1}{\rho}} & \text{if} \quad \rho > 0 \\ \exp -t\psi & \text{if} \quad \rho = 0. \end{cases}$$

and censoring times were distributed uniformly on $(0,\gamma)$. Five predictor variables were generated independently from the Uniform $(0,1)$distribution. We define $\eta = \log\psi$.Therefore, for individual i the models are

1. $\eta_i = I(Z_{1i} < .5)I(Z_{2i} < .5)$ and $\rho = 0$

2. $\eta_i = I(Z_{1i} < .5)I(Z_{2i} < .5) + .367$ and $\rho = 1$.

The first model is an exponential survival model with constant hazard ratios between individuals over time, and the second model corresponds to decreasing hazard ratios between individuals over time. Thus three covariates had no prognostic value, and the underlying structure for the other two covariates can be discovered with two splits of the data, resulting in three terminal nodes. Each simulated data set consisted of 300 observations; 100 data sets were simulated.

Figures 14 and 15 show the number of terminal nodes resulting from growing and pruning trees for both survival models, respectively, using Martingale residuals with 10-fold cross-validation (MRCV), and goodness of split complexity based on the logrank test and bias correction using the permutation method (PERM) and the bootstrap (BOOT). The predominant impression form these Figures is that the Martingale residual method tends to miss the structure much more frequently than the split complexity methods. The effects of censoring and nonproportionality are modest (for the two conditions sampled). A little censoring even seems to help the Martingale method. Overall, our impression is that split complexity with a penalty parameter α of 4 and bias correction using the bootstrap is the preferred of these approaches.

Accordingly, we grew and pruned a tree using the SWOG 8229 data using a minimum node size of 40. The result is given in Figure 16. Below each split in the tree is the logrank test statistic and the permutation p-value (as a descriptive statistic). Below each terminal node is the logarithm of the relative risk compared to the left most node, and the number of observations in the node. The first split is on serum β_2 microglobulin (sb2m) at a value of 5.4, as can be deduced from Figure 11. Patients with low serum β_2 are then split on calcium, while those with high serum β_2 are split on age, etc. Farther down the tree is a split on creatinine (creat) with a permutation p-value of .263 which would have been pruned using the permutation method but was retained by the bootstrap. The final tree has 7 terminal nodes.

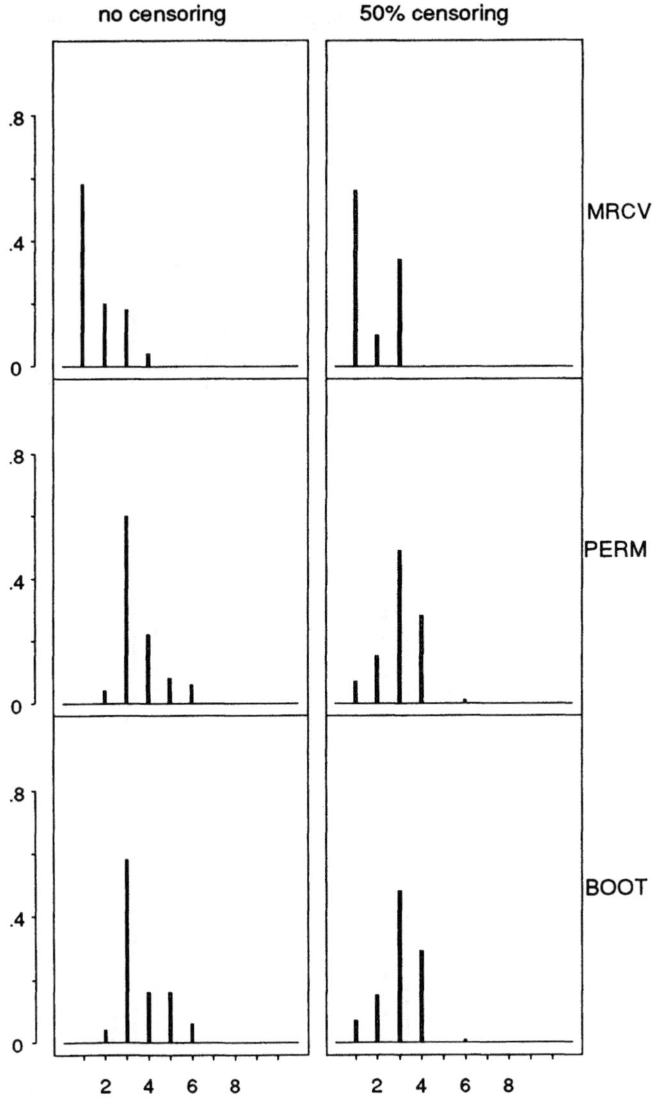

FIGURE 14. Number of terminal nodes for exponential survival model data

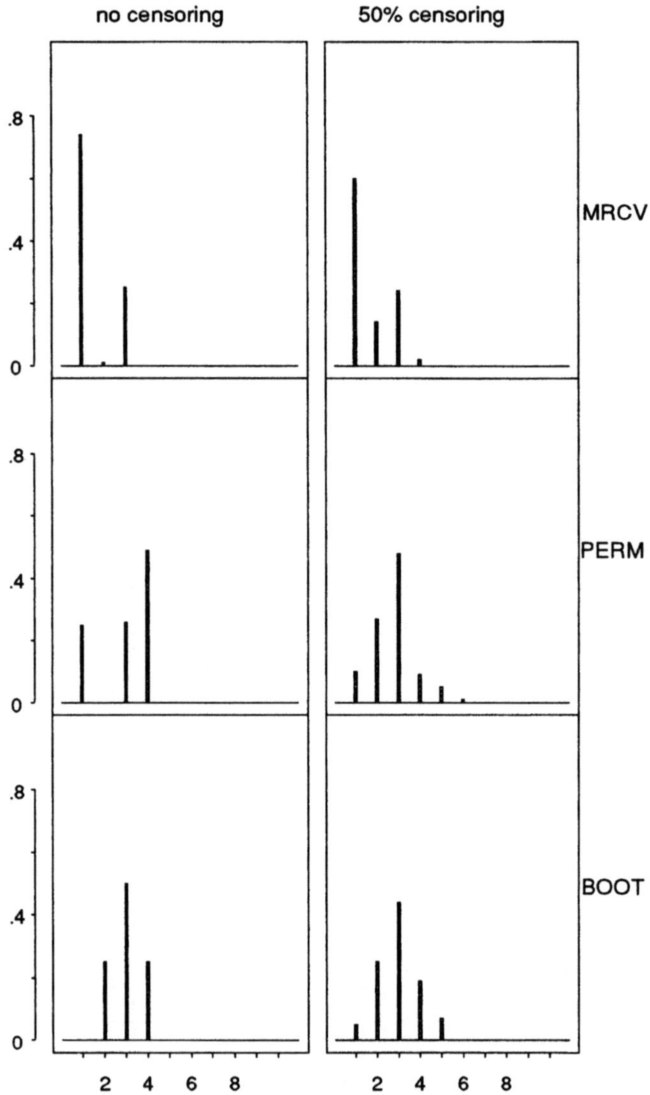

FIGURE 15. Number of terminal nodes for G^1 survival model data

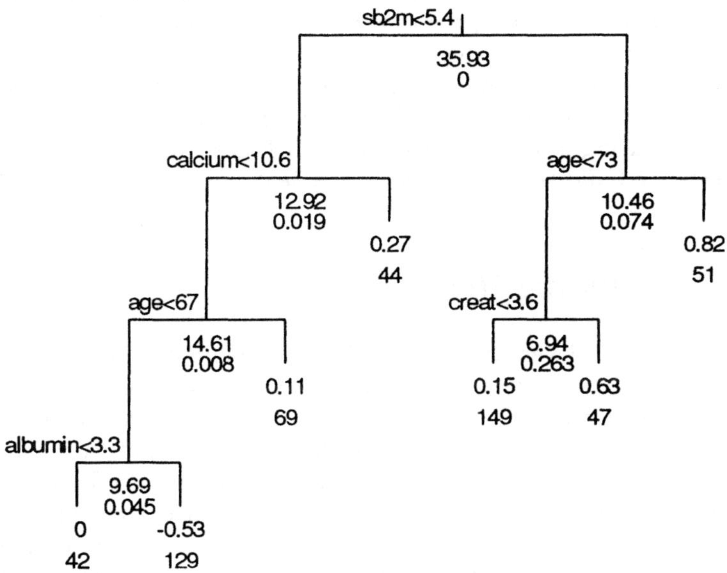

FIGURE 16. Pruned survival tree

Amalgamation

Some of the terminal nodes in Figure 16 from separate parts of the tree have a similar survival outcome, as can be seen by the log relative risks. A sensible step would thus be to combine or amalgamate these nodes to form the final prognostic groups. Ciampi, Hogg, McKinney and Thiffault (1988) use an algorithm based on minimal logrank tests to combine nodes of the tree into classes. An alternative we have proposed (LeBlanc and Crowley, 1993) is based on defining an ordered categorical variable from some measure (such as relative risk) on the terminal nodes of the tree, then subjecting this variable to a recursive partitioning procedure. For the purposes of comparing with the 3 stages of the Durie-Salmon scheme, we found the best 3 groups (defined by maximizing the partial likelihood) from the original 7 terminal nodes. Good prognosis patients are those with low serum β_2, low calcium, low age and high albumin (second node from the left in Figure 16). Poor prognosis patients are those with high serm $\beta_2$and high age, or those with high serum β_2, low age, and high creatinine (two rightmost nodes in Figure 16). Middle prognosis patients are the rest. This staging system is shown for the SWOG 8229 patients in Figure 17, and validated using SWOG 8624 patients in Figure 18. The groups are easily charaterized and the prediction is excellent.

4 Concluding Remarks

We are firm believers in the graphical and computer-intensive exploration of survival data. Several approaches have been presented here and implemented on a set of data on patients with multiple myeloma. While methods based on Martingale residuals have the advantage that they can be implemented with existing software, in most cases we have found more direct methods tailored to each situation that perform better. The three methods we used to develop prognostic groups all seem to do better than the Durie-Salmon staging system, and for this example result in quite similar results, most likely due to the fact that the covariates used have similar regression coefficients in the Seattle model. The regression approach using tertiles is simple to implement but does not result in groups which have an easy interpretation. Using the number of bad things is a more easily explainable procedure but is more arbitrary, and the resulting groups still lack simple interpretation. Recursive partitioning is easy to explain and simple to interpret, though the cutpoints chosen for splitting can be sensitive to small perturbations in the data and may not be reproducible from one data set to the next. The running logrank plots of Figure 11, along with some judgment, can be helpful in this regard. Many of our routines are implemented in the S language (Becker, Chambers and Wilks, 1988); these routines can be obtained from the second author.

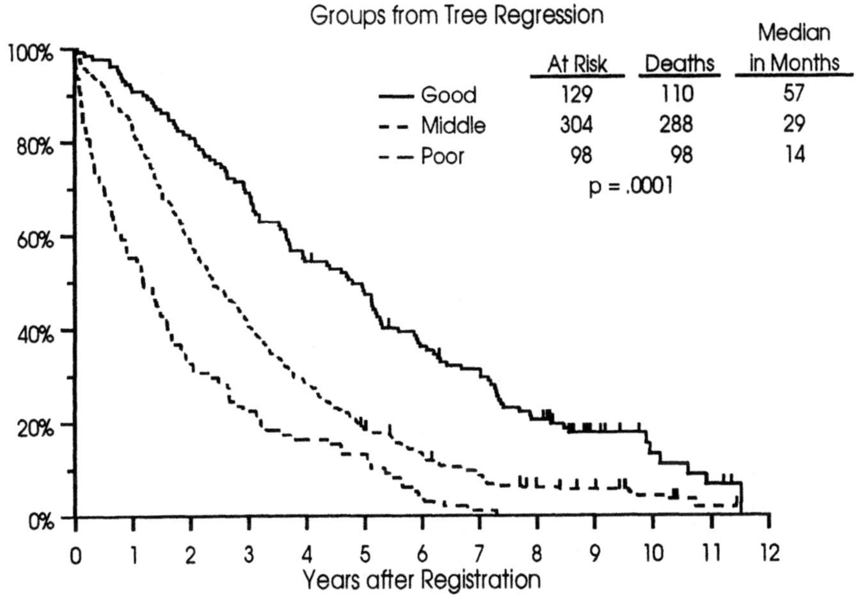

FIGURE 17. Prognosis Groups: SWOG 8229

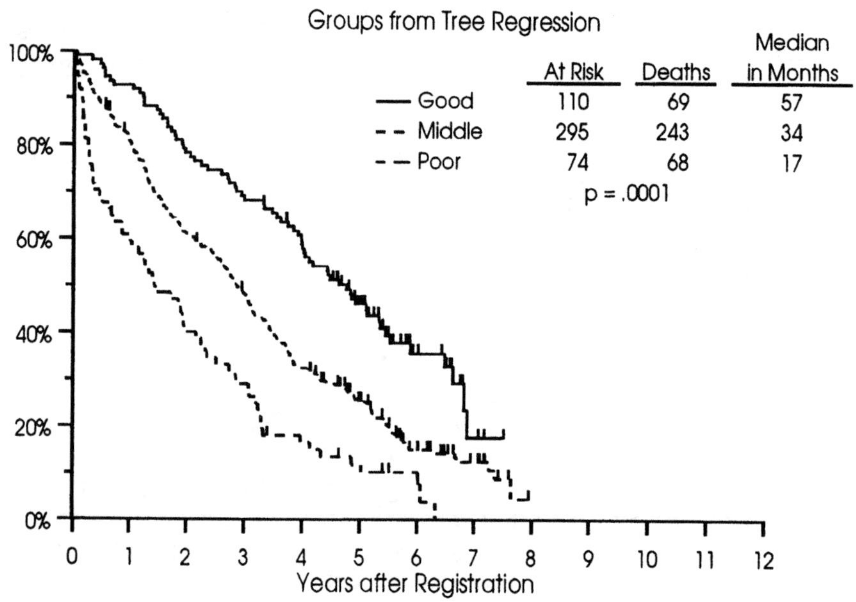

FIGURE 18. Prognosis Groups: SWOG 8624

References

Arjas, E. (1988). A graphical method for assessing goodness of fit in Cox's proportional hazards model. *Journal of the American Statistical Association,* **83**, 204-212.

Becker, R., Chambers, J. and Wilks, A. (1988). *The New S Language.* Wadsworth, Pacific Grove, CA.

Beran, R. (1981). Nonparametric regression with randomly censored survival data. Technical Report, Department of Statistics, University of California-Berkeley.

Breiman, L., Friedman, J. H., Olshen, R. A. and Stone, C. J. (1984). *Classification and regression trees.* Wadsworth, Belmont, CA.

Breslow, N. (1974). Covariance analysis of censored survival data. *Biometrics,* **30**, 89-99.

Ciampi, A., Thiffault, J., Nakache, J.-P. and Asselain, B. (1986). Stratification by stepwise regression, correspondence analysis and recursive partition. *Computational Statistics and Data Analysis*, **4**, 185-204.

Cox, D. R. and Snell E. J. (1968). A general definition of residuals (with discussion). *Journal of the Royal Statistical Society, Series B,* **30**, 248-275.

Cox, D. R. (1972). Regression models and life-tables (with discussion). *Journal of the Royal Statistical Society, Series B,* **34**, 187-220.

Cox, D. R. (1975). Partial likelihood. *Biometrika*, **62**, 269-276.

Crowley, J. and Hu, M. (1977). Covariance analysis of heart transplant survival data. *Journal of the American Statistical Association,* **72**, 27-36.

Crowley, J., LeBlanc, M., Gentleman, R. and Salmon, S. (1995). Exploratory methods in survival analysis. In *Analysis of Censored Data*, IMS Lecture Notes - Monograph Series 27, H. L. Koul and J. V. Deshpande, Eds, 55-77, Institute of Mathematical Statistics, Hayward, CA.

Crowley, J and Storer, B. E. (1983) Comment on Aitkin, M., Laird, N. and Francis, B. A reanalysis of the Stanford heart transplant data. *Journal of the American Statistical Association,* **78**, 277-281.

Davis, R. and Anderson, J. (1989). Exponential survival trees. *Statistics in Medicine,* **8**, 947-962.

Doksum, K. A. and Yandell, B. S. (1983). Properties of regression estimates based on censored survival data. In *Festschrift for Erich L. Lehmann*, P. J. Bickel, K. A. Doksum and J. L. Hodges, Eds, 140-156, Wadsworth, Belmont, CA.

Durie, B. G. M. and Salmon, S. E. (1975). A clinical system for multiple myeloma. Correlation of measured myeloma cell mass with presenting clinical features, response to treatment and survival. *Cancer,* **36**, 842-854.

Efron, B. and Tibshirani, R. (1993). *An Introduction to the Bootstrap.* Chapman and Hall, New York.

Fleming, T. R. and Harrington, D. P. (1991) *Counting Processes and Survival Analysis.* John Wiley and Sons, New York.

Gentleman, R. and Crowley, J. (1991a). Graphical methods for censored data. *Journal of the American Statistical Association,* **86**, 678-683.

Gentleman, R. and Crowley, J. (1991b). Local full likelihood estimation for the proportional hazards model. *Biometrics,* **47**, 1283-1296.

Gordon, L. and Olshen, R. A. (1985). Tree-structured survival analysis. *Cancer Treatment Reports*, **69**, 1065-1069.

Harrington, D. and Fleming, T. (1982). A class of rank test procedures for censored survival data. *Biometrika,* **69**, 553-556.

Kalbfleisch, J. D. and Prentice, R. L. (1980). *The Statistical Analysis of Failure Time Data.* John Wiley and Sons, New York.

Kaplan, E. L. and Meier, P. (1958). Nonparametric estimation from incomplete observations. *Journal of the American Statistical Association,* **53**, 457-481.

Kay, R. (1977). Proportional hazard regression models and the analysis of censored survival data. *Applied Statistics,* **26**, 227-237.

Lagakos, S. W. (1981). The graphical evaluation of explanatory variables in proportional hazards regression models. *Biometrika,* **40**, 1037-1048.

LeBlanc, M. and Crowley, J. (1993). Survival trees by goodness of split. *Journal of the American Statistical Association*, **88**, 457-467.

LeBlanc, M. and Crowley, J. (1995). Step-function covariate effects in the proportional-hazards model. *The Canadian Journal of Statistics,* **23**, 109-129.

Mantel, N. (1966). Evaluation of survival data and two new rank order statistics arising in its consideration. *Cancer Chemotherapy Reports*, **50**, 163-170.

Salmon, S. E., Tesh, D., Crowley, J., Saeed, S., Finley, P., Milder, M. S., Hutchins, L. F., Coltman, C. A. Jr., Bonnet, J. D., Cheson, B., Knost, J. A., Samhouri, A., Beckord, J. and Stock-Novack, D. (1990). Chemotherapy is superior to sequential hemibody irradiation for remission consolidation in multiple myeloma: A Southwest Oncology Group Study. *Journal of Clinical Oncology,* **8**, 1575-1584.

Salmon, S. E., Crowley, J., Grogan, T. M., Finley, P., Pugh, R. P. and Barlogie, B. (1994). Combination chemotherapy glucocorticoids, and interferon alpha in the treatment of multiple myeloma: A Southwest Oncology Group study. *Journal of Clinical Oncology,* **12**, 2405-2414.

Segal, M. R. (1988). Regression trees for censored data. *Biometrics*, **44**, 35-48.

Therneau, T., Grambsch, P. and Fleming, T. (1990). Martingale-based residuals for survival models. *Biometrika,* **77**, 147-160.

Tibshirani, R. J. and Hastie, T. (1987). Local likelihood estimation. *Journal of the American Statistical Association,* **82**, 559-567.

Survival Analysis in the Regulatory Setting

Susan S. Ellenberg
Jay P. Siegel

ABSTRACT The powerful methods for analyzing survival data (particularly censored survival data) that were introduced in the late 1960's and early 1970's represented major advances in the statistical assessment of such data. It is important to recognize, however, that these may not always be most appropriate to address the fundamental study question in any situation in which time-to-event techniques could be applied. In this paper, several such situations are considered, and the particular implications for regulatory decision-making are discussed.

1 Introduction

The seminal papers by Mantel (1966) and Cox (1972) ushered in a new era of data analysis in studies evaluating length of survival or time to some landmark event. The new methods described in these papers permitted extraction of substantially more information from a comparative study in which the primary objective was to prevent or delay the occurrence of an undesirable event than the then-prevailing practice of comparing simple proportions of each treatment group with and without the event at a specific point in time. The analytical approaches developed by Mantel and Cox so clearly represented major advances that their incorporation into the standard practice of biostatisticians involved in medical research was virtually immediate.

The issue we will address in this paper has little do with the specific methodologies or techniques of survival analysis, but rather the fundamental study question that generates the form of analysis. It sometimes appears that the desire to utilize powerful time-to-event techniques may lead investigators to pose study questions in such a way as to be able to make use of these techniques in situations where a different question (requiring alternative analytic approaches) might be more appropriate. This problem is of particular concern in a regulatory setting.

The end result of regulatory evaluation is a binary outcome. A new product is approved for marketing, or it is not. This is in contrast to clinical trials in other settings–for example, studies of new surgical techniques, or comparison of currently available intervention strategies–in which the study results will be analyzed, interpreted and (one hopes) published by the investigators, but each individual member of the relevant medical community will be able to make his or her own judgment concerning the appropriate interpretation of the results and to take action consistent with this individual interpretation. In the regulatory setting, the ability of individual physicians to act on their own interpretion of the study results will often depend on the binary outcome of the regulatory process noted above.

Pharmaceutical companies clearly have a great deal at stake in the interpretation of results of studies intended to serve as the primary basis for product approval. By the time such studies are designed, many years of effort and extensive financial resources have been invested in the product which will not be recouped if the product never reaches the marketplace. For this reason, it is essential to define as carefully as possible, prior to the beginning of the study, the basis for the final evaluation of the results. Up-front agreement on primary and secondary endpoints, interim monitoring procedures, approaches to adjustment for covariates, subset analyses of interest, and the analytic approaches to be used help minimize confusion, debate and concern about multiplicity issues at the time the results are reviewed by the Food and Drug Administration and its advisors. It is highly desirable, from the perspective of the regulatory agency as well as the companies, for the process leading to the ultimate regulatory decision to be as well-defined and transparent as possible.

2 Advantages of the Time-to-Event Approach

Three aspects of time-to-event analyses as compared with comparison of proportions are particularly attractive. First, as mentioned earlier, time-to-event incorporates more information into the analysis. It permits utilization of actual times, rather than a simple binary outcome for each subject; furthermore, it permits utilization of "censored" data, for those subjects whose experience is known for only a portion of the interval of interest. The additional power of these analyses in testing a hypothesis relating to prolongation of time to event as a function of hazard is readily demonstrated. Consider a disease context in which the annual mortality hazard λ is 0.4, implying a median time-to-event of about 1.7 years. To ensure power of 80% to detect a reduction in annual hazard to 0.3, assuming 2 years of accrual and 3 years further follow-up, and assuming that the hazards are proportional, would require entry of 262 patients per arm for a two-arm trial (Rubenstein, Gail and Santner, 1981). If we were to design our study instead with a plan to perform a simple comparison of proportion alive at 3 years, the hypothesized hazard rates would lead to proportions alive at 3 years of approximately 0.3 and 0.4 in the two arms; to detect this difference with 80% power using a binomial test of proportions would require

376 patients per arm (Fleiss, 1981), a 44% increase. For lower event rates, the difference in power is smaller, but the time-to-event analysis will always be more powerful if the proportional hazards assumption holds.

A second advantage is that, in long-term studies, the time-to-event approach provides important information that a simple binomial comparison does not, i.e. that is whether the outcome of interest, even if not prevented during the interval of observation, is delayed. In long-term mortality studies, for example, prolongation of life is generally the objective of the intervention under study; additional months or years of life is clearly beneficial even if the proportion surviving beyond some arbitrary time period (e.g., five years) is not changed.

A third advantage of the time-to-event approach is that it protects against a random fluctuation of the data at an arbitrary point in time, making that time point "unrepresentative" in some sense for evaluating the treatment difference. In the above example, for instance, the survival curves for the two treatments might be clearly separated throughout the time interval of interest, but might be closest together just at the 3 year point. Conversely, the curves might be generally overlapping or criss-crossing over the interval but show a difference at the 3 year point just large enough to achieve statistical significance. In either case, "three-year survival" would not have been an optimal choice of endpoint. While one could argue that this is not really a problem in the statistical sense–the error rates are unaffected since random highs and random lows are equally likely–the appearance in a particular case of data that seem to be telling a different story from that told by the statistical test applied to those data is troublesome. Use of an analysis that takes the experience over the entire interval into account reduces the concern about possibly anomalous data at any particular point.

Three types of situations in which the time-to-event approach to design and analysis of studies might be considered but might not optimally address the most critical question to be asked of the data will be discussed. These are studies of acute processes, studies with composite endpoints, and studies with repeating events. In each case, concerns can be raised regarding the appropriateness of the time-to-event approach as the primary basis for evaluating the investigational product.

3 Studies of Acute Processes

Consider a situation where the hazard of the event exists almost entirely over a short interval of time, a few weeks or even days. Because the time period of interest is short, the censoring is absent or minimal. The primary concern in such a situation is generally whether or not the event of concern occurs during the at-risk interval. Examples of such medical contexts include myocardial infarction, stroke, acute pulmonary embolism, and life-threatening infections.

A medical condition of this type for which numerous therapies have been investigated in recent years is sepsis syndrome. Sepsis syndrome, is an extreme complication of infection that involves serious systemic sequelae. The natural history of this syndrome has been well studied(Parrillo, Parker, Natanson, et al., 1990; Bone, Fisher, Clemmer, et al., 1989; Knaus, Harrell, Fisher, et al, 1993); mortality is substantial, exceeding 30% in many studies, and other serious complications such as organ failure are also common.

The primary objective of treatment for sepsis is to increase the probability of surviving the episode. Because the mortality hazard is highest near the time of diagnosis and decreases over time, the time intervals used for evaluating new therapeutic strategies have been short, typically one month or less. While there is no specific choice of interval that will be both long enough to ensure observation of all deaths due to sepsis (or drug toxicity), and short enough so that deaths due to sepsis or drug toxicity are the only deaths observed (many patients with sepsis have other serious underlying conditions such as traumatic injuries or advanced cancer), intervals of about one month have frequently been used. This short period of primary observation (during much of which the patient is hospitalized) permits complete follow-up on virtually all patients.

Pharmaceutical companies developing drugs for sepsis have sometimes proposed that the primary evaluation of results be based on time-to-event for the reasons described earlier. The question, however, of whether the time-to-event analysis addresses the objective of these studies directly depends on the validity of the proportional hazards assumption. In the absence of proportional hazards, a treatment might clearly prolong survival but not so clearly increase the overall probability of surviving the acute septic episode. In the context of a short time interval of interest, this could mean that deaths that might otherwise have occurred on day 2 or 3 might be delayed to day 10 or 11, but not prevented. (Alternatively, in a setting in which death may result from a variety of mechanisms, the earliest deaths might be reduced while later deaths due to other mechanisms might be increased.) It might be argued, of course, that any prolongation of survival, even if only a few days, is a worthwhile benefit; but it seems clear that the question of greatest interest would be whether the proportion of patients surviving the acute episode is increased. For this reason, the Food and Drug Administration has encouraged companies developing sepsis drugs to design studies to test the latter hypothesis, rather than one relating to time-to-event.

It should be noted that both types of analyses will inevitably be performed; if the proportional hazards assumption appears tenable both will be considered. In fact, in studies conducted over the past few years (Ziegler, Fisher, Sprung, et al., 1991; Greenman, Schein, Martin, et al, 1991; Fisher,

Dhainaut, Opal, et al., 1994; Abraham, Wunderink, Silverman, et al., 1995), a wide variety of comparisons have been performed and reported by sponsors as important primary or secondary analyses in sepsis studies: all-cause and cause-specific mortality at day 7 and day 14, in addition to one month; comparisons within specific subgroups; time-to-event analyses using different weighting systems (e.g., Wilcoxon, Cox proportional hazards model.); and all of the above both unadjusted and then adjusted for (multiple sets of) covariates. While there is substantial interdependence among these analyses, the inflation of Type I error would not be trivial if statistical significance on any of these analyses were to be considered adequate evidence of efficacy. The problem is therefore too complex for the "look at it both ways" approach, even if adjustment for multiple comparisons could be attempted. In addition, if the primary analysis is to be based on a comparison of proportions, the sample size will have to be larger than for a time-to-event approach specifying a constant hazard ratio, so that the decision about the primary endpoint must be made up front if adequate power is to be ensured.

Another medical condition with significant mortality is acute myocardial infarction. As with sepsis, most deaths occur within the first few days; 28 to 35 days from the event is the time window most often used for primary mortality analysis; and longer term outcomes are also of interest. Time-to-death analyses could increase power but, as in the first example, might not be the most appropriate survival measure if the proportional hazards assumption does not hold. A thrombolytic agent might improve patency of an obstructed coronary artery, thus diminishing early mortality, but risk (possibly even increased risk) would persist for later death from reocclusion and reinfarction, heart failure, arrhythmia, stroke or other causes. Unlike the case with sepsis, however, investigators and sponsors studying therapies for acute myocardial infarction have generally proposed as the primary endpoint the proportion surviving the episode.

There are circumstances in which time-to-event analyses do seem most appropriate even when the time interval of interest is short. An example would be a study of an agent to promote neutrophil engraftment following bone marrow transplantation. In this case, our interest is in reducing time-to-event rather than prolonging it. A reduction of the time during which the patient is at high risk of serious infection due to profound neutropenia by a week, or even a few days, would be viewed as a clinically important benefit.

4 Use of Composite Endpoints

When a disease has a variety of manifestations or complications, the primary endpoint may be an indicator variable that is positive if any one of a number of conditions arises. Such a variable is often referred to as a composite endpoint; the statistical properties of such endpoints have been the subject of recent methodological work (O'Brien, 1984; Tang, Geller and Pocock, 1993; Legler, Lefkopoulou and Ryan, 1995). An example of a composite endpoint in wide use is development of AIDS (acquired immunodeficiency syndrome). Individuals infected with the human immunodeficiency virus (HIV) typically progress over a period of years to a diagnosis of AIDS. This diagnosis may be made on the basis of a variety of different conditions, including occurrence of opportunistic infections, development of certain malignancies and reduction in immune system markers to below a threshold level (Centers for Disease Control, 1992).

The AIDS example is a good one to illustrate the problems associated with applying the time-to-event approach as the primary analysis in a study using a composite endpoint. All but one of the AIDS-defining events represents a serious clinical outcome requiring hospitalization and/or intensive therapy that may or may not be successful. The exception is the level of the immune marker, the CD4+ cell count, which correlates with the prognosis of the patient but often does not reflect the clinical status. (A decline in the CD4 count to below 200/mm^3 has been considered an AIDS-defining event since the last change in the AIDS definition was made in 1993.) While it is true that an individual with a CD4+ cell count of 180 is statistically somewhat more likely to become clinically ill than an individual with a CD4 + cell count of 220, those two individuals may present identical clinical pictures at a given point in time. It therefore might be inappropriate to give a CD4+ cell count decline the same weight as a life-threatening clinical event such as *pneumocystis carinii* pneumonia or cytomegalovirus infection in a study to assess the benefit of a new therapy. Further, most individuals will show a drop in CD4+ cell count to the AIDS-defining level (200/mm^3) before they experience any of the serious clinical manifestations of HIV infection; thus, a time-to-event analysis with AIDS diagnosis as the "event" would essentially be equivalent to a time-to-CD4-drop-to-under-200/mm^3 analysis. With this endpoint, a therapy that prevented or delayed serious infections but had a limited effect on CD4+ cell count might appear inferior to another therapy that had a greater effect on CD4+ cell count but had little effect on the more serious clinical sequelae of HIV. In recognition of this issue, clinical trials of antiretroviral therapies sponsored by the National Institutes of Health for patients who have not yet been diagnosed with AIDS have generally used composite endpoints composed only of major clinical events. With composite endpoints based on clinical events of comparable impact, time-to-event analyses may be much more appropriate.

Underlying the concern that "time-to-composite event" may place an undesirable emphasis on a single component of the composite is the concern that a therapy may act entirely or preferentially on only one or a small number of the components. While this is a concern relating to the use of composite endpoints in general, its implication for time-to-event analysis is even stronger: if the "responsive" component tends to occur earlier in the disease process, a time-to-event analysis that counts only the first event could be very misleading with regard to the clinical benefit of an investigational therapy (Neaton, Wentworth, Rhame, et al., 1994). Further, if the responsive event tends to occur later, a time-to-event analysis could miss the clinical effect entirely.

A recent example of a study that raised these issues was an investigation of a new antithrombotic agent in patients undergoing coronary angioplasty. The hope was that the new therapy would prevent a variety of adverse consequences, including death, myocardial infarction, need for emergency repeat angioplasty or other urgent interventions, need for nonurgent repeat angioplasty or other elective interventions, and recurrent myocardial ischemia. Clearly, these events vary greatly in clinical impact–the first problem. Another problem is that mortality outcomes would virtually be ignored, since relatively few deaths would occur without having been preceded by any of the other outcomes. Thus, a time-to-event analysis might select as the superior treatment one that prevented or delayed the most frequent and least serious of the components, but had no effect (or worse, an adverse effect) on the most important and most serious component–mortality. This issue is a concern generally with composite endpoints, but can be especially problematic with time-to-event analysis because the earliest events may be weighted more heavily. Ultimately, a composite endpoint including only the more serious clinical outcomes was used for primary evaluation of the product. Special attention was paid to the mortality component; had there been a suggestion of an adverse impact on survival in conjunction with benefit on the composite endpoint, additional studies could have been required.

There are many examples, of course, of appropriate use of composite endpoints in conjunction with time-to-event analyses. In addition to the AIDS and coronary heart disease examples discussed above, the composite endpoint of tumor recurrence or death is frequently used in studies of adjuvant cancer therapies, and the composite of death, kidney transplant or need for dialysis is a commonly used endpoint in renal disease. In these situations the endpoint components all reflect serious adverse outcomes and therapy would not generally be expected to have widely differing impacts on the individual components.

5 Prevention of Recurring Events

In many diseases, the natural history is characterized (entirely or in part) by a series of adverse clinical events, with intervals during which the patient is less symptomatic or even symptom-free. Examples include rheumatoid arthritis (flares), post-transplant recovery (transplant rejection episodes), chronic asthma (acute attacks), multiple sclerosis (exacerbations), pediatric AIDS (bacterial infections) and papillary bladder cancer (recurrence of tumor). An ideal therapy would, of course, prevent all adverse clinical outcomes. In general, however, it is more reasonable to hope that a new therapy will reduce the frequency of the adverse episodes and/or reduce their severity.

Developers of new agents for such diseases frequently propose use of a time-to-first event analysis as the primary approach to assessing efficacy. Certainly, the result of such an analysis is of interest, and delaying the time to the next observed event following initiation of therapy is clearly of benefit to the patient. Unless this delay were substantial, however (for example, a delay measured in years of an event that might be expected to occur every few months), a more clinically meaningful endpoint would seem to be the number of events occurring over some defined interval of time. An agent that delayed an initial episode by a short period of time but did not affect the natural history once an event had occurred (or worse, caused the frequency to increase) would not be as valuable as an agent that reduced the event's frequency over a longer period of time. An example might be an immunosuppressive agent given at time of organ transplantation to decrease the likelihood of a rejection episode. While the time to the first rejection episode might be an endpoint that would be relatively easily achieved and measured, it would not evaluate the very real possibility that, when the immune system recovers days or weeks later, the incidence of rejection episodes may return to baseline or even rebound above baseline.

Statistical techniques have been developed to permit time-to-event analyses when recurring events are of interest (Anderson and Gill, 1982; Prentice, William and Peterson, 1981; Wei, Lin and Weissfeld, 1989; Fleming and Harrington, 1991; Anderson, Borgan, Gill and Keiding, 1993). Determination of whether to use time to event analyses and which to use is highly dependent on knowledge and assumption regarding the disease process and the anticipated drug effect (Siegel, 1997). Of particular importance are assumptions about the correlations among events; if events are highly intercorrelated, and new therapies are unlikely to modify these correlations, analyses focused on the first event may be sufficiently informative and more efficient than accounting for successive events (Hughes, 1997). Such information, however, is usually not known at the time a study of a new therapy is designed. Current methods with robust variance estimators reduce the need for assumptions about the intercorrelation among consecutive events (Therneau, 1997). Given the practical concerns noted above, analytic approaches that takes account of the pattern of recurrences may be preferable in some instances.

6 Discussion

Despite their many advantages, time-to-event techniques may not always address the question of primary interest in trials of experimental therapies. In the examples described, one might be misled with regard to the actual clinical benefit of a treatment if one defined the primary endpoint in terms of the time to event rather than some other (perhaps less powerful) approach such as a comparison of proportions.

These considerations may be even more important for studies with active controls than for placebo-controlled studies. Active control trials subject the proportional hazards assumption to a greater challenge; each therapy might work in a different way and one would need to observe events over time to ensure that an early advantage conferred by one treatment was not outweighed by a later advantage of the other treatment.

It is a well-accepted practice in clinical trials to define the primary endpoint and the analytic approach to evaluating that endpoint in the study protocol, prior to initiating patient entry. When endpoints and analytic techniques not mentioned in the original study plan become the focus of final analyses, the possibility of a data-driven approach to analysis that would inflate the Type I error to an unknown extent is inevitably raised. The suggestion that analytic approaches may have been determined by the data limits the credibility of the study's conclusions.

Such concerns are particularly important in a regulatory setting, in which regulatory scientists must make decisions that have substantial and wide-ranging financial implications in addition to potentially important impact on public health. Under these circumstances, it is essential for all parties to the process to have as clear an understanding as possible of the basis on which the regulatory decision will be made. The selection of a primary endpoint that reflects most appropriately the clinical effect of a new product, and an approach to analysis that permits the clearest possible picture of the basis for decision-making are essential components of a regulatory process that is scientifically sound, and is fair both in actuality and in perception.

References

Abraham, E., Wunderink, R., Silverman, H., et al. (1995). Efficacy and safety of monoclonal antibody to human tumor necrosis factor and in patients with sepsis syndrome. *Journal of the American Medical Association*, **273**, 934-941.

Anderson, P. K., Borgan, O., Gill, R. D., and Keiding, N. (1993). *Statistical Models Based on Counting Processes.* New York: Springer-Verlag.

Bone, R. C., Fisher, C. J. Jr., Clemmer, T. P., et al. (1989). Sepsis syndrome: a valid clinical entity. *Critical Care Medicine*, **17**, 389-393.

Centers for Disease Control. (1992). 1993 Revised classification system for HIV infection and expanded surveillance case definition for AIDS among adolescents and adults. *Morbidity and Mortality Weekly Report*, **41**, 1-35.

Cox, D. R. (1972). Regression models and life tables (with discussion). *Journal of Royal Statistical Society*, Series B, **34**, 187-220.

Fisher, C. J. Jr., Dhainaut J. F. A., Opal, S. M., et al. (1994). Recombinant human interleukin 1 receptor antagonist in the treatment of patients with sepsis syndrome. *Journal of the American Medical Association*, **271**, 1836-1843.

Fleiss, J. L. (1981). *Statistical Methods for Rates and Proportions.* New York: Wiley.

Fleming, T. R. and Harrington D. P. (1991). *Counting Processes and Survival Analysis.* New York: Wiley.

Greenman, R. L., Schein, R. M. H., Martin, M. A., et al. (1991). A controlled clinical trial of E5 murine monoclonal IgM antibody to endotoxin in the treatment of gram-negative sepsis. *Journal of the American Medical Association*, **266**, 1097-1102.

Hughes, M. D. (1997). Power considerations for clinical trials using multivariate time-to-event data. *Statistics in Medicine*, in press.

Knaus, W. A., Harrell, F. E., Fisher, C. J. Jr., et al. (1993). The clinical evaluation of new drugs for sepsis: a prospective study design based on survival analysis. *Journal of the American Medical Association*, **35**, 861-867.

Legler, J. M., Lefkopoulou, M., Ryan, L. M. (1995). Efficiency and power of tests for multiple binary outcomes. *Journal of the American Statistical Association*, **90**, 680-693.

Mantel, N. (1966). Evaluation of survival data and two new rank order statistics arising in its consideration. *Cancer Chemotherapy Reports*, **50**, 163-170.

Neaton, J. D., Wentworth, D. H., Rhame, F., et al. (1994). Methods for studying interventions: considerations in choice of a clinical endpoint for AIDS clinical trials. *Statistics in Medicine*, **13**, 2107-2125.

O'Brien, P. C. (1984). Procedures for comparing samples with multiple endpoints. *Biometrics*, **40**, 1079-1087.

Parrillo SE, Parker MM, Natanson C, et al. (1990). Septic shock in humans. *Annals of Internal Medicine*, **113**, 227-242.

Prentice R.L., Williams B.J. and Peterson A.V. (1981). On the regression analysis of multivariate failure time data. *Biometrika*, **68**, 373-379.

Rubenstein, L. V., Gail, M. H., and Santner, T. J. (1981). Planning the duration of a comparative clinical trial with loss to follow-up and a period of continued observation. *Journal of Chronic Disease*, **34**, 469-479.

Siegel, J. P. (1997). Discussion of the paper by L. J. Wei and D. V. Glidden on "An overview of statistical methods for multiple failure time data in clinical trials". *Statistics in Medicine*, in press.

Tang, D. I., Geller, N. L., and Pocock, S. J. (1993). On the design and analysis of randomized clinical trials with multiple endpoints. *Biometrics*, **49**, 23-30.

Therneau, T. M. (1997). Extending the Cox model. In *Proceedings of the First Seattle Symposium in Biostatistics: Survival Analysis*, Eds. D. Y. Lin and T. R. Fleming, pp. xxx-xxx. New York: Springer-Verlag.

Wei, L. J., Lin, D. Y., and Weissfeld, L. (1989). Regression analysis of multivariate incomplete failure time data by modeling marginal distributions. *Journal of the American Statistical Association*, **84**, 1065-1073.

Ziegler, E. J., Fisher, C. J. Jr., Sprung, C. L., et al. (1991). Treatment of gram-negative bacteremia and septic shock with HA-1A human monoclonal antibody against endotoxin. *New England Journal of Medicine*, **324**, 429-436.

Proposed Strategies for Designing and Analysing Sepsis Trials

Peter A. Lachenbruch
Jawahar Tiwari

ABSTRACT We discuss the design and analysis of trials of drugs and biologics designed to combat sepsis. We consider the variety of trials that have been conducted, and comment on the problems that have occurred in their conduct. These include variation in populations, therapies, controls, and endpoints. Optimistic estimates of effect sizes have led to studies which have been underpowered. Analyses using the proportional hazards model may be underpowered if the data do not satisfy the assumption of proportionality.

1 Introduction

In recent years, design and analysis of sepsis trials are enjoying renewed interest. Given the complexity of definition, multitude of etiologies and heterogeneity of the patient population, it is a challenge to design a clinical trial that adequately but realistically address the development needs for a therapeutic agent. At present (March 1996), no clinical trials have been able to show significant clinical benefit over placebo of any agent. This, in turn, points to the need to readdress some basic issues in conducting sepsis trials. Sepsis and the related SIRS (Septic Inflammatory Response syndrome) affects approximately 500,000 patients annually (MMWR, 1990).

The terms sepsis, septic syndrome, septic shock, septicemia, bacteremia, and SIRS, have been used interchangeably. This paper discusses some of the problems encountered in clinical trials.

2 Definition and Sources of SIRS and Sepsis

The American College of Chest Physicians/Society of Critical Care Medicine (ACCP/SCCM) consensus committee considered definitions of sepsis to provide a uniform, consistent term to characterize a heterogeneous entity. Bone et al (1992) proposed the term SIRS, an acronym for Systemic Inflammatory Response Syndrome, defined as

- Temperature > 38° C. or < 36° C.
- Heart rate > 90 beats per minute
- Respiratory rate > 20 per minute (tachypnea)
- White Blood Count (WBC) > 12000/cu mm or < 4000/cu mm or more than 10% immature neutrophils (bands)

SIRS has about a 15% case-fatality rate (Bone, 1992). SIRS plus a culture-proven infection is denoted as sepsis and is associated with about a 25% case-fatality rate (Bone, 1992). Severe sepsis is defined as sepsis plus organ dysfunction, hypotension, or hyperperfusion (about a 40% case-fatality rate). Septic shock results from hypotension (despite fluid resuscitation) plus hyperperfusion abnormalities (with about a 55% case-fatality rate).

SIRS can arise from a variety of underlying causes. These include, but are not limited to, bacteremia, burns, fungal infections, pancreatitis, peritonitis, and trauma. Sources include high risk patients such as the elderly, the very young, and patients with serious diseases or conditions such as late stage cancer, AIDS, surgical wound infections, Adult Respiratory Distress Syndrome (ARDS), Urinary Tract Infections (UTI), and Acute Renal Failure (ARF). Because of multiple causes and facets of sepsis, there may be uncertainty of whether the infection is the cause of the inflammatory response, or is a symptom of more serious disease.

3 Dimensions of the Problem: Design Issues

More than 24 randomized clinical trials of therapeutic agents for sepsis have failed to show clear efficacy. There is a variety of possible causes for these problems. These are given in Table 1. Some trials have had problems involving several of these issues.

Table 1. *Why Have Sepsis Trials Not Shown Efficacy?*

1. Heterogeneity Issues:
   - a variety of causes of sepsis (e.g., ARDS, ARF, UTI, AIDS,cancer, surgery, trauma)
   - variety of pathogens (gram-negative, gram-positive, mixed)
2. Heterogeneity of Patient Population
   - Variety of ages (infants, middle-aged, elderly)
   - Duration of sepsis
   - Risk score

- Do Not Resuscitate orders (if many patients are in this category, the 28 day survival rate may not be useful)
- Organ failure - should this be an entry criterion, an exclusion criterion, an inclusion criterion?

3. Heterogeneity of Standards of Care
   - By country
   - By Hospital within country (e.g., primary, secondary, tertiary)
   - By Physician within Hospital
4. Specification of Treatment
   - Dosage
   - Route of administration
   - Schedule (frequency, duration, frequency)
5. Effectiveness of Treatment
   - Wrong endpoint
   - Treatment ineffective

The therapies that have been proposed attack the process at various stages and in a variety of ways. The trials that have been conducted have often taken all presenting patients with the variety of stages, therapies, and causes. Since little effort has been made to control these factors, either clinically or by statistical design, there may have been substantial dilution of treatment effect.

4 Variations in Study Endpoints Definition

Many endpoints have been proposed. The predominant one is twenty-eight day all-cause mortality. This endpoint is the one that the FDA has accepted as a standard. The reasoning is that the risk of mortality due to the acute sepsis incident is believed to be substantially diminished by that time, and that later mortality would be predominantly due to other causes.

A proposal has been to use 7 day or 14 day all-cause mortality. A potential problem with this endpoint is that early mortality may improve but there may be increased mortality over days 15 to 28. This translation of mortality by a few days results in little clinical benefit to the patient population. If another therapy would alleviate this later mortality, however, this early benefit would be useful.

A third endpoint, Days in Intensive Care Unit (ICU), raises questions of standardized criteria for entry and discharge from the ICU. A shortened

ICU stay without a corresponding survival increase or long term clinical benefit is of questionable value. Another problem arises if many patients die in the ICU. Then the ICU days would be reduced by an inferior treatment, and one would not wish to conclude that reduced ICU stay is beneficial. There are obvious statistical ways of handling this problem.

The time to organ failure might affect a variety and number of organs. A therapy might be designed to prevent a specific organ failure; in general this might be an appropriate secondary endpoint. If multiple organs are affected, resolution of one organ dysfunction is of little value if the others are still compromised. Thus, the clinical benefit may be limited. Survival models may be especially useful for secondary endpoints, possibly anticipating later trials of combination regimes.

Reduction of cytokine levels, resolution of fever, and reduction of mortality risk scores have been proposed as surrogates. Because of the short time to view the 28 day results, these surrogates may be more useful as secondary endpoints. That is, they are intermediate to showing clinical benefit to the patient.

Mortality risk scores are frequently proposed as endpoints of sepsis trials. The APACHE (Acute Physiology and Chronic Health Evaluation) system uses the log-normal distribution to predict 28-day mortality risk (Knaus et al, 1993). The MPM (mortality probability model) uses logistic regression to predict the 28-day risk adjusted mortality rate as does the SAPS (Simplified Acute Physiology Score). These models were developed by Lemeshow et al (1994) . The reader should note that these risk scores indicate the probability of survival to 28 days and do not necessarily indicate the chance for a good therapeutic response. They may allow removal of very low or very high risk patients from mortality studies. Such patients would contribute little to the clinical trial results since their outcome might be unaffected by therapy. These models typically have many predictor variables (over 15), and were developed on large populations. They have apparently not been calibrated among themselves: that is, they have not been correlated on the same data sets. Furthermore, the APACHE system is proprietary and the scoring system is not publicly available. A final problem is that an intervention might affect a clinical measure used in a risk score without affecting the outcome of the sepsis. For example, how does the change in the risk score at 7 days relate to 28 day mortality?

5 Design Issues Related to Trial Failures

As knowledge of sepsis grows, concerns are surfacing that the many failed sepsis trials may not be due entirely to ineffective agents. The hope that a single drug or biologic might exist for the variety of problems leading to end organ failure or death may be misplaced. It may be appropriate for products attacking a common pathway of the inflammatory response. However, different underlying causes (pathogens, physiologic responses) of sepsis suggest that different agents may need to be tailored for each cause. We examine statistical design practices that can help account for such heterogeneity.

If there is a lag between diagnosis and randomization, the most seriously ill patients may die, and the remaining patients will have a better prognosis. Patients randomized at different times in disease progression or treatment phase will experience differential survival effects. Thus, the design or analysis of the study must account for variation in time from diagnosis to treatment. Therapies may have their maximal effect at a specific time in the septic process. This effect may be diluted by cases in which there is little or no benefit. Likewise, covariates may change importantly between onset, diagnosis, and randomization. Randomization should balance these, but the effect of a covariate may be diluted if the baseline time is after it can have a substantial affect on the outcome.

There are several time scales that one could use as covariates: time in hospital prior to randomization, time in ICU prior to randomization, time from the onset of sepsis (in fact, a left censored variable), time from diagnosis of sepsis. Each of these has different inferential implications. The time in hospital and time in ICU prior to randomization might encompass diagnostic delays or the hardiness of the patient. Since these occur prior to randomization they are not affected by treatment.

Different hospital practices and care standards may have substantial effects. For example, a large tertiary care hospital may have equipment that a small primary care hospital does not, and this could affect survival. One approach might be to perform initial studies in tertiary care facilities and continue studies in smaller facilities for those agents that showed promise. Another alternative would be to stratify randomization by hospital type and include this factor (and its treatment by factor interaction) in the analytic model.

It is recognized that the causative pathogen may be unknown at the time of randomization unless rapid assays are available, a problem that needs consideration in calculation of sample size. Patients found to have pathogens known to be resistant to the study agent are generally excluded from the efficacy analyses, since in principle it was established at baseline that the patient could not benefit from the agent. All such patients must be evaluated for safety. Indeed, detrimental effects might ensue in patients receiving inappropriate treatment. An example of this would be a therapy designed to treat gram-negative sepsis. The antibody would be ineffective in patients with gram-positive sepsis (Ziegler et al, 1991).

Undersized trials are another reason for failure. If a trial is designed to detect a mortality reduction from 40% to 20%, it requires 82 patients per arm in a two- sided 0.05-level test with power 0.8. If the product really reduces the rate to 30%, the power will be 0.267. It is well-known that smaller differences require larger sample sizes. The effect on power is equally evident, but perhaps not as well appreciated. The following table gives some simple results:

Table 2. *Sample Size and Power for a Two-Sample Test to Detect the Alternative (Type I error=0.05, Power =0.8)*

P_{control}	$P_{\text{experimental}}$	$n_1 = n_2$	Power under $n_1 = n_2 = 82$
0.4	0.20	82	0.80
0.4	0.25	152	0.54
0.4	0.30	357	0.267
0.4	0.35	1471	0.097

Thus, the chance of detecting a product that affects sepsis can be drastically reduced if the effect is less than the trial is designed for.

The design might specify a large trial with several interim analyses for efficacy. If the estimates of the event rates in placebo and treatment groups are not based on substantial phase II data, then it may be advisable to plan in advance to reconsider the sample size at an interim analysis.

Different therapies may be effective at different times. Thus, a sequence of therapies might be "just what the doctor ordered". An interesting question which must be addressed is, if the therapeutic intervention consists of a sequence of three new therapies (possibly from different companies), "how should the trials be conducted?" Would a 28-day survival benefit imply benefit for all three products? A problem that might arise would be if one product appeared to provide a benefit in the first three days and the other products showed no difference from standard care, yet the 28-day results showed a clear benefit. These issues will be addressed on a case-by-case basis. Referees have expressed concern that the logistics of such trials would be extremely difficult (how would companies work together, would the timing in drug development permit a joint clinical trial, etc.). We recognize this and hope that the sequence of therapies model could be made feasible.

A final reason for failed trials is that some products do not work. That is, the null hypothesis (that there is no difference between 28 day survival on standard of care and the experimental agent) is true.

6 Problems and Possible Solutions in Analysis

Have the alternative hypotheses been constructed realistically? If the Phase II studies show a possible 20% benefit, the Phase III studies might be designed for a smaller difference. In sepsis, there are many potential endpoints as noted earlier. If this 20% benefit is the maximum benefit observed among several endpoints in the phase II trial, the expectation of the maximum difference is an upwardly biased estimate of the effect size of the endpoint corresponding to the maximum. Thus, in replicating a study, we would expect to find a smaller treatment effect on the best endpoint observed in the first study, and hence a sample size computation should account for this phenomenon. A referee notes that this can occur due to regression to the mean, since only positive phase II studies are likely to progress to phase III. Indeed, any process which selects extreme values will have less extreme values in the next study. Table 2 (sample sizes and power) gives an indication of the effect of over-optimistic guesses at the level of benefit: the test will be underpowered, and there is a substantially lower chance of finding a specified difference.

A second problem arises from non-proportional hazards, making a log-rank test or proportional hazards model inappropriate. Conceivably, a time-segmented proportional hazards model with different underlying hazards in various intervals would be appropriate. Parametric models such as the log-normal have been proposed. Similarly, a mixture of distributions model with high hazard rates occurring early could be a reasonable approach. Taking this one step further, a compound distribution model might be appropriate. For example, if each subject had an exponential survival model with parameter $\lambda \exp(z\beta)$, and λ has a gamma distribution with parameters γ and α, the resulting distribution for T is

$$f(t) = \gamma\alpha \exp(z\beta)/(1 + \exp(z\beta) * t)^{\gamma+1}.$$

It is necessary to validate such a model in phase II studies. In addition, providing model validation tests on the phase III data can be helpful. If the hazards are not proportional, then the power of the test may be reduced. The log-normal distribution has been proposed as an alternative distribution which would emphasize early differences (Knaus, et al, 1995). Other alternative analysis methods would be the Prentice-Wilcoxon test or the $G^{\rho,\gamma}$ test family adjusted to be sensitive to early differences in the distributions. The Prentice-Wilcoxon test is sensitive to early differences, while the $G^{\rho,\gamma}$ can be tailored for sensitivity to crossing hazards (See, e.g., Fleming and Harrington, 1991). Other distributions (such as the compound of an exponential with a gamma mentioned earlier) might be considered. Again, it is important that these methods be specified in the protocol rather than developed by examining the data set.

Segmented statistical models may be useful. The product of logistic models or survival models would be useful if different covariates affected survival in different time intervals. Similarly, if a sequence of therapies is proposed based on achieving various landmarks (e.g., resolution of fever, discharge from ICU, etc.), a sequence of survival models might prove useful. The simplest of these would assume a constant hazard within time segments.

Mixture models might also be considered. These, and compound distribution models have already been mentioned. A possible extension might be to model a 'cured' group separately, together with the mixture of early failures.

Thus, the model would have the form $\alpha \Pr(T > t|\text{not cured}) + (1-\alpha)\Pr(T > t|\text{cured})$. The specific identification would not be necessary.

Different covariates could have important effects at different times. These lead to parametric survival models, proportional hazards models, and logistic models in which there is a time × covariate interaction. Our *a priori* bias is that the probability of survival should be a continuous smooth function. These models may give us better insight to the mechanisms of action of the products. These covariates and models should be determined before the start of the phase III trial. An alternative would be to determine the covariates from the control group or the pooled data without an term expressing treatment effect in the model. If the covariates are selected on the basis of the data, the null hypothesis has not been specified *a priori*, and the inference is uncertain. (A similar line of reasoning would apply to the model: if the distribution is determined from the data, is the observed result possibly a function of the data set instead of a particular benefit of the product?) A particularly interesting covariate might be the interaction between the risk score and the treatment indicator. This would reveal if there are different treatment effects for different risk levels.

One dilemma that faces analysts is handling missing values. An important missing value problem in sepsis trials relates to computing a risk index when some variables are not available at the time of randomization for all patients. If these are treated as missing, and replaced by a published value observed in non-diseased patients, the risk score may underestimate the "true" risk. This can create too optimistic a value or risk index. An alternative would be to replace the value by the mean in the placebo group. This, of course, leads to underestimates of variability for that value. A more complex approach would be to use a multiple imputation procedure for the missing values. For each imputation an estimate of the treatment effect would be obtained and these could be used to obtain an estimate of the variability due to the missing value replacement procedure. A fourth procedure would develop a risk score for each pattern of missing values. This might be prohibitively expensive if there were many such patterns. At any rate, using the value from the normal population seems to be fraught with problems, particularly in estimating the baseline risk of mortality.

Ignoring center effects can affect significance levels and reduce power in a survival analysis context. Specifically, the proportional hazards model may fail to hold if center effects are present and not included in the model. In an analysis of variance context, the F test becomes meaningless since the denominator expectation is the wrong one for testing the treatment effect.

7 Discussion and Conclusions

We believe there are many serious problems that have led to failure of Phase III sepsis trials. SIRS, sepsis, severe sepsis, and septic shock represent the spectrum of increasing disease severity formerly called sepsis. A plethora of unknown and uncontrolled variables have increased the variability of the primary outcome variables, leading to trials which are insensitive to any treatment effect. Thus,

> 1. It is important to identify sources of variation not due to the treatment. The number of covariates should be relatively small. In our experience, not more than 5 or 10 are used, but this is not a hard and fast rule.

The heterogeneity of the disease process suggests the presence of important subgroups which may benefit from specific therapies. Similarly, dividing patients into subgroups based on risk indices may help identify groups with a good chance of gaining from treatment. Hence,

> 2. An important contribution of phase II studies is to identify subgroups that will benefit (or not benefit) from the therapy being studied.

Additionally, high or low risk groups may be identified in phase II. In the preliminary studies, characteristic times when a therapy is effective (e.g., in the first three days following diagnosis of sepsis) may be identified. This information may be used to construct sequences of therapies which may be studied. Consequently,

> 3. Phase II studies may be used to identify therapies that can be used sequentially or in combination.

Because trials usually are conducted in many centers, it can be important to control for center by treatment (or investigator) interactions. Ignoring such factors can lead to questions about the generalizability of the trial. Accordingly,

> 4. Center and Treatment by Center effects may be studied for both phase II and phase III studies. Failure to control for these effects and interactions in phase II studies may mislead the investigator in the design of the phase III studies. Inadequate control of these interactions in phase III trials may lead to a failure to demonstrate an effective treatment.

The preceding comments suggest that there are many important issues that could profitably be studied prior to embarking on definitive trials of new agents. Since the target group for most sepsis therapies is not usually precisely defined at the onset of phase II studies, these studies typically should be large and entry relatively unrestricted. Then, exploratory, hypothesis-generating analyses may be performed to identify groups that have a high potential of deriving benefit from the therapy. Phase II studies generally do not show such strong positive results that phase III testing is curtailed. The phase II studies may be used to identify important risk factors or subgroups that may benefit from the product In designing these studies, it is useful to account for potential shrinkage in effect size when moving from phase II to phase III studies. One rule of thumb would be to select an effect size equal to some fraction (e.g.,0.5) of that seen in the phase II study.

Acknowledgements

The views expressed in this paper are those of the authors and not necessarily those of the Food and Drug Administration. We thank Dr. Ralph Harkins and Dr. Aloka Chakravarty for many useful, helpful discussions of problems and solutions seen in sepsis. Dr. Susan Ellenberg and Dr. William Schwieterman made many helpful suggestions on the manuscript. The referees comments have added immeasurably to this paper.

References

Bone, R. C., Balk, R. A., Cerra, F. C., et al. (1992). Definitions for sepsis and organ failure and guidelines for the use of innovative therapies in sepsis. *Chest*, **101**, 1656–1662.

FDA (1987). *Guidelines for the Format and Content of the Clinical and Statistical Sections of New Drug Applications.*

Fleming, T. R. and Harrington, D. P. (1991). *Counting Processes and Survival Analysis.* New York: John Wiley and Sons.

Knaus, W. A., Harrell, F. E., Fisher, C. J., et al. (1993). The clinical evaluation of new drugs for sepsis: a prospective study design based on survival analysis. *Journal of the American Medical Association.*, **270** (10), 1233–1241.

Knaus, W. A., Harrell, F. E., LaBreque, J. F., et al. (1996). Use of predicted risk of mortality to evaluate the efficacy of anticytokine therapy in sepsis. *Critical Care Medicine*, **24**, 46–56.

Lemeshow, S., Teres, D., Klar, J., et al. (1993). Mortality Probability Models (MPM II) based on an international cohort of intensive care unit patients. *Journal of the American Medical Association.* **270** (20), 2478–2486.

MMWR (1990). *Increase in National Hospital Discharge Survey Rates for Septicemia - United States, 1979-1987.* **39**, 31–34.

Siegel, J. P. (1994). Letter to Brussels Roundtable Consensus Conference. Personal communication.

Ziegler, E. J., Fisher, C. J., Sprung, C. L., et al. (1991). Treatment of gram-Negative bacteremia and septic shock with HA-1A human monoclonal antibody against endotoxin. *New England Journal of Medicine*, **324**, 429–436.

Coarsening at Random: Characterizations, Conjectures, Counter-Examples

Richard D. Gill
Mark J. van der Laan
James M. Robins

ABSTRACT The notion of *coarsening at random* (CAR) was introduced by Heitjan and Rubin (1991) to describe the most general form of randomly grouped, censored, or missing data, for which the coarsening mechanism can be ignored when making likelihood-based inference about the parameters of the distribution of the variable of interest. The CAR assumption is popular, and applications abound. However the full implications of the assumption have not been realized. Moreover a satisfactory theory of CAR for continuously distributed data—which is needed in many applications, particularly in survival analysis—hardly exists as yet. This paper gives a detailed study of CAR. We show that grouped data from a finite sample space *always* fit a CAR model: a nonparametric model for the variable of interest together with the assumption of an arbitrary CAR mechanism puts no restriction at all on the distribution of the observed data. In a slogan, `CAR is everything`. We describe what would seem to be the most general way CAR data could occur in practice, a sequential procedure called *randomized monotone coarsening.* We show that CAR mechanisms exist which are not of this type. Such a coarsening mechanism uses information about the underlying data which is not revealed to the observer, without this affecting the observer's conclusions. In a second slogan, `CAR is more than it seems`. This implies that if the analyst can argue from subject-matter considerations that coarsened data is CAR, he or she has knowledge about the structure of the coarsening mechanism which can be put to good use in non-likelihood-based inference procedures. We argue that this is a valuable option in multivariate survival analysis. We give a new definition of CAR in general sample spaces, criticising earlier proposals, and we establish parallel results to the discrete case. The new definition focusses on the distribution rather than the density of the data. It allows us to generalise the theory of CAR to the important situation where coarsening variables (e.g., censoring times) are partially observed as well as the variables of interest.

0. Overview

The phenomena of missing data in multivariate analysis (some components of a multivariate vector not observed), censoring in survival analysis, and grouped data in general, have in common that rather than observing a random variable (or vector) X of interest, one is only able to observe that X

takes a value in some possibly randomly determined set of values; by 'randomly determined' we mean that the set not only depends on X itself but also possibly on auxiliary random variables. The notion of '*coarsening at random*' was introduced by Heitjan and Rubin (1991) to single out exactly those situations in which the coarsening mechanism can be ignored when making inference on the distribution of X.

The same notion, but restricted in application to missing observations in a multivariate vector, goes back to Rubin (1976) and Little and Rubin (1987); in this context it is called 'missing at random'. These papers, and those of Heitjan (1993, 1994) have studied the statistical consequences of MAR and CAR in parametric models. Heitjan (1993) gives significant biomedical examples of CAR. Robins and Rotnitzky (1992), van der Laan (1993, 1995), Robins, Rotnitzky and Zhao (1994), and Robins (1996a) study statistical consequences in non- and semi-parametric models (with positive probability of complete observations). In survival analysis, with right-censored observations, coarsening at random is intimately connected to the central notion of 'independent censoring'; see Jacobsen and Keiding (1995).

'Coarsening at random' has clearly become an important topic in survival analysis, in biostatistics; in general, in applied statistics. Practitioners are keen to be able to assume that coarsened data is 'coarsened at random'. Yet in our opinion the notion is still poorly understood and this has dangers in uncritical application. We want to to find out what it really means, from a modelling point of view, to make the CAR assumption. To underline the problem, we note that CAR has only been defined for discrete data but the mere definition of CAR for continuous data—e.g., censored survival times—is not obvious at all. This is not just a question of measure-theoretic technicalities. Jacobsen and Keiding (1995) offer a definition, but though we learnt a great deal from their results, in our opinion their (rather complex and at the same time restrictive) definition does not capture the intended content of the notion.

The paper has a large number of sections, falling into two main parts. Sections 1 to 5 concentrate on the discrete case; 6 to 10 on the general; section 11 concludes. In section 1 we set out the necessary preliminaries. We give the (discrete case) definition of CAR in terms of the conditional distribution of the coarsened data given the underlying data: the conditional probability of a particular outcome depends on the underlying data value only through what the observed data tells us about this value. We show that this is equivalent to a factorization of the likelihood into separate parts corresponding neatly to the underlying data and the coarsening mechanism, and that it is equivalent to a condition concerning the probability law of the underlying data given the observed: it is the same as if the possible coarsenings had been fixed in advance, independently of the underlying data; so-called Coarsening Completely at Random.

In section 2 we give one of our main new results, 'CAR is everything': if we assume nothing about the distribution of the underlying complete data, and nothing about the coarsening mechanism except that it is CAR, then we are in effect assuming nothing at all about the distribution of the data. Any coarsened data from a finite sample space fits exactly to a CAR model. Without adding further assumptions either on the complete-data generating mechanism, or on the coarsening mechanism, the CAR assumption is untestable. This result generalises the well-known result in survival analysis known as the unidentifiabilty of the independent competing risks assumption. That our result is a reasonable conjecture follows by counting parameters; see Rubin, Stern and Vehovar (1995; section 4) for the case of Missing at Random. Our proof combines likelihood and convexity theory; in a sense, we use statistics to prove a theorem of algebra.

A counterexample shows that extending to infinite sample spaces, the result is not always true. However, every distribution of coarsened data can still be arbitrarily well approximated by a CAR model, so no amount of real data will ever be able to rule out a CAR model.

In section 3 we look at the CAR assumption as a modelling assumption, asking ourselves what kind of coarsening mechanisms could arise in nature, satisfying CAR. We argue that the most general realistic (physical) mechanism which produces CAR data is a sequential procedure we call 'sequential randomised coarsening'. A rather natural conjecture is then that *all* CAR mechanisms can be realised in this way. If the conjecture were true, CAR would not just be an attractive assumption to make (because of its data-analytic consequences) but also an assumption with physical or subject-matter content. If the conjecture is false, then CAR may be convenient, but in itself difficult to justify. Put yet another way, if you *can* justify the CAR assumption on subject-matter grounds, you actually know more about the coarsening mechanism than just the fact that it is CAR. It turns out that CAR mechanisms exist which cannot be represented sequentially, so that CAR is indeed much more than it seems.

Section 4 is a kind of interlude, establishing further nice properties of a rather special sequential coarsening mechanism, called 'monotone coarsening'. In this situation nonparametric maximum likelihood estimation can be done explicitly (without iteration), as in the case of the Kaplan-Meier estimator for right-censored survival data.

If the data are CAR, additional knowledge about the coarsening mechanism is irrelevant for likelihood-based inference. However we argue in section 5 that there are many situations, and we give an example in multivariate survival analysis, where restricting oneself to likelihood based methods may leave one with no practically useful methods at all. It can be better to make use of asymptotically irrelevant additional information about the coarsening mechanism and construct ad hoc (frequency based) methods, which are not asymptotically efficient, but do actually work in realistically sized samples.

The mathematical results so far were restricted to a finite sample space though we argue that their broad implications hold generally, for instance in survival analysis. In section 5 we already drew such implications. However we also want to develop the mathematical theory of CAR in this direction. The second main group of sections, 6 to 10, is devoted to the generalisation of the earlier results to arbitrary sample spaces. To begin with, we need a general definition of CAR.

CAR is usually stated as a condition on the (discrete) density function of the observed given the underlying data. One then is led to define CAR in general also in terms of densities. This leads immediately into technical problems, since coarsened data is a random set, and it is not clear how to introduce density functions into the picture. Jacobsen and Keiding (1995) took this route, defining CAR in terms of the densities of the variables in the model under consideration, relative to a 'reference model'. Our philosophy is different. We remark that in discrete models, a discrete density is just a probability. We read the definition of CAR as a statement about certain conditional *distributions*, not conditional densities. Now it is more or less immediate how one should generalise such a statement to a general case. Our general definition of CAR, in section 6, is that certain conditional distributions should coincide on certain parts of the sample space: 'the distribution of the observed data given the underlying variable of interest only depends on that variable through the information given to us about it, in the data'. There are minor measure-theoretic issues in making this definition mathematically rigorous, since conditional distributions can be changed at will on conditioning events of probability zero. For the more practically motivated reader it is enough to know that it is possible to make the definition precise in such a way that it both can be applied to the cases of interest in practice, and that it has the expected statistical consequences. The section shows that from our general definition of CAR do indeed follow the expected factorization of the likelihood, and the expected property of the conditional distribution of the underlying data given the observed. However these properties though implied by CAR are no longer equivalent to CAR, so here an important difference with the discrete case emerges.

Section 7 contrasts our 'absolute' definition of CAR (for general sample spaces) with the 'relative' definition of Jacobsen and Keiding (1995), and establishes the connections.

In section 8 we consider an important general issue: suppose we also observe to some extent some aspects of the coarsening mechanism. Is there still a natural definition of coarsening at random? For example, often in survival analysis one observes part or all of the censoring variables, even for uncensored observations. The original notion of CAR is only applicable when the actual data is strictly a coarsening of the underlying survival time. No further information is supposed to be available. We show that there are no problems in extending our definition and results to this more general case still. Here again our philosophy of thinking in terms of (conditional) distributions, not densities, pays off.

In sections 9 and 10 we attempt to extend our results on existence and uniqueness of a CAR model for arbitrarily coarsened data to the general case. Recall that in section 2 we show that discrete coarsened data always fits exactly to a CAR model, and that the underlying distribution and the CAR mechanism can be essentially uniquely reconstructed from the law of the data. However this breaks down in general sample spaces (even in countable sample spaces), though there is a good sense in which it is true for practical purposes: CAR is 'almost everything'. Any coarsened data whatsoever can be fit arbitrarily well by a CAR model, if not exactly. Anyway, in section 9 we obtain another 'next best' result stating, in the language of semiparametric models and information bounds, that the CAR model places no restrictions on the distribution of the data. In the neighbourhood of a CAR model one has so much freedom (although subject to the CAR assumption) that the set of possible score-functions is everything, and estimation is as difficult as in a completely non-parametric model. Section 10 gives a uniqueness result on the factorization of the likelihood in a CAR model, generalising the uniqueness of the CAR decomposition established in the discrete case.

In section 11 we conclude and in particular survey the many open problems which remain. We see that Coarsening at Random is not only a topic full of importance and interest from an applied point of view, but also full of challenges to theoreticians, opening a view to a rich and delicate theory. Survival analysis and mathematical statistics continue to enrich one another over the years.

A companion paper in this volume, Gill and Robins (1997), studies coarsening mechanisms with a sequential structure. Following Robins (1996b) we consider a generalisation of CAR in which coarsening occurs in a number of phases, each of which separately is CAR while the overall result is not. We also further investigate sequential mechanisms for producing MAR data. How could one 'physically' realise a general MAR mechanism? Is the MAR assumption an assumption which on its own can be supported by subject matter knowledge, or is it the case that if one can argue for MAR, one actually knows more (and therefore, outside of likelihood-based inference, has more options in data-analysis)?

1. Preliminaries

Suppose X is a random variable taking values x in a finite set E. Let $\mathcal{E}$ denote the set of all subsets of E, and let $\mathcal{X}$ denote a random nonempty subset of E: so $\mathcal{X}$ takes values A in $\mathcal{E} \setminus \{\emptyset\}$. We say that $\mathcal{X}$ is a coarsening of X if, with probability 1, $X \in \mathcal{X}$. The observed data, the random set $\mathcal{X}$, is usually denoted by Y in the literature on CAR. However later we will make a distinction between the random set $\mathcal{X}$ and its representation in the data Y as a list of coordinates, coefficients, or types.

If $\mathcal{X}$ is a coarsening of X, and one observes $\mathcal{X}$ but not X itself, one may ask if the observation '$\mathcal{X} = A$' can be treated for statistical purposes as the observation '$X \in A$'; i.e., as if the value of $\mathcal{X}$ instead of being random, had been provided in advance. Heitjan and Rubin (1991) show that this is the case if the conditional distribution of $\mathcal{X}$ given $X = x$ satisfies the following *coarsened at random* (CAR) assumption:

$$\text{for all } A \in \mathcal{E}, \quad \Pr(\mathcal{X} = A | X = x) \quad \text{is constant in } x \in A. \tag{1}$$

Obviously $\Pr(\mathcal{X} = A | X = x) = 0$ if $x \notin A$, if $\mathcal{X}$ is a coarsening of X. In a moment we derive their main result on ignorability of the coarsening mechanisms under CAR, but first we note that the CAR assumption intuitively seems to say that the observation of $\mathcal{X} = A$ is not influenced by the specific value of X in A which was taken, only by the fact that X *did* take a value in A. In fact CAR is obviously equivalent to

$$\Pr(\mathcal{X} = A | X = x) \quad = \quad \Pr(\mathcal{X} = A | X \in A) \quad \forall A, x \in A. \tag{2}$$

The CAR assumption is an assumption on the coarsening mechanism leading from X to $\mathcal{X}$, by which we emphasize that coarsening is seen as occuring in two stages: firstly the random variable X of interest is realised; secondly, a conceptually different process (usually associated with features of measurement or observational restrictions, rather than the scientific phenomenon under study itself), given the value x taken by X, replaces this value by a set $\mathcal{X} = A \ni x$.

However, having observed $\mathcal{X}$, we are free to consider the conditional distribution of X given $\mathcal{X} = A$, even though this compounds two quite different processes. Since (2) can be rewritten as (for all $x \in A$)

$$\Pr(\mathcal{X} = A | X = x \text{ and } X \in A) \quad = \quad \Pr(\mathcal{X} = A | X \in A)$$

we can recognise it as a conditional independence assumption: given $X \in A$, the events $X = x$ and $\mathcal{X} = A$ are independent. By symmetry of (conditional) independence, we therefore equivalently have:

$$\Pr(X = x | \mathcal{X} = A \text{ and } X \in A) \quad = \quad \Pr(X = x | X \in A)$$

But since the former is equal to $\Pr(X = x | \mathcal{X} = A)$ we have that CAR is equivalent to:

$$\Pr(X = x | \mathcal{X} = A) \quad = \quad \Pr(X = x | X \in A) \quad \text{for all } x \in A. \tag{3}$$

Thus the observation of $\mathcal{X} = A$ tells us no more, in the sense of what is now the conditional distribution of X, than the obvious '$X \in A$'.

So far we have only discussed the (probabilistic) interpretation of the CAR assumption. Now we give Heitjan and Rubin's statistical consequence. Suppose the distribution of X depends on a parameter θ, while the coarsening mechanism, supposed CAR, depends on a distinct, variation independent, parameter γ. We suppose CAR holds, for each γ. Write

$$p_x^\theta \quad = \quad \Pr{}^\theta(X = x); \qquad p_A^\theta = \Pr{}^\theta(X \in A);$$

$$\begin{aligned}\pi_A^\gamma &= \Pr^\gamma(\mathcal{X} = A | X = x) \qquad (x \in A) \\ &= \Pr^\gamma(\mathcal{X} = A | X \in A).\end{aligned}$$

The marginal distribution of $\mathcal{X}$ is

$$\begin{aligned}f_A^{\theta,\gamma} = \Pr^{\theta,\gamma}(\mathcal{X} = A) &= \Pr^{\theta,\gamma}(\mathcal{X} = A \text{ and } X \in A) \\ &= \Pr^\theta(X \in A) \Pr^\gamma(\mathcal{X} = A | X \in A) \\ &= p_A^\theta \pi_A^\gamma. \end{aligned} \tag{4}$$

So under CAR, the joint likelihood for θ and γ *factors* and the θ part can be written down *without knowledge of the coarsening mechanism*: as far as θ is concerned, the observation '$\mathcal{X} = A$' can be treated like an observation '$X \in A$'; i.e., as if the coarsening mechanism had been Coarsening Completely at Random: the sample space is partitioned in advance, independently of X, and one just looks to see which element of the partition X falls into. At the same time, the likelihood for γ can be written down without knowing the distribution of X, and moreover the likelihood for γ based on the data $\mathcal{X}$ is the same as the likelihood for γ based on the conditional distribution of $\mathcal{X}$ given X (even though X itself cannot be observed).

2. CAR is everything

Suppose $\mathcal{X}$ is a coarsening of X (in the discrete set-up of the previous section). We observe $\mathcal{X}$ only. If we assume nothing about the distribution of X, but we do assume CAR, does this imply anything about the distribution of the observable $\mathcal{X}$? Put another way, given a random non-empty set $\mathcal{X}$, can we construct a random variable X such that $\mathcal{X}$ is a coarsening of X and CAR holds?

Mathematically we have the following

Question. *Given a probability distribution $(f_A : A \in \mathcal{E}, A \neq \emptyset)$ of a random non-empty set $\mathcal{X}$, can we write*

$$f_A = p_A \pi_A \tag{5}$$

where $(p_x : x \in E)$ is a probability distribution on the finite set E, p_A is defined by $p_A = \sum_{x \in A} p_x$, and $(\pi_A : A \in \mathcal{E}\backslash\{\emptyset\})$ is a set of probabilities such that, for each $x \in E$,

$$\sum_{A \ni x} \pi_A = 1 \ \ ?$$

For given that the distribution of $\mathcal{X}$ factors as in (5), construct a joint distribution of X and $\mathcal{X}$ by letting

$$\begin{aligned}&\Pr(X = x | \mathcal{X} = A) = p_x / p_A \quad && x \in A,\ f_A > 0, \\ &\Pr(X = x | \mathcal{X} = A) = 0 && x \notin A.\end{aligned}$$

Under (5), if $f_A > 0$ then $p_A > 0$ too so the construction is well-defined. The construction forces $X \in \mathcal{X}$ to hold with probability 1. Moreover

$$\Pr(X = x \text{ and } \mathcal{X} = A) = \frac{p_x}{p_A} p_A \pi_A = p_x \pi_A \text{ for } A \neq \emptyset,\ x \in A,\ f_A > 0$$

and trivially

$$\Pr(X = x \text{ and } \mathcal{X} = A) = p_x \pi_A \qquad A \neq \emptyset,\ x \in A$$

if $f_A = 0$ and hence $p_A = 0$ or $\pi_A = 0$. Adding over $A \ni x$ shows that the marginal distribution of X is (p_x). Dividing by p_x shows

$$\Pr(\mathcal{X} = A | X = x) = \pi_A \qquad A \ni x,\ p_x > 0$$

which doesn't depend on x, so CAR holds.

This argument, together with the conditional independence arguments of Section 1, shows that CAR can equivalently be formulated as:

CAR($\mathcal{X}|X$): the conditional law of $\mathcal{X}$ given X satisfies (1) or (2);

CAR($X|\mathcal{X}$): the conditional law of X given $\mathcal{X}$ satisfies (3); and

FACTOR($\mathcal{X}$): the marginal law of $\mathcal{X}$ factors as in (5).

Now we return to the question posed at the beginning of this section. Just counting equations and unknowns suggests, but does not prove, that the answer is *yes*. The equations are non-linear and the unknowns have to be probabilities. We will prove that the answer is yes and moreover that the factorization (5) is unique for those A with $f_A > 0$. Our proof uses statistical ideas; we consider the expected log likelihood for a CAR model for the given observed data distribution. First we state the result formally:

Theorem. *Let $\mathcal{X}$ be a random non-empty set with distribution $(f_A : A \in \mathcal{E}\backslash\{\emptyset\})$. Then there exist CAR probabilities (π_A) and a distribution (p_x) on E such that $f_A = \pi_A p_A$ for all A, where $p_A = \sum_{x \in A} p_x$. For each A with $f_A > 0$, π_A and p_A are unique.*

Proof. Consider the problem of maximization of

$$\sum f_A \log(p_A \pi_A) = \sum f_A \log p_A + \sum f_A \log \pi_A$$

$$\text{over} \quad p_A = \sum_{x \in A} p_x,\ p_x \geq 0,\ p_E = 1,\ \sum_{A \ni x} \pi_A = 1\ \forall x,\ \pi_A \geq 0.$$

Considered as a function of the p_A and π_A for A with $f_A > 0$, with all these variables varying unrestrictedly in $[0, \infty)$, $\sum f_A \log p_A + \sum f_A \log \pi_A$ is continuous and strictly concave, and takes values in $[-\infty, \infty)$. The subset of p_A and π_A satisfying all the further listed constraints (to be interpreted as 'there exist p_x, π_A such that ...', as far as these variables were not yet involved) is convex and compact. So the supremum is attained uniquely as far as the original restricted set of variables are concerned, but may not be unique when we take account of the supplementary variables as well. We want to show that at least one of the solutions not only solves the maximization problem but also provides a factorization $f_A = \pi_A p_A$.

We study a solution for the p_A separately, in more detail. The analysis would be much simplified if we knew in advance that each $p_x > 0$ (which is for instance the case if each $f_{\{x\}} > 0$) and the reader could first restrict attention to this case. Consider the maximization of $\sum_A f_A \log p_A$, now over variables $p_x \geq 0$, subject to the constraint $p_E = 1$, where $p_A = \sum_{x \in A} p_x$. There exists a solution, and by concavity of $\sum f_A \log(\sum_{x\in A} p_x)$ we know, see for instance Whittle (1961), that there exists a Lagrange multiplier λ such that any solution is also solution of the problem: maximize $\sum_A f_A \log p_A - \lambda p_E$ over $p_x \geq 0$. At a given solution, for those x satisfying $p_x > 0$ differentiating with respect to p_x shows

$$\sum_{A \ni x} \frac{f_A}{p_A} - \lambda = 0. \tag{6}$$

For other x such that $p_x = 0$ we only have

$$\sum_{A \ni x} \frac{f_A}{p_A} - \lambda \leq 0.$$

If at this solution $p_A = 0$, then we must have $f_A = 0$ (otherwise $f_A \log p_A = -\infty$) and $p_x = 0$ for all $x \in A$. Multiplying (6) by p_x and adding over x such that $p_x > 0$ gives

$$\begin{aligned}
0 &= \sum_{x:p_x>0} p_x \sum_{A\ni x} \frac{f_A}{p_A} - \lambda \sum_{x:p_x>0} p_x \\
&= \sum_{x} \sum_{A\ni x} p_x \frac{f_A}{p_A} - \lambda \qquad \text{where } 0/0 = 0 \\
&= \sum_{A} \sum_{x\in A} p_x \frac{f_A}{p_A} - \lambda \\
&= \sum_{A} p_A \frac{f_A}{p_A} - \lambda = \sum_A f_A - \lambda \qquad \text{since } p_A = 0 \Rightarrow f_A = 0 \\
&= 1 - \lambda.
\end{aligned}$$

So $\lambda = 1$ and we have

$$\begin{aligned}
\sum_{A\ni x} \frac{f_A}{p_A} &= 1 \quad \text{if} \quad p_x > 0 \\
\sum_{A\ni x} \frac{f_A}{p_A} &\leq 1 \quad \text{if} \quad p_x = 0.
\end{aligned}$$

Define now $\pi_A = f_A/p_A$ *except* that if $p_x = 0$, so also $f_{\{x\}} = 0$, define $\pi_{\{x\}} = 1 - \sum_{A \ni x} f_A/p_A$ (with $0/0 = 0$ throughout). We then have:

$$\sum_{A \ni x} \pi_A = 1 \quad \text{for all } x;$$

and $f_A = p_A \pi_A$ for all A (also A with $p_A = 0$, also singletons).

Thus a factorization $f_A = \pi_A p_A$ exists. Since $\sum f_A \log \widehat{f}_A \leq \sum f_A \log f_A$ for all $\widehat{f}_A = \pi_A p_A$, the factorization we have found must also be a solution of the maximization problem we considered at the outset. As we remarked, this uniquely determines the p_A and π_A for A with $f_A > 0$. □

It is difficult to give necessary and sufficient conditions for uniqueness of *all* p_x and π_A in the factorization $f_A = p_A \pi_A$. If $f_{\{x\}} > 0$ for all x, then $p_x > 0$ and is uniquely determined for all x, hence $p_A > 0$ for all A and $\pi_A = f_A/p_A$ is uniquely determined for all A. Consider the incidence matrix with rows corresponding to A with $f_A > 0$ and augmented if necessary with a row for $A = E$; columns corresponding to $x \in E$; and the (A, x) element equal to the indicator of $x \in A$. The vector of p_A's with $f_A > 0$ augmented with p_E, equals this matrix times the vector of (p_x); so if the matrix has rank equal to the number of elements of E, (p_x) is uniquely determined. This rank condition is however not necessary, since the inequalities $p_x \geq 0$ might also help to uniquely determine (p_x) from $(p_A : f_A > 0 \text{ or } A = E)$.

One would hope that the result 'car is everything' remains true in general sample spaces but the following counter-example, due to Ya'akov Ritov, shows that this hope fails already in a countable sample space. The example may seem artificial, but actually it corresponds to a survival analysis situation where every observation is censored and arbitrarily large censored observations can occur. Let E be the natural numbers $\{0, 1, \ldots\}$ and suppose the only subsets of E which get positive probability are $\{n, n+1, \ldots\}$ for $n = 1, 2, \ldots$. If we try to factor $f_A = p_A \pi_A$ by maximizing the log likelihood $\sum f_A \log p_A$ we see that the likelihood is always increased by moving probability from the left to the right. The maximizer would like to put all the probability mass at $+\infty$ but there is no such point in E so the maximizer does not exist. Hence there can be no factorization, since if there were one, it would maximize the log likelihood by fitting the f_A exactly. One could try to save the situation by adding a point $+\infty$ to E but this only helps if one also adds the same point to all the sets $A = \{n, n+1, \ldots\}$. In other words, this example can be repaired by compactifying both the sample space and all the observed random sets in a careful way. That might be true in general.

Even if CAR is not everything in the strict sense, we do argue that it is 'almost everything'. At least from a practical point of view, every coarsened data model can be arbitrarily well approximated by a model for discrete coarsened data, and a CAR model fits that exactly. Combining these two facts, a CAR model fits arbitrarily well, even if not exactly. Similarly, the empirical distribution of any finite sample of coarsened data can be exactly fit to a CAR model: computation of the non-parametric maximum likelihood estimator does exactly this job, where without loss of generality one can assume that the support of the underlying variable X is finite.

3. Sequential representations of CAR

So far a CAR mechanism is described in an algebraic way: just a collection of probabilities π_A satisfying

$$\sum_{A \ni x} \pi_A = 1$$

for each $x \in E$. Is there a more appealing way to describe all CAR mechanisms? Is there a convenient way to simulate any CAR mechanism?

The direct way to simulate the random set $\mathcal{X}$ is first to generate X according to the law $(p_x : x \in E)$, then $\mathcal{X}$ according to the conditional law $(\pi_A : A \ni x)$. This makes no use of the fact that the coarsening mechanism is actually CAR. Moreover in the course of the simulation we have to look at the specific value taken by X, even though this value is not later revealed by $\mathcal{X}$. Another way is to directly generate $\mathcal{X}$ from its marginal distribution $(f_A = p_A \pi_A : A \subseteq E)$. Again, once the probabilities f_A have been calculated, no use is made of the fact that coarsening is CAR.

A rather special kind of CAR does allow an appealing simulation construction: so-called *monotone coarsening* (or *monotone missingness*). Consider the collection of subsets A with $\pi_A > 0$. Suppose no two of these subsets overlap non-trivially. Consider the directed graph on $\{A : \pi_A > 0\} \cup \{E\}$ where there is an edge from A to A' if and only if $A' \subset A$ and no A'' exists with $A' \subset A'' \subset A$ (and $\pi_{A''} > 0$). This graph forms a tree with root at E.

The leaves of the tree form a partition of E; and in fact the branches leading from any node A form a partition of A. For suppose the contrary were true: there exists A, $x \in A$, $A' \subset A$ with $\pi_{A'} > 0$, and $x \notin A'$. Moreover $x \notin A''$ for any $A'' \subset A$ with $\pi_{A''} > 0$. Choose $x' \in A'$. We have $1 = \sum_{A'' \ni x} \pi_{A''} < \pi_{A'} + \sum_{A'' \ni x} \pi_{A''} \leq \sum_{A'' \ni x'} \pi_{A''} = 1$, which is impossible.

The random set $\mathcal{X}$ can be generated by a random walk up the tree, starting at the root E, and stopping somewhere, $\mathcal{X} = A$, on the way up. Suppose at some stage we have just moved into the node A. We then stop in A with probability $\pi_A / (1 - \sum_{A' \supset A} \pi_{A'})$. (This expression is a probability, since for any $x \in A$ we have $1 = \sum_{A' \ni x} \pi_{A'} \geq \pi_A + \sum_{A' \supset A} \pi_{A'}$.) Conditionally on *not* stopping in A, we choose a branch A' with probability $p_{A'}/p_A$ and move into A'. Since the branches A' from A form a partition of A, the branching probabilities add to 1. An equivalent description of this step is that, knowing now that $X \in A$ and that we do not stop here, we look to see which element of the partition of A contains X, and move to that element. Note that we use the CAR probabilities π_A to decide whether to stop or not; we use the underlying probabilities p_A to choose a branch, given that we did not stop.

A direct calculation shows that this procedure generates $\mathcal{X}$ with probability distribution $p_A \pi_A$. To see what is going on more intuitively, consider the pair X, $\mathcal{X}$. If the value $X = x$ were known in advance, only one path through the tree would be relevant, the path starting at E and ending at the leaf containing x. Call this path $E = A_0 \supset A_1 \supset \ldots \supset A_k \ni x$. The probabilities π_{A_i} along this path form the distribution of $\mathcal{X}$ given $X = x$, and the 'stopping probability' $\pi_A/(1 - \sum_{A' \supset A} \pi'_A)$, for $A = A_j$, equals $\pi_{A_j}/(1 - \sum_{i<j} \pi_{A_i})$. In fact in our simulation we do not generate X in advance but at a given step, when we know already that $X \in A$ and that we do not stop here, we decide which branch A' from A contains X, according to the conditional distribution of X given $X \in A$.

Of course CAR probabilities such that the sets A with $\pi_A > 0$ lie on a tree are rather special. However the idea of generating $\mathcal{X}$ by successively partitioning a set in which X is known to lie, and observing in which element of the partition X lies, seems to us the most general way conceivable to physically realize a CAR mechanism. In monotone coarsening the partitions are given in advance. Now we will allow the partitions to be chosen at random; in principle the choice could depend on what happened at previous stages. However it should not depend on future choices since that would require foreknowledge of X, in conflict with the required CAR property.

Partitioning a set into say k subsets and observing in which X lies can also be carried out by a series of partitions in 2 subsets. Thus we arrive at the following definition of a *randomized monotone coarsening* scheme. Initially, $n = 1$ and $\mathcal{A}_0 = E$. By step n we have generated a sequence of nonempty subsets $\mathcal{A}_0 \supset \mathcal{A}_1 \supset \ldots \supset \mathcal{A}_{n-1}$ and we know $X \in \mathcal{A}_{n-1}$. We may now decide to terminate, and set $\mathcal{X} = \mathcal{A}_{n-1}$, or we decide to continue. In the latter case we choose at random a subset B_n of $\mathcal{A}_{n-1}$. We observe whether X lies in B_n or in $\mathcal{A}_{n-1} \backslash B_n$, and set $\mathcal{A}_n$ equal to B_n or $\mathcal{A}_{n-1} \backslash B_n$ accordingly. Increment n by one, and repeat. The probability of stopping and the probability distribution of B_n, given we do not stop, may depend in an arbitrary way on the past sequence $\mathcal{A}_0, \mathcal{A}_1, \ldots, \mathcal{A}_{n-1}$. If these probabilities only depend on the current data $\mathcal{A}_{n-1}$ (and on n) then we call the scheme a *Markov coarsening*.

A very natural conjecture is: *any set of CAR probabilities can be represented by a randomized monotone coarsening scheme.* In a simulation of randomized monotone coarsening, as a separate procedure taking place after the generation of X, we do not use more information about X than that which is finally revealed in the value of $\mathcal{X}$. Conversely, given X, to simulate a CAR mechanism which is *not* randomized monotone, the computer program will require in the course of the procedure information about X—perhaps even its precise value—which is ultimately not revealed in the value of $\mathcal{X}$ output by the computer in its final `print` statement. But the fact that the computer has had to hide information from us does not affect our face-value inference, '$\mathcal{X} = A$ tells us no more than that $X \in A$'.

The conjecture is easily found to be true when $\#E = 2$. However already when $\#E = 3$ there are counter-examples. Let $E = \{1, 2, 3\}$; the list of possible A is $\{1\}$, $\{2\}$, $\{3\}$, $\{1,2\}$, $\{1,3\}$, $\{2,3\}$, $\{1,2,3\}$. The only way the procedure can stop with the set $\{2,3\}$ is when at the very first stage, the partition $\{1\}, \{2,3\}$ is chosen (and $X = 2$ or 3). Note that the very first partition is chosen independently of X. So if $X = 1$, we would also have stopped immediately if this had been the first partition chosen. However one can also arrive, when $X = 1$, at $\mathcal{X} = \{1\}$ after various other first stage choices. So the probability to terminate with $\mathcal{X} = \{1\}$, when $X = 1$, is at least as large as the probability to terminate with $\mathcal{X} = \{2,3\}$ when $X = 2$, or when $X = 3$; i.e., $\pi_{\{1\}} \geq \pi_{\{2,3\}}$. However for instance

$$\begin{aligned} \pi_{\{1\}} = \pi_{\{2\}} = \pi_{\{3\}} &= 0.1 \\ \pi_{\{1,2\}} = \pi_{\{1,3\}} = \pi_{\{2,3\}} &= 0.4 \\ \pi_{\{1,2,3\}} &= 0.1 \end{aligned}$$

satisfies $\sum_{A \ni x} \pi_A = 1 \; \forall x = 1, 2, 3$ but $\pi_{\{2,3\}} > \pi_{\{1\}}$.

Obviously the counter-example can be extended to $E = \{1, 2, \ldots, n\}$ for any $n \geq 3$, comparing $\pi_{\{1,2,\ldots,n-1\}}$ with $\pi_{\{n\}}$.

The probabilities in our counter-example are pretty arbitrary. A more extreme example is obtained by letting the probability of each duplet $\{i, j\}$, $i \neq j$, be equal to 0.5, while letting the probabilities of the singeletons and the triplet be equal to zero. Now it is more clear why these probabilities cannot be realized sequentially: when $X = 1$ the computer ultimately has to choose between reporting '$X \in \{1,2\}$' and '$X \in \{1,3\}$'; for the other values of X it has to make similar choices. How can it make its choice without observing X completely? (The reader familiar with the so-called quiz-master problem might like to ponder if that is an example of CAR. A prize car is hidden between one of three doors. You choose one door. The quizmaster, knowing the location of the car, opens another, showing that no car is behind it. He then asks if you would like to revise your choice. Do you? We come back to this example in section 8.)

This unsatisfactory state of affairs leaves many questions open. The authors cannot conceive of a more general mechanism than a randomized monotone coarsening scheme for constructing the CAR mechanisms which one would expect to meet with in practice, but is this just a lack of imagination? Can one easily recognise if a given CAR mechanism has a randomized monotone representation or not? As the set E gets larger, do 'most' CAR mechanisms admit a representation?

4. Computing the monotone CAR model

In general, the problem of computing the decomposition $f_A = p_A \pi_A$ of a given (f_A) does not have an explicit solution. As we saw, it can be phrased as a constrained maximization problem $(\max \sum f_A \log p_A)$, which can be solved by various numerical procedures including the iterative EM or Turnbull algorithm, $p_x^{\text{new}} = \sum_{A \ni x} f_A p_x^{\text{old}} / p_A^{\text{old}}$.

One case (and one case only?) does have an explicit solution. This is when (f_A) is *monotone* by which, similarly to the definition of a monotone coarsening mechanism in the previous section, we mean that for $A \neq A'$ such that $f_A > 0$ and $f_{A'} > 0$, either $A \subset A'$, or $A' \subset A$, or $A' \cap A = \emptyset$. In other words: no two A with positive probability in the data overlap non-trivially. Before, the same statement was supposed to be true conditionally on $X = x$, for any x.

We saw in the last section that a monotone coarsening mechanism could be simulated on the computer as a random walk on a tree. This representation can also be used to compute explicitly the decomposition $f_A = p_A \pi_A$ from given (f_A), using a 'tree'-version of the Kaplan-Meier estimator from ordinary survival analysis—right censored survival times supplying a particularly simple example of monotone data.

Consider the collection of A such that $f_A > 0$, augmented (if they are not already included), with E and all the singletons. This collection also has no non-trivially overlapping members. Define a directed graph on this collection of nodes, with a branch (directed edge) from A to A' if $A \supset A'$ but there is no A'' with $A \supset A'' \supset A$. The symbol $\supset$ means strict inclusion. The graph is easily seen to be a tree with root in E and with the singletons $\{x\}$, $x \in E$, as its leaves. The branches from any node A partition A.

We know that a decomposition $f_A = p_A \pi_A$ exists. Moreover it may be chosen so that $f_A = 0 \Rightarrow \pi_A = 0$ except possibly if A is a singleton. So all A with $\pi_A > 0$ are on our tree: the coarsening mechanism is also monotone.

Consider a point x and in imagination hold the tree at the root E and leaf $\{x\}$, pulling it tight, and so forming a straight path from the ground to the sky with side branches on the way up. Consider the steps up the main path of the tree from E to $\{x\}$ as discrete time-steps, with the leaf at time infinity. Consider all the f_A as empirical (relative) frequencies of a large sample of data. Each observation A, together with the underlying true value of $X = x' \in A$, represents a path on the tree, starting from E, going through A, and ending at $\{x'\}$. At some point this path must branch off from the main route up to $\{x\}$ (unless $x' = x$); either before, at, or after A. Consider the branching time as an underlying survival time (you die when you leave the straight and narrow path; you live for ever if $x' = x$). The underlying survival time is observed exactly if A is off the main path or if $A = \{x\}$; but is unknown if A is on the main path, before $\{x\}$. If A lies on the main path the observation is censored just *before* this time point. The underlying survival function between consecutive time steps $A \supset A'$

going up the main path is $p_{A'}$. Hence $p_{\{x\}}$ is the estimated probability to be still alive just after the last branching before $\{x\}$, or as one could say, the probability of eternal life. This Kaplan-Meier estimated survival function just before time ∞ is as usual undefined if the 'last observation is censored'. In this context that occurs when the last node before $\{x\}$, say A, which necessarily satisfies $f_A > 0$, is such that $f_{A'} = 0$ for all $A' \subset A$. Then we can only calculate p_A itself, the probability of surviving till just before A. The estimated survival curve actually tells us all the p_A for A on the main route to $\{x\}$ and by division we can recover the π_A also.

5. Examples in survival analysis

Ordinary censored survival times $(\widetilde{T}, \Delta)$ provide a classic example of monotone data. The underlying variable of interest will be denoted by T instead of X. A coarsened observation is the interval $(\widetilde{T}, \infty)$ if $\Delta = 0$, and the singleton $\{\widetilde{T}\}$ if $\Delta = 1$. Of course this example needs a continuous sample space to be treated properly, and we will do that in the next section. However our main points on CAR and survival analysis can be made while pretending the survival times are discrete. So let us suppose that all time points, denoted t or $\widetilde{t}$, are integers between 0 and N say. In ordinary survival analysis (one-dimensional right-censored survival times) the coarsened data is represented by half intervals $\{\widetilde{t} + 1, \ldots, N\}$ in the case of censored observations, and by singletons $\{t\}$ in the case of uncensored. Two outcomes are either disjoint or one contains the other: the data is monotone.

Suppose we assume nothing about the distribution of the underlying survival time, and nothing about the coarsening mechanism except that it is CAR. Since CAR is everything we are assuming nothing at all about the distribution of the observations. Computing the non-parametric maximum likelihood estimator comes down to maximising $\sum f_A \log p_A$ over underlying distributions (p_t), where (f_A) is the empirical distribution of observed sets A. If each of n observations is different, then each observation yields one set A with f_A equal to $1/n$. The maximization can be done explicitly (because of monotonicity), yielding the Kaplan-Meier estimator for (p_t). The CAR probabilities can be computed explicitly too. The CAR mechanism is actually a random (independent) censoring model: in other words, one can generate the observed coarsening by choosing a C independent of $T \sim (p_t)$, and then reporting the set $\{T\}$ if $T \leq C$, and $\{C+1, \ldots, N\}$ if $T > C$. By CAR we have that for each $\widetilde{t}$, $\Pr\{\widetilde{T} = \widetilde{t} | T = t\}$ is the same for each $t > \widetilde{t}$. These probabilities, for $\widetilde{t} = 0, 1, \ldots$, supply the claimed distribution of C. The fact 'CAR is everything' is well known, for censored survival data, as *the unidentifiability of the independent competing risks assumption*: any pair $\widetilde{T}, \Delta$ whose first element is a random time and whose second element is a zero/one variable can be written, in distribution, as $\min(T, C), 1\{T \leq C\}$, for an *independent* couple T, C (each possibly taking the value $+\infty$ with positive probability). Note that the possibility of mass at infinity is crucial to the nonidentifiability of dependent competing risks. Specifically it follows from Ritov's counterexample to 'CAR is everything' (end of section 2) that we could rule out independence of competing risks (i.e., independence of censoring and failure time) if we knew that with probability one every observation will be censored and we knew the failure time distribution has no mass at infinity.

Turning to multivariate survival analysis, the situation becomes much more complex. The problem of nonparametric estimation for multivariate censored survival times has remained open for a long time and only recently was a lot of striking progress made. The difficulty is directly connected with CAR. Let us represent one observation again as $(\widetilde{T}, \Delta)$ where $\widetilde{T}$ is now a *vector* of censored survival times and Δ a vector of censoring indicators. If the observation takes the value $(\widetilde{t}, \delta)$ then we know that the underlying survival vector T lies in the set A formed by taking the Cartesian product of singletons or half intervals defined precisely as in univariate censoring from each pair of components $(\widetilde{t}_i, \delta_i)$.

Let us assume nothing about the distribution of T. Our aim is to estimate its (multivariate) survival function, let us call it S, based on n censored observations. Ignoring again the fact that the sample space should be continuous, an assumption of CAR together with no assumption on the distribution of T means that we are not assuming anything about the distribution of the data at all. Computing the non-parametric maximum likelihood estimator of S by maximising the sum (over the n observations) of logs of probabilities of observed sets is no more than computing the reparametrisation from observed data probabilities (f_A) to underlying (p_t), (π_A) where we plug in as (f_A) the empirical distribution of the observed data: probability $1/n$ for each observed set A. Since our model is completely nonparametric there are no other reasonable estimators. In fact, as $n \to \infty$, in a situation where root n consistent estimation is possible at all—roughly speaking, when the CAR probabilities to get an exact observation are all positive–the nonparametric maximum likelihood estimator for S should be asymptotically efficient and moreover any other asymptotically regular estimator of S will be asymptotically equivalent with it. (Sometimes the NPMLE itself may fail to have good asymptotic behaviour, but still all asymptotically efficient estimators will be asymptotically equivalent to one another. Typically a simple modification of the NPMLE turns it into one of these good estimators. See van der Laan (1993, 1995) for general theory and many applications).

In the univariate case these facts are true and nowadays quite well-known. Assuming CAR (with otherwise completely unknown coarsening mechanism) is in fact equivalent to assuming Random Censorship with unknown censoring distribution. If both survival and censoring distribution are completely unknown, the model for the data is completely nonparametric. There is no essential alternative for nonparametric estimation of the survival function to the Kaplan-Meier estimator. Apparent alternatives such as nonparametric Bayes estimators, or the negative exponential of the

Nelson estimator of cumulative hazard, are asymptotically equivalent to the Kaplan-Meier estimator. Only if one assumes some knowledge of the censoring distribution (or is also able to observe censoring times of uncensored observations) do inefficient, strictly different estimators become available such as the reduced sample estimator (see Kaplan and Meier, 1958), or the reweighted (according to the censoring survival function) empirical distribution of the uncensored observations.

With multivariate censored data one can also consider the natural analogue of the univariate Random Censoring model. This says that there exists a vector of censoring times C, independent of T, and with completely unknown distribution, such that $(\widetilde{T}, \Delta)$ is formed componentwise from the components of T and C as in the univariate case. This does define a CAR mechanism. But there are many CAR mechanisms producing multivariate censored observations which cannot be represented in this way. The nonparametric multivariate random censoring model really is a model.

Here is a simple example of bivariate censoring which is coarsened at random but not random censored. Let $T = (T_1, T_2)$ be a bivariate survival time. Suppose T_1 and T_2 are actually consecutive durations between events in the lifetime of one individual, starting at time 0. Let C be an independent censoring time in the one-dimensional 'calendar time' time scale at which there are two events of two different types at times T_1 and $T_1 + T_2$. Thus $(T_1, T_1 + T_2)$ is randomly censored by (C, C). The data may still be represented, in 'two-dimensional duration time', as $(\widetilde{T}, \Delta) = \big((\min(T_1, C), \min(T_2, \max(0, C - T_1))),\ (1\{T_1 \le C\}, 1\{T_1 + T_2 \le C\})\big)$.

Assuming only CAR (a correct assumption in this example), and assuming nothing about S, we have no option than to compute the NPMLE of S, or an asymptotically equivalent version of it. Such estimators have been studied by van der Laan (1996). In fact, in general the NPMLE of bivariate censored survival data does not work correctly as it stands; one has to modify it slightly by an asymptotically negligeable further coarsening of the data. In the two dimensional case the possible observations are points, half-lines, and quadrants. The half-lines cause problems because we have to put probability mass in these lines, but have no information about how to do that since there will typically be no point-observations within the lines. The half-lines should be slightly expanded to thin strips, containing a few uncensored observations, and then the NPMLE makes sense and can be made asymptotically efficient. Its computation is time-consuming and its mathematical analysis very delicate. These problems are associated with the curse of dimensionality: under the completely nonparametric model we are forced to use the NPMLE or a modification thereof, and that forces us into binning or smoothing high-dimensional data in order to estimate conditional densities of some components given others. This only makes sense with huge data sets.

Suppose however we know that multivariate censored data is not just

Coarsened (in this case, censored) at Random but actually Randomly Censored. Although assuming nothing about survival or censoring distributions, we are now making identifiable assumptions; we have a real (restrictive) model. Going for full asymptotic efficiency gives us again no options: the same, delicate, NPMLE. However we can use our information on the censoring mechanism to generate a multitude of inefficient estimators. Some of these—the beautiful Kaplan-Meier generalisations of Dabrowska (1988) and of Prentice and Cai (1992)—do not lose much efficiency, are easy to calculate, and work very well already with quite small sample sizes. The other side of the coin is that there are CAR mechanisms which are not Random Censoring under which those estimators are inconsistent. They truly need the 'nuisance assumption' to work.

The likelihood factored; information about nuisance parameters should be irrelevant; yet we have made use of such information to generate alternative and practically valuable estimators. But a major justification for likelihood methods is their good large sample properties. The curse of dimensionality may prohibit their practical use, and one can be better off using asymptotically irrelevant information to construct well-behaved though (asymptotically) inefficient statistical procedures; see Robins and Ritov (1996) for an in depth study of this phenomenon.

Our example with calendar and duration time mix-up illustrates again the pitfalls. A sensible statistician would represent the data as censored times of events $(T_1, T_1 + T_2)$, knowing that from the joint distribution of these two times one can easily compute the joint distribution of (T_1, T_2). The data is actually monotone. The NPMLE can be computed explicitly. It is based on combining the marginal Kaplan-Meier estimator of the distribution of T_1 with conditional Kaplan-Meier estimators of the distribution of $T_1 + T_2$ given T_1, for each observation for which T_1 is uncensored. Actually, because we will be using one observation to estimate each conditional survival function for each observed value t_1, one can expect this estimator to make nonsense. But binning of the observations according to values of t_1 solves that problem if the sample size is large enough, and yields an asymptotically efficient estimator. One could alternatively use the Dabrowska or the Prentice-Cai estimators: they do not require any artifical grouping or smoothing of the data, but are asymptotically inefficient.

A less sensible statistician will treat the data precisely as bivariate censored observations of the durations (T_1, T_2). If he or she sticks to NPMLE (or modifications thereof) nothing will go wrong; the data is CAR and likelihood based methods do not need further information about the coarsening mechanism. However the Dabrowska or Prentice-Cai estimators will now be inconsistent since the Random Censoring model is not true.

Section 3 considered sequential mechanisms for generating CAR data. We showed that not all CAR mechanisms can be represented in this way. An important direction for future reseach is to define and study similar mechanisms in the context of multivariate censored data. This will lead to a class of censoring models intermediate between Random Censoring and Censoring (Coarsening) at Random.

6. CAR in general sample spaces

In a discrete sample space, equivalent definitions of CAR and important consequences of it were easy to obtain. In a general sample space, the various possible definitions may not be easy to formulate any more; moreover, even if they can be formulated in a natural way, they may no longer be equivalent. In that case, which definition one takes as primary should be influenced by which desirable results can be obtained from it.

In the first section we defined CAR in terms of the conditional distribution of a coarsening $\mathcal{X}$ given the coarsened variable X, CAR($\mathcal{X}|X$). We showed in section 2 that the definition was equivalent to a condition on the conditional distribution of X given $\mathcal{X}$, CAR($X|\mathcal{X}$), and to a specific factorization of the marginal distribution of $\mathcal{X}$, FACTOR($\mathcal{X}$).

The original definition—in terms of the distribution of $\mathcal{X}$ given X—respects the idea that *after* the random variable X has been generated, it is coarsened to the observation $\mathcal{X}$ by a conceptually distinct process. The condition on the conditional distribution of X given $\mathcal{X}$ describes in an appealing way that under CAR, knowing $\mathcal{X} = A$ tells us no more about X than the obvious fact $X \in A$. It is moreover useful in statistical inference—e.g., in the E step of the EM algorithm, in running the Gibbs' sampler, in calculating score functions; in all cases using exactly this conditional distribution. Finally factorization of the marginal distribution allows likelihood based inference on the distribution of X to be carried out completely ignoring the coarsening mechanism.

Before giving a general definition we must set up the measure-theoretic background to be able to talk about all these conditional distributions. If X is, say, a k-dimensional vector, our random set $\mathcal{X}$ takes values in the set of *all* subsets of $\mathbb{R}^k$. There is no natural topology on this very large space, no natural Borel σ-algebra. The space is so large that conditional distributions of X given $\mathcal{X} = A$ are not guaranteed to exist. In practice however, the range space of $\mathcal{X}$ can be taken to be quite small (e.g., rectangles only). Each possible value can typically be described by a short list of types, coefficients, coordinates or whatever. So we suppose that $\mathcal{X}$ can be described in a 1–1 way as function of some, say real vector, Y; $\mathcal{X} = \alpha(Y)$. In fact if we just suppose that X and Y take values in *Polish* spaces (separable, metric spaces) then sets of regular conditional distributions of X given Y and of Y given X both exist (see, e.g., Chang and Pollard, 1997). We also want the values of $\mathcal{X}$ to be measurable sets for X, and the set of values of Y consistent with a given value of X, to be measurable too. This is taken care of by assuming that the mapping $(x, y) \mapsto 1\{x \in \alpha(y)\}$ is jointly measurable in x and y, where the domain of the mapping is given the Borel σ-algebra corresponding to the topologies on the spaces where X and Y lie.

From now on, we assume this bare minimum of regularity without comment and also, when it is not relevant in the present context, drop the distinction between the set $\mathcal{X}$ and its description Y. Suppose then Y is a coarsening of X so, abusing our notation as announced already, $X \in Y$ with probability 1. The natural generalisation of CAR is

CAR($Y|X$). *The conditional distributions of Y given X do not depend on the values x taken by X, except for the restriction implied by Y being a coarsening of X, namely that given $X = x$, the random set Y takes values in $\{y : y \ni x\}$. More precisely, taking account of the fact that conditional distributions are not uniquely defined on sets of probability zero, we suppose that versions of $P_{Y|X=x}(\mathrm{d}y) = \Pr(Y \in \mathrm{d}y|X = x)$ can be chosen for P_X-almost all x, such that for x, x' not in the exceptional set,*

$$P_{Y|X=x}(\mathrm{d}y) = P_{Y|X=x'}(\mathrm{d}y) \text{ on } \{y : y \ni x\} \cap \{y : y \ni x'\}. \tag{7}$$

One might hope that if CAR is true according to this definition, then versions of $P_{Y|X=x}$ can be chosen making (7) hold *everywhere*. The following recipe (also used in the proof of 'car is everything' in section 2) might work: for the bad x, redefine $P_{Y|X=x}$ on $\{y \ni x\}$ to be equal, for each good x', to $P_{Y|X=x'}$ on $\{y \ni x'\}$. If probability mass still remains to be assigned, put all the remainder as an atom on the singleton $\{x\}$. One must check that this pasting together of bits of many other probability distributions does not entail using *more* than total probability 1. This problem is open. Alternatively, one could simply delete all bad x from the original sample space, and merge corresponding y (with and without bad x), arriving at a new coarsening model in which (7) holds without exception, and only differing from the original in indistinguishable events.

The precise formulation we have taken of CAR allows us to establish the following property of the Radon-Nikodym derivative (likelihood ratio) between *two* distinct coarsening mechanisms P, P' *each* satisfying CAR separately:

$$\frac{\mathrm{d}P'_{Y|X=x}}{\mathrm{d}P_{Y|X=x}}(y) \quad \text{does not depend on } x \in y, \tag{8}$$

it only depends on y itself. We call this derived property CAR(REL), REL standing for relative, in contrast with (7) which can be called CAR(ABS). In other words, we show that CAR(ABS) for P and CAR(ABS) for P' implies CAR(REL) for P' with respect to P. From CAR(REL) will follow a factorization and a result of the type CAR($X|Y$) concerning conditional distributions in the reverse direction.

Lemma. *Suppose that* (7) *holds for each of two coarsening mechanisms P, P', with the* same *marginal distribution for X, thus $P_X = P'_X$, and (without loss of generality) with the same exceptional set. Then* (8) *holds: i.e., versions of $\mathrm{d}P'_{Y|X=x}/\mathrm{d}P_{Y|X=x}(y)$, $y \ni x$, can be chosen which only depend on y, for P_X-almost all x.*

Note that we do not assume any dominatedness, so Radon-Nikodym derivatives may be zero or infinite on non-null sets. With the natural conventions $1/0 = \infty$, $1/\infty = 0$, (8) is symmetric with respect to P and P': it does not make any difference which is placed in numerator and which in denominator.

Proof. We will prove the lemma by establishing that 'not (8)' implies '(7) cannot hold for both P and P''. This is equivalent to showing that if (8) is not true while P does satisfy (7), then P' does not satisfy (7).

Now the negation of (8) implies that for each of a P_X-positive set of points x one may find at least one, and possibly many, points x' with

$$\frac{\mathrm{d}P'_{Y|X=x}}{\mathrm{d}P_{Y|X=x}}(y) \neq \frac{\mathrm{d}P'_{Y|X=x'}}{\mathrm{d}P_{Y|X=x'}}(y)$$

on a $P_{Y|X=x}$- or a $P'_{Y|X=x}$-positive set of points y in $\{y \ni x\} \cap \{y \ni x'\}$. Also all the points x' so involved must together have positive P_X probability, for otherwise we could also simply put them in the exceptional set.

For each such pair (x, x'), either we must have '$<$' or '$>$' on either a $P_{Y|X=x}$- or a $P'_{Y|X=x}$-positive set of points y. The resulting four combinations define four (possibly overlapping) sets of pairs (x, x'). At least one of these four sets must involve both a P_X-positive set of points x and a P_X-positive set of points x': otherwise (8) is saved by simply augmenting the exceptional (but still null) set. Almost without loss of generality we suppose the surviving combination has '$<$'; more obviously, we must consider the cases of a $P_{Y|X=x}$-positive or a $P'_{Y|X=x}$-positive set of y separately.

Suppose first that for each of a P_X-positive set of points x one can find one or more points x', altogether also making up a P_X-positive set, with

$$\frac{\mathrm{d}P'_{Y|X=x}}{\mathrm{d}P_{Y|X=x}}(y) < \frac{\mathrm{d}P'_{Y|X=x'}}{\mathrm{d}P_{Y|X=x'}}(y) \tag{9}$$

on a $P_{Y|X=x}$-positive set of points in $\{y \ni x\} \cap \{y \ni x'\}$. Since we have strict inequality, the left-hand side is finite *everywhere* on this set. Integrating over the set with respect to $P_{Y|X=x}$, by (7) equal to $P_{Y|X=x'}$, gives (both integrals of course finite)

$$P'_{Y|X=x} < P'_{Y|X=x'} \tag{10}$$

on some set of points y, for a collection of pairs (x, x'), each coordinate covering a P_X-non null set. Thus (7) fails for P'. Though integrating '$+\infty$' over a null set may give a positive result, this can only happen on the right-hand side, not harming the inequality. Note that the argument does not depend on the direction of the inequality in (9) and (10), since by our assumption (7) for P one can interchange the roles of x and x' if desired.

Suppose on the other hand (9) holds, now with positive $P'_{Y|X=x}$ probability, for each of the usual collection of pairs (x, x'). Again integrate with respect to $P_{Y|X=x} = P_{Y|X=x'}$ over the indicated set. The integrand on the left-hand side must have been finite *everywhere* (since we had *strict* inequality) and the result is strictly positive since it is just the $P'_{Y|X=x}$ probability of the set over which we integrate. The set must therefore also have had $P_{Y|X=x} = P_{Y|X=x'}$ positive probability and hence the result on the right-hand side is strictly larger giving us (10) again; thus again (7) fails for P'.

This last argument relied on our having '<' rather than '>' in the inequality (9). With the reverse inequality, now the right-hand side is finite everywhere. If the $P_{Y|X=x} = P_{Y|X=x'}$ measure of our set of points y is zero, the right-hand side integrates to zero, which must also be its $P'_{Y|X=x'}$ measure, and consequently less than its a priori known positive $P'_{Y|X=x}$ measure. But if we start with positive $P_{Y|X=x}$ measure, then finiteness of the integrand on the right-hand side everywhere preserves the strict inequality on integration. In either case we get the desired (reversal) of (10); and again (7) fails for P'. □

We now want to establish a factorization of the marginal distribution of Y, FACTOR(Y), and a property of the conditional distribution of X given Y, CAR($X|Y$). For the latter, we would like to derive:

$$P_{X|Y=y}(\mathrm{d}x) \overset{\text{(def.)}}{=} \Pr(X \in \mathrm{d}x|Y=y) \;=\; \Pr(X \in \mathrm{d}x|X \in y) \tag{11}$$

However, the last expression here is *not* well-defined in general. If y is a singleton, or y is a set of positive probability for X, then we know how $\Pr(X \in \mathrm{d}x|X \in y)$ should be interpreted; in the other specific examples we also may be able to guess a reasonable definition. But the general set-up so far does not permit a unique interpretation (more on that later).

However, an important feature of (11) is that the very right hand side should be computable from the marginal distribution of X, without knowledge of the conditional distribution of Y given X (the coarsening mechanism); in fact, it should be the *same* whatever the distribution of Y given X. So just as when discussing factorization of a likelihood function, we consider a family of joint distributions of X and Y, each satisfying the probabilistic CAR assumption (7), and the property we derive is a property of the resulting *statistical model.*

To respect the distinction between the underlying variable of interest X and the coarsening mechanism leading to Y we suppose their joint distribution depends on variation independent parameters belonging separately to these two aspects, say θ and γ respectively;

$$P^{\theta,\gamma}_{X,Y}(\mathrm{d}x,\mathrm{d}y) = P^{\theta}_{X}(\mathrm{d}x)P^{\gamma}_{Y|X=x}(\mathrm{d}y). \tag{12}$$

We assume $P^{\gamma}_{Y|X=x}(\mathrm{d}y)$ satisfies CAR, (7), for each value of γ. Moreover we assume that $P^{\theta}_{X} \ll P^{\theta_0}_{X}$; and for $P^{\theta_0}_{X}$ almost all x, $P^{\gamma}_{Y|X=x} \ll P^{\gamma_0}_{Y|X=x}$.

Theorem. *Under CAR, the likelihood for θ, γ based on observation of Y factors:*

$$\frac{\mathrm{d}P_Y^{\theta,\gamma}}{\mathrm{d}P_Y^{\theta_0,\gamma_0}}(y) = \frac{\mathrm{d}P_{Y|X=x}^{\gamma}}{\mathrm{d}P_{Y|X=x}^{\gamma_0}}(y) \cdot E_{\theta_0,\gamma_0}\left(\frac{\mathrm{d}P_X^{\theta}}{\mathrm{d}P_X^{\theta_0}}(X) \,\middle|\, Y = y \right) \tag{13}$$

(for arbitrary $x \in y$ not in the null set for x, this being possible for $P_Y^{\theta_0,\gamma_0}$-almost all y). Moreover, $P_{X|Y=y}^{\theta,\gamma}(\mathrm{d}x)$ does not depend on γ.

Proof. Consider, for $x \in y$,

$$\begin{aligned} \frac{\mathrm{d}P_{X,Y}^{\theta,\gamma'}}{\mathrm{d}P_{X,Y}^{\theta,\gamma}}(x,y) &= \frac{\mathrm{d}P_X^{\theta}}{\mathrm{d}P_X^{\theta}}(x)\,\frac{\mathrm{d}P_{Y|X=x}^{\gamma'}}{\mathrm{d}P_{Y|X=x}^{\gamma}}(y) \\ &= 1 \cdot k(y;\gamma',\gamma) \end{aligned} \tag{14}$$

for some function k, by the Lemma.

Since the right hand side of (14) does not depend on x, we must have

$$\frac{\mathrm{d}P_Y^{\theta,\gamma'}}{\mathrm{d}P_Y^{\theta,\gamma}}(y) = k(y;\gamma',\gamma).$$

Thus

$$\begin{aligned} \frac{\mathrm{d}P_Y^{\theta,\gamma}}{\mathrm{d}P_Y^{\theta_0,\gamma_0}}(y) &= \frac{\mathrm{d}P_Y^{\theta,\gamma}}{\mathrm{d}P_Y^{\theta,\gamma_0}}(y)\,\frac{\mathrm{d}P_Y^{\theta,\gamma_0}}{\mathrm{d}P_Y^{\theta_0,\gamma_0}}(y) \\ &= k(y;\gamma,\gamma_0)\,E_{\theta_0,\gamma_0}\left(\frac{dP_{X,Y}^{\theta,\gamma_0}}{dP_{X,Y}^{\theta_0,\gamma_0}}(X,Y) \,\middle|\, Y = y \right) \\ &= k(y;\gamma,\gamma_0)\,E_{\theta_0,\gamma_0}\left(\frac{dP_X^{\theta}}{dP_X^{\theta_0}}(X) \,\middle|\, Y = y \right), \end{aligned}$$

which by (14) is the claimed factorization (13).

We also have

$$\frac{\mathrm{d}P_{X,Y}^{\theta,\gamma'}}{\mathrm{d}P_{X,Y}^{\theta,\gamma}}(x,y) = \frac{\mathrm{d}P_Y^{\theta,\gamma'}}{\mathrm{d}P_Y^{\theta,\gamma}}(y)\,\frac{\mathrm{d}P_{X|Y=y}^{\theta,\gamma'}}{\mathrm{d}P_{X|Y=y}^{\theta,\gamma}}(x)$$

hence

$$k(y;\gamma',\gamma) = k(y;\gamma',\gamma)\,\frac{\mathrm{d}P_{X|Y=y}^{\theta,\gamma'}}{\mathrm{d}P_{X|Y=y}^{\theta,\gamma}}(x).$$

Since $k(y;\gamma,\gamma_0) = (\mathrm{d}P_Y^{\theta,\gamma}/\mathrm{d}P_Y^{\theta,\gamma_0})(y)$, it is positive for $P_Y^{\theta,\gamma}$ almost all y. So for such y, and all $x \in y$,

$$\frac{\mathrm{d}P_{X|Y=y}^{\theta,\gamma}}{\mathrm{d}P_{X|Y=y}^{\theta,\gamma_0}}(x) = 1$$

or $P^{\theta,\gamma}_{X|Y=y} = P^{\theta,\gamma_0}_{X|Y=y}$ for $P^{\theta,\gamma}_Y$ almost all y. □

The theorem not only shows there is a factorization in the likelihood for θ, γ but also fairly explicitly tells what the two factors are.

The γ-part (CAR mechanism) is the same as the likelihood for γ based on the conditional distribution of Y given $X = x$, even though X itself is not observed, only Y. In other words, inference about the coarsening mechanism can be done as if X had also been observed.

The θ-part (underlying variable of interest) can be written down without knowing the CAR parameter γ: just pick any value, say γ_0, and compute the second factor of the right hand side of (13). It does seem that we do need to know the *structure* of the coarsening mechanism. However, even if we believe in a particularly complex mechanism—so a particular family $P^{\gamma}_{Y|X=x}$—we can calculate the likelihood from the second factor of (13) using a set of conditional distributions $P^*_{Y|X=x}$ *outside* this family, and perhaps of much simpler structure. We only need to have $P^{\gamma}_{Y|X=x} \ll P^*_{Y|X=x}$, in other words: the 'reference' coarsening mechanism $P^*_{Y|X=x}$ used in the calculations can generate all the sets y which can occur 'in reality'.

Put another way: the likelihood for θ, under CAR, is the same as under any specific CAR mechanism which, given $X = x$, can generate the same (or more) random sets. We illustrate this and further points with a succession of examples. Examples 1 and 2 concern censored data (univariate and multivariate respectively). Example 3 is a classical paradox from the theory of conditional distributions. We concentrate on the main features of the examples, leaving more technical details to the interested reader.

Example 1. *Univariate right censoring.* Suppose $X \in [0,\infty)$ and $\mathcal{X} = \alpha(Y)$ is either a singleton $\{x\}$ or a half-line $(\widetilde{x},\infty)$. The data Y is traditionally the pair $(\widetilde{X},\Delta)$ where $\widetilde{X}$ takes values in $[0,\infty)$, Δ in $\{0,1\}$, and $\Delta = 1$ if $\mathcal{X} = \{\widetilde{X}\} = \{X\}$, $\Delta = 0$ if $\mathcal{X} = (\widetilde{X},\infty) \ni X$.

Suppose such data arose from a CAR coarsening of X. The data is monotone and we expect to be able to retrieve the coarsening mechanism and the underlying law of X fairly explicitly from the law of the data. We also expect that such data in general fits a CAR model. We will see that for practical purposes, that is true. Every outcome is either a singleton or a set of positive probability for X, so we can guess in advance the likelihood for the interest part of the model.

The CAR property is that for any $x < x'$,

$$\text{law}((\widetilde{X},\Delta)|X = x) = \text{law}((\widetilde{X},\Delta)|X = x')$$

restricted to values of $(\widetilde{X},\Delta)$ such that $\mathcal{X}$ contains both x and x'. This means $\Delta = 0$ and $\widetilde{X} < x$. So CAR is equivalently the assumption $\Pr\{\widetilde{X} \in d\widetilde{x}, \Delta = 0|X = x\}$ does not depend on x, for $\widetilde{x} < x$. We may therefore define $G(t) = \Pr\{\widetilde{X} \le t, \Delta = 0|X = x\}$ for arbitrary $x > t$ and G is a (sub)-distribution function defined at least up to the right hand end-point of the support of $\widetilde{X}$. Letting t increase up to x from below, we obtain $G(x-) = \Pr\{\Delta = 0|X = x\}$ and hence $1 - G(x-) = \Pr\{\Delta = 1|X = x\}$. We now have described the law of the coarsened data given X completely in terms of a distribution G. It could be generated on a computer by drawing a random variable C from the distribution G, independently of X, and reporting $\Delta = 1, \widetilde{X} = X$ if $C \geq X$, $\Delta = 0, \widetilde{X} = C$ if $C < X$.

Suppose X has distribution function F and survival function S. We compute the marginal distribution of the data as $\Pr\{\widetilde{X} \in \mathrm{d}\widetilde{x}, \Delta = 0\} = G(\mathrm{d}\widetilde{x})(1 - F(\widetilde{x}))$, $\Pr\{\widetilde{X} \in \mathrm{d}\widetilde{x}, \Delta = 1\} = F(\mathrm{d}\widetilde{x})(1 - G(\widetilde{x}-))$. Moreover $\Pr\{\widetilde{X} \geq \widetilde{x}\} = (1 - F(\widetilde{x}-))(1 - G(\widetilde{x}-))$ so that the hazard measure of X can be calculated from the distribution of the data as $\Lambda(\mathrm{d}x) = F(\mathrm{d}x)/(1 - F(x-)) = \Pr\{\widetilde{X} \in \mathrm{d}x, \Delta = 1\}/\Pr\{\widetilde{X} \geq x\}$, at least, for x such that $\Pr\{\widetilde{X} \geq x\} > 0$. Product-integration of the hazard yields the survival function $S = \prod(1 - \mathrm{d}\Lambda)$, for the values of x just mentioned. Once S has been recovered we can also recover G from the distribution of the data.

Even if we did not know that $\mathcal{X}$ was CAR, we could use these calculations to produce a CAR model exactly fitting the data. This would succeed unless the calculations implied a defective distribution of X, suggesting that $X = \infty$ has positive probability. For instance, if in reality the data was generated by a censoring time C identically equal to one minus the failure time X, and arbitrary large values of X can occur, then knowing that X is actually finite we could conclude from the distribution of the data that the coarsening was not CAR. □

Example 2. *Multivariate right censoring.* Suppose $X = (X_1, \ldots, X_k)$ and the possible realisations of $\mathcal{X} = \alpha(Y)$ are Cartesian products of singletons $\{x_i\}$ and half-lines (x_i, ∞). Such sets are generated by the *random-censoring* model: $C = (C_1, \ldots, C_k)$ is independent of X and for each i we observe X_i if $X_i \leq C_i$, C_i if $X_i > C_i$. Write $\widetilde{X}_i = X_i \wedge C_i$, $\Delta_i = 1\{X_i \leq C_i\}$.

It is easily checked that this specific model is CAR. For a point $(\widetilde{x}, \delta)$ and vector x let $x_\delta = (x_i : \delta_i = 1)$, $x_{\overline{\delta}} = (x_i : \delta_i = 0)$. The distribution of the data $\Pr(\widetilde{X} \in \mathrm{d}\widetilde{x}, \Delta = \delta)$ can be expressed as

$$P_{X_\delta}(\mathrm{d}\widetilde{x}_\delta) P_{X_{\overline{\delta}}|X_\delta = \widetilde{x}_\delta}((\widetilde{x}_{\overline{\delta}}, \infty_{\overline{\delta}})) \cdot P_{C_{\overline{\delta}}}(\mathrm{d}\widetilde{x}_{\overline{\delta}}) P_{C_\delta|C_{\overline{\delta}} = \widetilde{x}_{\overline{\delta}}}([\widetilde{x}_\delta, \infty_\delta)).$$

Intuitively, for a particular value of the vector of censoring indicators δ, we first write down the probability for the exactly observed variables to take values in tiny intervals around those values, and multiply by the conditional probability for the other variables to exceed the relevant values. We use the independence of X and C to further split each term in two. Finally we regroup to exhibit the CAR factorization.

Therefore under *any* CAR model producing the same sets the likelihood for parameters of the distribution of X is

$$P_{X_\delta}(\mathrm{d}\widetilde{x}_\delta) P_{X_{\overline{\delta}}|X_\delta = \widetilde{x}_\delta}((\widetilde{x}_{\overline{\delta}}, \infty_{\overline{\delta}})).$$

This example allows sets y having probability zero which are not singletons. One must check that any CAR mechanism which produces right-censored data is dominated by some random censorship model, as required to apply the theorem. □

Example 3. *Borel's paradox.* This is a famous example of two different ways in which one could observe that a point, uniformly distributed on the surface of a sphere, actually lies on a particular great circle. The two data generating mechanisms are such that the conditional distribution of the point, given it lies on the circle, is uniform in the one case, and non-uniform in the other case. The example can be 'fixed' by changing conditional distributions on an event of probability zero, but it is also a genuine example of the fact that the law of X given $X \in y$ cannot be talked about without reference to a 'reference model' generating the same random sets y as occur in the CAR model under consideration.

Suppose X is uniformly distributed on the surface of the unit sphere, and let Θ be its longitude $\in [0, 2\pi)$ and Φ its latitude $\in [-\pi/2, \pi/2]$. If $\Phi = -\pi/2$ or $+\pi/2$ (South or North pole) then Θ can be defined arbitrarily: this case has probability zero anyway.

One easily computes that Θ and Φ are independent, Θ is uniformly distributed while Φ has density $\frac{1}{2}\cos\phi$. Consequently, given $\Theta \bmod \pi = \theta$, the point X is distributed on the great circle through the poles on longitudes θ and $\theta + \pi$ with probability $\frac{1}{2}$ to be on each side of the globe, and its latitude having density $\frac{1}{2}\cos\phi$. On the other hand, given $\Phi = \phi$, the point X is uniformly distributed on the circle of constant latitude ϕ.

Taking $Y = \Theta \bmod \pi$ or $Y = \Phi$ is in both cases a coarsening at random of X. In the second case it is possible that $\Phi = 0$, conditional on which X is uniformly distributed on the equator. In the first case it is possible that $\Theta \bmod \pi = 0$ and then X is *non*-uniformly distributed on the great circle through North pole and Greenwich (England).

We took the coordinate system as being the same in the two cases. But we could have used different coordinate systems, so that the equator in one case was the same as the Greenwich meridian in the other. Then we have two CAR coarsenings which can produce exactly the same set y, but such that the distribution of X given $X \in y$ depends on which coarsening was involved.

There is no conflict with our main theorem. The theorem tells us that two coarsening at random mechanisms *which produce the same sets* have the same conditional distributions of X given the set. In the present example the sets produced by the two coarsening mechanisms are completely disjoint except for the single case of the equator which has zero probability under both mechanisms. □

If y is a singleton, then the distribution of X given $Y = y$ is degenerate at this point, so (13) gives the usual 'complete data' likelihood for θ. Suppose on the other hand sets y can occur with positive P_X probability. Construct a reference CAR coarsening model by choosing one of the sets at random, independently of X, and observing whether or not X lies in the set. For this model, $P_{X|Y=y} = P_{X|X\in y}$ and (13) gives the 'right' answer $P_X^\theta(y)$.

Our theorem satisfactorily shows that our general notion of CAR($Y|X$) has the required consequences CAR($X|Y$) and FACTOR(Y). It is an open question as to whether (and how) these can be made actually *equivalent* to CAR($Y|X$). A special case in which that can be shown, generalizing the discrete case with which we started, is when the distributions of Y given $X = x$ are dominated (over x) by a single, σ-finite, measure. One may check that it then conversely holds that the distributions of X given Y are also dominated; in fact, DOM($Y|X$) $\iff$ DOM($X|Y$), and under this condition, CAR($Y|X$) $\iff$ CAR($X|Y$). However this special case hardly has interesting applications beyond the discrete case.

Our definition of CAR($Y|X$) was an absolute or probabilistic definition for each coarsening mechanism in the model separately. In the lemma we derived a relative or statistical consequence concerning the likelihood ratios between different coarsening mechanisms, and that was all we used in our theorem. The theorem did not show *what* the factorisation was, nor *what* is the, for each θ fixed in γ, distribution of X given $Y = y$. In order to formulate necessary and sufficient conditions for CAR we must further specify these ingredients.

As we saw, only CAR(REL) and not CAR(ABS) was needed to prove the theorem. It is easy to show that if a statistical model satisfies CAR(REL) (with respect to γ, for each θ), and one point in it satisfies CAR(ABS) (for each θ), then so do the rest. One may say: if CAR(REL) holds, then CAR(ABS) either holds at all points in the model or none. It is possible that interesting statistical models can be found which satisfy CAR(REL) without CAR(ABS) holding anywhere. In fact CAR(REL) is simply a classical sufficency condition: assuming domination, it is the factorization criterion, in the model when *both* X and Y are observed, for Y to be sufficient for γ for each fixed θ. Consequently (and equivalently) we have sufficiency according to the definition in terms of conditional distributions: for each θ, the distribution of X given $Y = y$ does not depend on γ. This is just the second part of our theorem. (See Chang and Pollard (1997) for a modern proof of this equivalence). The first part of our theorem is the classical result that the likelihood function based on the sufficient statistic is the same as the likelihood function based on the original data.

7. CAR according to Jacobsen and Keiding

Jacobsen and Keiding (1995) have a somewhat different definition of CAR in general sample spaces. Their definition assumes much more structure on X and Y, which allows rather concrete representations of the various conditional distributions of interest, without making our regularity conditions. Their definition of CAR becomes less transparent since it is stated in terms of a density with respect to a particular reference experiment. Their conclusions are partly more strong, since more explicit, but on the other hand do not reveal so explicitly as ours the practical interpretation and calculation of the factors in the likelihood. We summarize their results below, first describing the main features of their set-up. Further analysis of the difference between our and their approach is given by Nielsen (1996).

Suppose besides X there is a non-observable random variable G generating the coarsening of X. Thus for each value g of G there is a partition of the sample space, and we observe the element Y of the partition in which X lies. G may be dependent on X. The coarsening mechanism is described by the distribution of G given $X = x$, for each x, and the partition generated by $G = g$, for each g. Write $Y = \Phi(X, G)$; because of the partitioning structure we have, for every $y = \Phi(x, g)$ for some x, g, that $x \in y$ and, if also $x' \in y$, then $\Phi(x', g) = y$ too. In fact $\{(x, g) : \Phi(x, g) = y\} = y \times \{g : \Phi(x, g) = y \text{ for some } x\}$.

Jacobsen and Keiding assume that there exists a reference model, which we shall call P^*, under which: X and G are *independent*, X with distribution μ, G with distribution ν. (They give P^* the name ρ.) Then they consider families of distributions $P^{\theta,\gamma}$ such that X has marginal distribution $P^\theta \ll \mu$, and, for each x, $G|X = x$ has distribution $P^\gamma_{G|X=x} \ll \nu$. Call the corresponding densities $f(x; \theta)$ and $h(g; x, \gamma)$. Define

$$k(y; x, \gamma) = E^*(h(G; x, \gamma)|Y(x, G) = y);$$

i.e., write the conditional expectation $E^*(h(G; x, \gamma)|Y(x, G))$ as a function of $Y(x, G)$. They note that k is the density of the conditional $P^{\theta,\gamma}$ distribution of Y given $X = x$, with respect to its distribution under P^*. The P^*-independence of X and G plays a crucial role in these calculations. Finally they define CAR as: $k(y; x, \gamma)$ does not depend on $x \in y$, for each γ. They prove that under CAR, the likelihood based on Y, for θ and γ, factors, and the θ part is $E^*(f(X; \theta)|Y = y)$. When y is a singleton, or a set of positive μ-probability, the likelihood becomes as one would hope $f(X; \theta)$ or $P^\theta_X(y)$ respectively.

It can be checked that if both Jacobsen and Keiding's structure, and our regularity conditions, are present, then CAR according to Jacobsen and Keiding implies CAR according to us, thereby giving more interpretability to their condition and giving further conclusions. In the terminology of the last section, their 'reference model' satistifies CAR(ABS), while the rest of the model satisfies CAR(REL). We do not know if their set-up is essentially more restrictive than ours. Starting from a given law of (X, Y) as we do, one might take for G a collection of uniform random variables used in a computer simulation of the law of Y given X, and let Φ be the deterministic part of the algorithm. However then different CAR mechanisms will correspond to different functions Φ, not to different distributions of G.

Jacobsen and Keiding's definition of CAR works for a given statistical model and is relative to a specific 'reference model'. In fact, the reference model P^* itself satisfies CAR according to their own definitions (take the densities f and h identically equal to 1, then also k is identically equal to 1). Their reference model is also CAR according to our, absolute (non-relative) definition.

One could say that Jacobsen and Keiding define CAR precisely through *assuming* the factorization holds of likelihoods, *with respect* to a specific reference model, which itself in our broader sense is CAR. Their results therefore do give a context in which it is true that CAR (absolutely—in terms of $P_{Y|X=x}$) and the factorization are *equivalent*, certainly a very nice theorem to have. This works by having a reference model through which it is defined what the θ part of the likelihood factorization should be, and how the distribution of X given $X \in y$ should be defined.

8. More data

In many applications the coarsening mechanism depends on an underlying random variable G which may be observed, or partially observed, along with the coarsening of X. For instance in survival analysis, potential censoring times are sometimes known even for uncensored observations.

Let us represent the data by some random variable Y. Suppose there is a function α from the sample space for Y to the non-empty subsets of E such that $\mathcal{X} = \alpha(Y)$ is a coarsening of X: thus $X \in \alpha(Y)$ with probability one. Typically we will have $Y = \Phi(X, G)$ for some known function Φ, and $\alpha(y) = \{x : \exists g \text{ with } \Phi(x, g) = y\}$.

The only difference with the set-up of Section 6 is that we do not suppose the function α is one-to-one: two different points y, y' could give the same functional information about X, namely $X \in \alpha(y) = \alpha(y')$. So Y is not just a parametrisation of $\mathcal{X}$.

The coarsening mechanism is fixed by describing the conditional laws of Y given $X = x$ together with the function α. We will define Coarsening at Random in terms of these two ingredients. If actually $Y = \Phi(X, G)$ for some Φ, G, and corresponding α, the CAR property can be rephrased as a more clumsy looking condition on the laws of G given $X = x$ and the function Φ.

In Section 6, Y was thought of as a convenient encoding of the observed coarsening of X, $\mathcal{X} = \alpha(Y)$. But the implicit fact that α is one-to-one was not used anywhere at all! Therefore we may *maintain the definition* (7) of CAR($Y|X$) in our new context, and *all the results of Section 6 remain valid*. Everywhere, the statement '$x \in y$' should just be read as shorthand for '$x \in \alpha(y)$'.

The discrete-case results of Section 1 can also be copied. Let X, Y be discrete random variables and $\mathcal{X} = \alpha(Y)$ a coarsening of X. Our definition of coarsening at random becomes:

$$\text{CAR}(Y|X): \quad \Pr\{Y = y \mid X = x\} \quad \text{does not depend on } x \in \alpha(y).$$

Consequently $\Pr\{Y = y \mid X = x\} = \Pr\{Y = y \mid X \in \alpha(y)\}$ for $x \in \alpha(y)$;

of course $\Pr\{Y = y \mid X = x\} = 0$ for $x \notin \alpha(y)$.

We compute the marginal distribution of Y: it is

$$\begin{aligned} \Pr\{Y = y\} &= \sum_{x \in \alpha(y)} \Pr\{Y = y \mid X = x\} \Pr\{X = x\} \\ &= \Pr\{Y = y \mid X = x\} \Pr\{X \in \alpha(y)\} \end{aligned}$$

for arbitrary $x \in \alpha(y)$; this is the factorisation property FACTOR(Y).

Finally, for $x \in \alpha(y)$,

$$\begin{aligned} \Pr\{X = x \mid Y = y\} &= \frac{\Pr\{X = x\} \Pr\{Y = y \mid X = x\}}{\Pr\{X \in \alpha(y)\} \Pr\{Y = y \mid X = x\}} \\ &= \Pr\{X = x \mid X \in \alpha(y)\}, \end{aligned}$$

so CAR($X|Y$) holds.

Sometimes one has in mind a model $Y = \Phi(X, G)$ involving a specific grouping or censoring variable G. Take

$$\alpha(y) = \{x : \exists g, \Phi(x, g) = y\}.$$

CAR($Y|X$) (in the discrete case) is immediately rewritten as

$$\Pr\{\Phi(x, G) = y \mid X = x\} \quad \text{does not depend on } x \in \alpha(y).$$

This is hard to rephrase in a more attractive way (without reverting to Y) and hard as it stands to generalise to arbitrary sample spaces. A little progress can be made in an important special case when Φ is *Cartesian* by which we mean that $\Phi^{-1}(y) = \{(x, g) : \Phi(x, g) = y\}$ is a Cartesian product, say $\Phi_X^{-1}(y) \times \Phi_G^{-1}(y)$. The data Y is equivalent to simultaneous coarsenings $\mathcal{X}$ and $\mathcal{G}$ of X and G. In particular, $\mathcal{X} = \alpha(y) = \Phi_X^{-1}(y)$. Then

$$\Pr\{\Phi(x, G) = y \mid X = x\} = \Pr\{G \in \Phi_G^{-1}(y) \mid X = x\}.$$

We see from this that under Cartesian coarsening, CAR is *implied* by the assumption that $\Pr\{G = g \mid X = x\}$ does not depend on $x \in \Phi_X^{-1}(y)$, for each $g \in \Phi_G^{-1}(y)$. This more pleasant looking sufficient condition, which one could call CAR($G|X$), could be reformulated in general sample spaces analogously to (7). If the coarsening is both Cartesian and CAR and moreover G is observed exactly—i.e., $\Phi_G^{-1}(y)$ is always a singleton—then the condition is also necessary.

On the other hand, it is easy to construct toy examples where CAR($Y|X$) is true but the coarsening is not Cartesian or CAR($G|X$) does not hold.

Under Cartesian coarsening, one can pretend that G is actually completely observed, as follows. The data Y is equivalent to coarsenings $\mathcal{X}, \mathcal{G}$ of X and G. Define G^* to be a function of Y which picks out in a deterministic, measurable, way an arbitrary element g^* of $\mathcal{G}$. Now $Y^* = (\mathcal{X}, G^*)$ is not only a function of Y, but actually equivalent to the data Y since one can recompute $Y = \Phi(X, G) = \Phi(x, G^*)$ by picking any x in $\mathcal{X}$. This trick is sometimes useful in calculations, because CAR($G^*|X$) does now hold, even if CAR($G|X$) didn't; see Robins and Rotnitzky (1992), Robins (1996a).

Nielsen (1996), following Jacobsen and Keiding (1995) has interesting further results in this area.

We conclude this section with a discussion of the quizmaster problem, mentioned at the end of section 3. Let us avoid modelling the psychology of the quizmaster and the player (us) by an initial randomization by which the three doors get secretly relabelled by us with the numbers $1, 2, 3$ uniformly and independently of the location of the car behind one of the doors. We choose one of the numbers at random and indicate that door. The quizmaster opens one of the other doors and shows that there is no car hidden behind it. (It is an important model assumption that he *always* does this; of course, he always *can*). He asks us to reconsider our choice.

Because of our initial randomisation, the location of the car X is uniformly distributed over $\{1, 2, 3\}$. Our choice of door is also independently and uniformly distributed. Given our choice of door and the location of the car, the quizmaster either has chosen one of the two possible doors with equal probability if we are right, and has made the only possible choice if we are wrong. Represent our data Y with the ordered pair whose first component is the number of the door we chose and whose second is the number of the remaining closed door. The coarsening $\alpha(Y)$ of X is the set whose unordered elements are the components of the ordered pair Y. We know that the car lies behind one of these two doors. It is easy to calculate that conditional on the data, the car is twice as likely to lie behind the other closed door than the door we chose. The law of X given $Y = y$ is therefore different from the law of X given $X \in \alpha(y)$, so this data cannot have been coarsened at random. If however we reduce the data to the coarsened value of X—for example $(1, 2)$ and $(2, 1)$ are reduced to $\{1, 2\}$—then we do have CAR, as is also easy to verify!

This perhaps explains why most people on first being told this story, say that the information given by the quizmaster is no use and that one might just as well keep to the same door. The coarsening of X on its own was coarsening at random, so not informative of the value of X. But the complete data is is not CAR, and tells us something nontrivial about X.

9. Locally, CAR is everything

We saw that 'car is everything' breaks down, strictly speaking, in general sample spaces. In this section we show that something very close is true, in a sense connected to asymptotic efficiency theory. Starting from a given CAR model, and then allowing the CAR mechanism to vary arbitrarily and the underlying distribution to vary arbitrarily, together allow the distribution of the data to vary away from the starting point in any direction whatsoever, just as if we had assumed nothing at all from the start. This has important consequences in asymptotic optimality theory. If we assume just that the coarsening is CAR, and assume nothing about the distribution of the variable of interest, all regular estimators of the underlying distribution of interest are asymptotically equivalent. Esentially there is only one estimator available: the NPMLE. Other (more technical) consequences are given below.

Suppose we have one observation of a random vector X and assume nothing whatsoever about its distribution. If X actually has distribution P^0 then for every bounded function h of X, such that $E^0 h(X) = 0$, $P^{\theta,h}$ defined by $P^{\theta,h}(\mathrm{d}x) = (1+\theta h(x))P^0(\mathrm{d}x)$ is for small enough $|\theta|$ also a probability distribution of X. In fact $(P^{\theta,h} : |\theta| \leq \varepsilon)$ is a one-dimensional parametric submodel for X with score function, at $\theta = 0$, equal to $h(X)$. The *tangent space* is by definition the closure (in the $\mathcal{L}^2(P^0)$ sense) of the linear span of all score-functions, at P^0, of regular one-dimensional parametric submodels for the distribution of X, passing through P^0. We see that if we assume nothing about X, then the tangent-space at P_0 is $\mathcal{L}_0^2(P^0)$, the space of all square-integrable, mean-zero functions of $X \sim P^0$. We also write $\mathcal{L}_0^2(X)$ for the same space when the distribution of X under which we work is clear from the context.

In fact, any element of $\mathcal{L}_0^2(X)$, not just the bounded ones, are score-functions of submodels: define alternatively

$$P^{\theta,h}(\mathrm{d}x) = \Big(1+\tfrac{1}{2}\theta h(x)\Big)^2 P^0(\mathrm{d}x) \,/\, \Big(1+\tfrac{1}{4}\theta^2 E^0(h(X)^2)\Big).$$

The tangent-space plays a central role in the theory of semi-parametric models. In particular, the asymptotic Cramèr-Rao lower bound for estimation of functionals of the distribution of X based on i.i.d. replicates is calculated via a calculation of the tangent space. The larger the tangent space, the harder is estimation and the larger is the Cramèr-Rao bound. As we have just seen, assuming nothing about the distribution of X leads to the largest possible tangent space: $\mathcal{L}_0^2(X)$.

Suppose now Y is a coarsening of X satisfying the CAR assumption. Our model for Y is built up of a model for P_X, the distribution of X, and for $P_{Y|X=x}$, the family of distributions of Y given X. We show here that: *if nothing is assumed about P_X, and nothing is assumed about $P_{Y|X=x}$ beyond the CAR assumption, then the tangent space at a particular point P_Y in the resulting model for the distribution of Y is $\mathcal{L}_0^2(Y)$*. Locally, we are not assuming anything about the distribution of Y.

Let P_X, $P_{Y|X=x}$ be given, the latter satisfying CAR, and define for given functions $h(x)$ and $k(y;x)$

$$\begin{aligned} P_X^\theta(\mathrm{d}x) &= \big(1+\theta h(x)\big)P_X(\mathrm{d}x) \\ P_{Y|X=x}^\gamma(\mathrm{d}y) &= \big(1+\gamma k(y;x)\big)P_{Y|X=x}(\mathrm{d}y). \end{aligned}$$

If h is bounded and $E(h(X)) = 0$ this defines a one-dimensional parametric submodel for the distribution of X with parameter θ (sufficiently close to zero). Similarly if k is bounded and $E\big(k(Y;x) \mid X = x\big) = 0$ we have a model for the distribution of Y given X with parameter γ. In order that the CAR assumption holds under $P^{\theta,\gamma}$ we require that $k(y;x)$, $y \ni x$, does not depend on x; so in fact $k(y;x) = k(y)$.

If we had observed X and Y the score functions (at $\theta = 0$, $\gamma = 0$) for θ and γ would have been $h(X)$ and $k(Y)$ respectively. When we observe only Y, the score functions are transformed to their conditional expectations given Y : $E(h(X)|Y)$ and $k(Y)$ respectively; see Gill(1989, §3, Ex. 2) for a heuristic derivation of this result and Bickel, Klaassen, Ritov and Wellner (1993, Prop. A5.5) for a rigorous one. (We refer in the sequel to this work as BKRW.)

Write E^X, E^Y for conditional expectation operators given X and Y respectively, considered as mappings on the following Hilbert spaces:

$$E^X : \mathcal{L}_0^2(Y) \to \mathcal{L}_0^2(X)$$
$$E^Y : \mathcal{L}_0^2(X) \to \mathcal{L}_0^2(Y).$$

Write $\|\cdot\|_X$, $\langle\cdot,\cdot\rangle_X$ etc. for the corresponding norms and inner products. E^X and E^Y are one-another's adjoint: writing $A = E^Y$ and defining $A^\top$ by $\langle g, Ah\rangle_Y = \langle A^\top g, h\rangle_X$ for all $g \in \mathcal{L}_0^2(Y)$ and $h \in \mathcal{L}_0^2(X)$ we have:

$$\begin{array}{rclclcl} \langle g, Ah\rangle_Y & = & \langle g, E^Y h\rangle_Y & = & E(g(Y)E(h(X)|Y)) & = & E(g(Y)h(X)) \\ & & & = & E(E(g(Y)|X)h(X)) & = & \langle E^X g, h\rangle_X \end{array}$$

proving that if $A = E^Y$, then $A^\top = E^X$.

We have shown that for each bounded function h of X, with mean zero, $E^Y h$ is a score function of a one-dimensional parametric submodel for the distribution of X. Similarly, for each bounded function $k(Y)$, with conditional mean given X zero, k is a score function of a one-dimensional parametric CAR submodel for the distribution of Y given X. Since taking $\theta \equiv \gamma$ gives a score function equal to the sum of the scores for θ and γ separately, we find by taking closures that our tangent space based on observation of Y contains $\overline{\mathcal{R}(A)} + \mathcal{N}(A^\top)$ where $\mathcal{R}$ and $\mathcal{N}$ denote range and null-space respectively. However, it is a well-known (and easily proved) fact from the theory of Hilbert spaces that for any bounded linear operator A from one Hilbert space H to another H', $\overline{\mathcal{R}(A)} + \mathcal{N}(A^\top)$ is a decomposition of the range space H' into two orthogonal components: for suppose g is orthogonal to $\overline{\mathcal{R}(A)}$. Then $\langle g, Ah\rangle_{H'} = 0$ for all h, thus $\langle A^\top g, h\rangle_H = 0$ for all h, thus $A^\top g = 0$ or $g \in \mathcal{N}(A^\top)$. So the tangent space is $\overline{\mathcal{R}(A)} + \mathcal{N}(A^\top) = \mathcal{L}_0^2(Y)$, the largest possible tangent space, corresponding globally to making no assumptions whatever on the distribution of Y.

We next prove, under an assumption concerning the probability to get a *complete observation*, $y = \{x\}$, *that the distribution of* X *is locally identified* under the completely nonparametric CAR model described above. The result was already given in van der Laan (1993, Lemma 3.3) and Robins and Rotnitzky (1992).

Under CAR, we obtained the factorization (13):

$$\frac{\mathrm{d}P_Y^{\theta,\gamma}}{\mathrm{d}P_Y^{\theta_0,\gamma_0}}(y) = \frac{\mathrm{d}P_{Y|X=x}^{\gamma}}{\mathrm{d}P_{Y|X=x}^{\gamma_0}}(y) \cdot E_{\theta_0,\gamma_0}\left(\frac{\mathrm{d}P_X^{\theta}}{\mathrm{d}P_X^{\theta_0}}(x) \,\middle|\, Y = y\right),$$

where the first factor depends only on y (not on $x \in y$). Fixing $\theta = \theta_0$ we see that the space of score functions of one-dimensional parametric submodels for the coarsening mechanism $P_{Y|X=x}$ not only contains but is actually exactly equal to $\mathcal{N}(A^\top)$, the space of zero-mean, square integrable functions of Y with conditional mean given X identically zero. Similarly, the space of score-functions of one-dimensional parametric submodels for the distribution of interest P_X is exactly equal to $\overline{\mathcal{R}(A)}$. Since $\overline{\mathcal{R}(A)}$ and $\mathcal{N}(A^\top)$ are orthogonal, we find from the theory of semiparametric models that the asymptotic Cramèr-Rao lower bound for estimation of functionals of P_X is the same when $P_{Y|X=x}$ is known and fixed, as when it is completely unknown (subject in both cases to CAR). Suppose we want to estimate $\kappa(P_X) = \int \kappa(x) P_X(\mathrm{d}x)$ for some bounded function $\kappa(x)$, e.g. $\kappa(x) = 1_A(x)$ corresponding to estimation of $\Pr(X \in A)$ for a given set A. Define $\widetilde{\kappa} = \kappa - E_X(\kappa)$. Then by BKRW or by van der Vaart (1991), we have: if $I = A^\top A$ has (at $\widetilde{\kappa}$) an inverse I^{-1}, then the asymptotic information bound for estimation of κ is

$$\|A(A^\top A)^{-}\widetilde{\kappa}\|^2 < \infty.$$

In fact $g(X) = (A^\top A)^{-}\widetilde{\kappa}(X)$ generates a 'hardest' one-dimensional submodel for estimating κ at P_X (maximizes the Cramèr-Rao bound over all parametric submodels). A slightly weaker condition for a finite asymptotic information bound is just that $\widetilde{\kappa}$ lies in the range of $A^\top$; this is obviously implied by $\widetilde{\kappa} = A^\top A g$ for some g. We will later argue that a finite information bound means in some sense local identifiability. But first the result:

Theorem. *(van der Laan, 1993). Suppose for each x, $P_{Y|X=x}(\{x\}) \geq \delta > 0$; i.e., the conditional probability of a complete observation is bounded away from zero. Then $I = A^\top A : \mathcal{L}_0^2(X) \to \mathcal{L}_0^2(X)$ is onto and has a bounded inverse; in fact $\|I^{-1}h\| \leq \delta^{-1/2}\|h\|$ for all h. Consequently $\|A(A^\top A)^{-1}\widetilde{\kappa}\|^2 \leq \delta^{-1}\|\widetilde{\kappa}\|$ or: the information bound for estimating κ based on Y is not more than $1/\delta$ times its bound based on observing X.*

Proof. The argument is based on van der Laan (1993; Lemma 2.2 and Lemma 3.3) with a minor supplement. To start with (cf. Lemma 3.3, van der Laan 1993), consider

$$\begin{aligned}
\|Ah\|^2 &= E\Big(E(h(X) \mid Y)^2\Big) \\
&\geq E\Big(h(X)^2\, 1\{Y = \{X\}\}\Big) \\
&= E\Big(h(X)^2 \Pr(Y = \{X\} \mid X)\Big) \;\geq\; \delta\|h\|^2.
\end{aligned}$$

So $0 < \delta \le \|A\|^2 \le 1$ and, if $\|h\| = 1$,

$$\begin{aligned} \|A^\top Ah\| &= \|A^\top Ah\|\,\|h\| \\ &\ge \langle A^\top Ah, h\rangle \qquad \text{(by Cauchy-Schwartz)} \\ &= \|Ah\|^2 \ \ge\ \delta. \end{aligned}$$

Thus for any h, $\|A^\top Ah\| \ge \delta\|h\|$. This shows in particular that $A^\top A$ is 1–1 since, if $A^\top Ah = A^\top Ah'$, then $A^\top A(h-h') = 0$ and $\delta\|h-h'\| \le \|A^\top A(h-h')\| = 0$, implying $h = h'$. Now (following van der Laan's Lemma 2.2) let us consider the operator $1 - A^\top A$, where 1 is the identity. This operator is self-adjoint. It is also bounded, since A, $A^\top$ and 1 are bounded. Therefore from Hilbert space theory (see, e.g., Kress, 1989, Theorem 15.9),

$$\begin{aligned} \|1 - A^\top A\| &= \sup_{h:\|h\|=1} |\langle h, (1 - A^\top A)h\rangle| \\ &= \sup_{h:\|h\|=1} |1 - \|Ah\|^2| \ \le\ 1 - \delta \ <\ 1. \end{aligned}$$

Consequently we have that $(A^\top A)^{-1} = (1 - (1 - A^\top A))^{-1}$ exists and is in fact given by $\sum_{n=0}^{\infty}(1 - A^\top A)^n$. The squared norm of the inverse is bounded by $(\sum_{n=0}^{\infty}(1-\delta)^n) = \delta^{-1}$. □

Remark. If we know $P_{Y|X=x}$ and moreover $P_{Y|X=x}(\{x\}) \ge \delta > 0$ for all x, one could estimate $\kappa(P_X) = \int \kappa dP_X$ based on n observations of Y by

$$\frac{1}{n}\sum_{i=1}^{n} \frac{1\{Y_i = \{X_i\}\}}{P_{Y|X=x}(\{x\})|_{x=X_i}}\kappa(X_i).$$

This estimator is unbiased and its variance is easily seen not to exceed $\|\widetilde{\kappa}\|^2/(n\delta)$. This shows directly that the information bound for estimation of κ is finite and not more than $(1/\delta)$ times the bound when X is observed, when $P_{Y|X=x}$ is known. By orthogonality the same bound applies even when $P_{Y|X=x}$ is unknown.

Now we discuss the interpretation of this result as a kind of local identifiability. Suppose we have n i.i.d. observations Y_i of Y and consider any parametric model $P_Y^{\theta,\gamma}$ constructed from P_X^θ and $P_{Y|X=x}^\gamma$. Consider the local models $\theta = \theta_0 + n^{-\frac{1}{2}}\eta$, $\gamma = \gamma_0 + n^{-\frac{1}{2}}\psi$. Define the optimal influence curve $\mathrm{IC}_{\mathrm{opt}} = A(A^\top A)^{-1}\widetilde{\kappa}$, working at the point $\theta = \theta_0$, $\gamma = \gamma_0$. Then

$$\widehat{\kappa} = \kappa(P_X^{\theta_0}) + \frac{1}{n}\sum_{i=1}^{n} \mathrm{IC}_{\mathrm{opt}}(Y_i)$$

is an estimator of $\kappa(P_X^\theta)$ based on $Y_1, \ldots, Y_n$ such that $n^{1/2}(\widehat{\kappa} - \kappa(P_X^{\theta_0})) \xrightarrow{\mathcal{D}} \mathcal{N}(\mu, \sigma^2)$ as $n \to \infty$, under $P_Y^{\theta_0 + n^{-1/2}\eta, \gamma_0 + n^{-1/2}\psi}$, where the limiting mean

$\mu = \lim_{n\to\infty} n^{1/2}(\kappa(P_X^{\theta_0+n^{-1/2}\eta}) - \kappa(P_X^{\theta_0}))$, and $\sigma^2 < \infty$. Thus asymptotically we can recover $\kappa(P_X^{\theta_0+n^{-1/2}\eta})$ from $(P_Y^{\theta_0+n^{-1/2}\eta,\gamma_0+n^{-1/2}\psi})^n$.

This holds separately for every parametric model passing through the same given point P^0, i.e., $P_X^0 = P_X^{\theta_0}$, $P_{Y|X=x}^0 = P_{Y|X=x}^{\gamma_0}$. Since even under CAR the tangent space at P^0 is everything, any P_Y close to P_Y^0 lies to a close approximation on one of these submodels. Thus in a local asymptotic sense, for P_Y close to a given model P_Y^0 determined by $P_X^0, P_{Y|X=x}^0$, one can recover P_X from P_Y.

10. Global identifiability of CAR

Suppose the triple X, Y, α is such that Y, α is a coarsening at random of X. The question we study here is: given the distribution of the data Y, and the coarsening $\mathcal{X} = \alpha(Y)$, are the marginal distribution of X and the conditionals of Y given $X = x$ uniquely determined? In other words, if a factorization of the distribution of Y exists, is it unique?

In Section 2 we saw that in the discrete case the factorisation $f_A = p_A\pi_A$ (which was always possible) was uniquely determined for A with $f_A > 0$. The (p_A) and (π_A) of the factorisation might not be hereby completely determined for A with $f_A = 0$. There might be some free choice between having $p_A = 0$ or $\pi_A = 0$, and consequently some free choice in the value given to the non-zero member of the pair.

In general sample spaces there is a similar non-uniqueness (if a factorisation exists at all). Let the function α and the marginal law of Y be fixed. Let P and P' denote two CAR models, such that the possible P-null exceptions x for the CAR property of $P_{Y|X=x}$ also form a P'-null set and vice-versa. We assume $P_Y = P'_Y$.

Define $Q_X = \frac{1}{2}(\mathrm{P}_X + P'_X)$ and $Q_{Y|X=x} = \frac{1}{2}(P_{Y|X=x} + P'_{Y|X=x})$. Then Q is also CAR, and P and P' are dominated by Q. Let $P^{\theta,\gamma}$ be defined by

$$
\begin{aligned}
P_X^\theta &= (1-\theta)P_X + \theta P'_X, \\
P_{Y|X=x}^\gamma &= (1-\gamma)P_{Y|X=x} + \gamma P'_{Y|X=x}.
\end{aligned}
$$

Thus $Q = P^{0.5,0.5}$. By the theorem of Section 7,

$$
\frac{\mathrm{d}P_Y^{\theta,\gamma}}{\mathrm{d}Q_Y}(y) = E_Q\Big(\frac{\mathrm{d}P_X^\theta}{\mathrm{d}Q_X}(X)\,\Big|\,Y=y\Big).\frac{\mathrm{d}P_{Y|X=x}^\gamma}{\mathrm{d}Q_{Y|X=x}}(y), \quad x \in \alpha(y).
$$

Thus

$$
\begin{aligned}
E_P \log \frac{\mathrm{d}P_Y^{\theta,\gamma}}{\mathrm{d}Q_Y}(Y) = {} & E_P \log\Bigg(E_Q\Big(\frac{\mathrm{d}P_X}{\mathrm{d}Q_X}(X)\,\Big|\,Y\Big) \\
& +\theta\Big(E_Q\Big(\frac{\mathrm{d}P'_X}{\mathrm{d}Q_X}(X) - \frac{\mathrm{d}P_X}{\mathrm{d}Q_X}(X)\,\Big|\,Y\Big)\Big)\Bigg) \\
& + E_P \log\Bigg(\frac{\mathrm{d}P_{Y|X}}{\mathrm{d}Q_{Y|X}}(Y) + \gamma\Big(\frac{\mathrm{d}P'_{Y|X}}{\mathrm{d}Q_{Y|X}}(Y) - \frac{\mathrm{d}P_{Y|X}}{\mathrm{d}Q_{Y|X}}(Y)\Big)\Bigg)
\end{aligned}
$$

where $(\mathrm{d}P_{Y|X}/\mathrm{d}Q_{Y|X})(Y)$ is defined on $Y = y$ as $(\mathrm{d}P_{Y|X=x}/\mathrm{d}Q_{Y|X=x})(y)$ for any $x \in \alpha(y)$.

Now the above function of θ and γ is concave in both arguments, and maximal both at $\theta = 0$, $\gamma = 0$, and at $\theta = 1$, $\gamma = 1$. Therefore it must be constant in θ and γ, or:

$$E_Q\Big(\frac{\mathrm{d}P'_X}{\mathrm{d}Q_X}(X)\,\Big|\,Y=y\Big) \;=\; E_Q\Big(\frac{\mathrm{d}P_X}{\mathrm{d}Q_X}(X)\,\Big|\,Y=y\Big)$$

for P almost all y,

$$\frac{\mathrm{d}P'_{Y|X=x}}{\mathrm{d}Q_{Y|X=x}}(y) \;=\; \frac{\mathrm{d}P_{Y|X=x}}{\mathrm{d}Q_{Y|X=x}}(y)$$

for P almost all y. Now the particular choice of Q dominating P was not important so we have that all CAR models reproducing P_Y have the same decomposition

$$\frac{\mathrm{d}P_Y}{\mathrm{d}Q_Y}(y) \;=\; E_Q\Big(\frac{\mathrm{d}P_X}{\mathrm{d}Q_X}(X)\,\Big|\,Y=y\Big).\frac{\mathrm{d}P_{Y|X=x}}{\mathrm{d}Q_{Y|X=x}}(y), \qquad x \in \alpha(y)$$

provided the same exceptional points x are involved; in particular, if actually CAR($Y|X$) holds without any exceptional points.

11. Open questions

We have shown that, in fairly general sample spaces, a certain definition of CAR in terms of $P_{Y|X=x}$ has desired consequences for $P_{X|Y=y}$ and for factorization of P_Y. In discrete sample spaces, these three properties are actually equivalent. The equivalence does not hold in general. But one can look for supplementary conditions to make this hold.

Part of this problem is the wish to be able to have from CAR (or even equivalent to CAR): $P_{X|Y=y} = P_{X|X\in y}$. In a general set-up however, there is not a unique way to interpret $P_{X|X\in y}$. Perhaps one should restrict attention to cases where Y has further special structure. The following covers all specific examples of which we are aware: it has features both from missing observations in a multivariate vector and from grouped (including censored) observations. Suppose observation of $Y = y$ is equivalent to observation of a discrete 'type' K, and, when $K = k$, observation that $\alpha_k(X) = a_k$ and $\beta_k(X) \in B_k$ for certain measurable functions α_k and β_k where furthermore $\Pr(\beta_k(X) \in B_k|\alpha_k(X) = a_k) > 0$ for all possible values a_k and sets B_k. Conditional on $Y = y$, we would now want $P_{X|Y=y}$ to coincide with the conditional distribution of X given $\alpha_k(X) = a_k$ and $\beta_k(X) \in B_k$ which, for each k, a_k and B_k is unambiguously defined, and which we may justly call $P_{X|X\in y}$. Now one could try to construct a CAR mechanism which produces observations of this form only, and which is generated by an underlying independent 'typing and grouping' variable G as in Keiding and

Jacobsen's reference experiment. In this reference model one should be able to compute $P_{X|Y=y}$ and show that it equals $P_{X|X \in y}$. Then by our result that $P_{X|Y=y}$ does not depend on the specific CAR mechanism at hand, it remains equal to $P_{X|X \in y}$ for all CAR mechanisms.

We showed that in discrete sample spaces, `CAR is everything`, but that this result breaks down in general. Is there a repair-job which will make it generally true? We conjecture that even if in general CAR is not everything, that MAR is: in other words, any model for the distribution of a random subset of the components of a multivariate vector can be reproduced exactly with a Coarsening (Missing) at random model.

Our negative results on sequential representations of CAR in this paper and on MAR in our companion paper need further study. `CAR is more than it seems!` CAR and MAR mechanisms exist, whose computer implementation has the following property: the computer needs to know more about X than it's willing to output in its final `print` statement, yet this fact does not affect our face-value inference.

We cannot conceive of more general mechanisms for generating CAR and MAR in an honest way, but is this just a lack of imagination? Can one easily recognise if a given CAR or MAR mechanism has a sequential representation? In 'large' spaces, do 'most' CAR and MAR mechanisms admit a representation?

We showed that in finite sample spaces, monotone coarsened data could be modelled by monotone coarsening rules, and that the CAR factorisation could be explicitly recovered from the observed data distribution. Robins and Rotnitzky (1992) and Robins (1996a) have obtained analogous results when monotone coarsened data results from right censoring by a continuously distributed censoring variable. It is a challenge to extend this to the general monotone case. Some kind of product-integration technique should be possible to mimic the Kaplan-Meier method we used. The branching structure of a tree should be replaced by the continuous branching of histories of a continuous time stochastic process, where the total history of the process 'up to and including time infinity' tells us the complete data. We conjecture that all monotone CAR data can be represented as a stopped stochastic process with the unstopped process corresponding to the underlying data X and the stopping time corresponding to a generalised censoring time.

Bibliography

P.J. Bickel, C.A.J. Klaassen, Y. Ritov and J.A. Wellner (1993), *Efficient and Adaptive Inference in Semi-parametric Models*, John Hopkins University Press, Baltimore.

J.T. Chang and D. Pollard (1997), Conditioning as disintegration, *Statistica Neerlandica* **51** (to appear).

D.M. Dabrowska (1988), Kaplan-Meier estimation on the plane, *Ann. Statist.* **16**, 1475–1489.

R.D. Gill (1989), Non- and semi-parametric maximum likelihood estimators and the von Mises method, Part 1, *Scand. J. Statist.* **16**, 97–128.

R.D. Gill and J.M. Robins (1997), Sequential models for coarsening and missingness, *Proc. First Seattle Symposium on Biostatistics: Survival Analysis*, ed. D.Y. Lin, Springer-Verlag.

D.F. Heitjan (1993), Ignorability and coarse data: some biomedical examples, *Biometrics* **49**, 1099–1109.

D.F. Heitjan (1994), Ignorability in general incomplete-data models, *Biometrika* **81**, 701–708.

D.F. Heitjan and D.B. Rubin (1991), Ignorability and coarse data, *Ann. Statist.* **19**, 2244–2253.

M. Jacobsen and N. Keiding (1995), Coarsening at random in general sample spaces and random censoring in continuous time, *Ann. Statist.* **23**, 774–786.

R. Kress (1989), *Linear Integral Equations*, Springer-Verlag, Berlin.

M.J. van der Laan (1993), *Efficient and Inefficient Estimation in Semiparametric Models*, Ph.D. Thesis, Dept. Mathematics, University Utrecht; reprinted (1995) as CWI tract 114, Centre for Mathematics and Computer Science, Amsterdam.

M.J. van der Laan (1996), Efficient estimation in the bivariate censoring model and repairing NPMLE, *Ann. Statist.* **24**, 596–627.

R.J.A. Little and D.B. Rubin (1987), *Statistical Analysis with Missing Data*, Wiley, New York.

S.F. Nielsen (1996), *Incomplete Observations and Coarsening at Random*, preprint, Institute of Mathematical Statistics, Univ. of Copenhagen.

R.L. Prentice and J. Cai (1992), Covariance and survivor function estimation using censored multivariate failure time data, *Biometrika* **79**, 495–512.

J.M. Robins (1996a), Locally efficient median regression with random censoring and surrogate markers, pp. 263–274 in: *Lifetime Data: Models in Reliability and Survival Analysis*, N.P. Jewell, A.C. Kimber, M.L. Ting Lee, G.A. Whitmore (eds), Kluwer, Dordrecht.

J.M. Robins (1996b), Non-response models for the analysis of non-monotone non-ignorable missing data, *Statististics in Medicine*, Special Issue, to appear.

J.M. Robins and R.D. Gill (1996), Non-response models for the analysis of non-monotone ignorable missing data, *Statistics in Medicine*, to appear.

J.M. Robins and Y. Ritov (1996), Towards a curse of dimensionality appropriate (CODA) asymptotic theory for semiparametric models, *Statistics in Medicine*, to appear.

J.M. Robins and A. Rotnitzky (1992), Recovery of information and adjustment for dependent censoring using surrogate markers, pp. 297–331 in: *AIDS Epidemiology—Methodological Issues*, N. Jewell, K. Dietz, V. Farewell (eds), Birkhäuser, Boston.

J.M. Robins, A. Rotnitzky and L.P. Zhao (1994), Estimation of regression coefficients when some regressors are not always observed, *J. Amer. Statist. Assoc.* **89**, 846–866.

D.B. Rubin (1976), Inference and missing data, *Biometrika* **63**, 581–592.

D.B. Rubin, H.S. Stern and V. Vehovar (1995), Handling "Don't Know" survey responses: the case of the Slovenian plebiscite, *J. Amer. Statist. Assoc.* **90**, 822–828.

A.W. van der Vaart (1991), On differentiable functionals, *Ann. Statist.* **19**, 178-204.

P. Whittle (1971), *Optimization under Constraints*, Wiley, New York.

Authors:

R.D.Gill, Mathematical Institute, University Utrecht, PO Box 80 010, 3508 TA Utrecht, Netherlands; email: gill@math.ruu.nl

M.J. van der Laan, Dept. of Biostatistics, University of California, Berkeley, Ca. 94720, USA; email: laan@stat.berkeley.edu

J.M. Robins, Depts of Epidemiology and Biostatistics, Harvard School of Public Health, 677 Huntington Road, Boston Ma. 02115, USA; email: robins@hsph.harvard.edu

Sequential Models for Coarsening and Missingness

Richard D. Gill
James M. Robins

ABSTRACT In a companion paper we described what intuitively would seem to be the most general possible way to generate Coarsening at Random mechanisms, a sequential procedure called *randomized monotone coarsening*. Counter-examples showed that CAR mechanisms exist which cannot be represented in this way. Here, we further develop these results in two directions. Firstly, we consider what happens when data is coarsened at random in two or more phases. We show that the resulting coarsening mechanism is not CAR anymore, but under suitable assumptions is identified and can provide interesting alternative analysis of data under a non-CAR model. Secondly, we look at sequential mechanisms for generating MAR data, missing components of a multivariate random vector. Randomised monotone missingness schemes, in which one variable at a time is observed and depending on its value, another variable is chosen or the procedure is terminated, supply in our opinion the broadest class of physically interpretable MAR mechanisms. We show that every randomised monotone missingness scheme can be represented by a Markov monotone missingness scheme, in which the choice of which variable to observe next only depends on the set of previously observed variables and their values, not on the sequence in which they were measured. We also show that MAR mechanisms exist which cannot be represented sequentially.

1. Sequential CAR is not CAR but it is everything

We use the notation of our companion paper Gill, van der Laan and Robins (1997). In particular $\mathcal{X}$ is a coarsening of an underlying random variable X. We suppose that X takes values in a finite set E. Its power set (the set of all subsets of E) is denoted by $\mathcal{E}$. So $\mathcal{X}$ takes values in $\mathcal{E} \setminus \{\emptyset\}$ and $X \in \mathcal{X}$ with probability one.

In this section, we study k-sequential coarsening, by which we mean that the observed random set $\mathcal{X}$ is formed by first coarsening the underlying random variable X to a random set $\mathcal{X}_1$, then further coarsening $\mathcal{X}_1$ to a new random set $\mathcal{X}_2, \ldots$, and finally coarsening $\mathcal{X}_k$ to $\mathcal{X}$. Only the final random set $\mathcal{X}$ is observed. We say the coarsening is k-sequentially CAR if each of the $k+1$-sequential coarsenings are individually CAR. Robins (1996) introduced sequential CAR in the special case of missing components of a multivariate random vector which he called generalised permutation missingness and we call sequential MAR. We will show that $\mathcal{X}_1, \ldots, \mathcal{X}_k, \mathcal{X}$ being k-sequentially CAR does not imply that the overall coarsening of X to $\mathcal{X}$ is

CAR. Thus, modeling the observed data $\mathcal{X}$ as CAR versus k-sequentially CAR implies different marginal distributions for the underlying variable X. Furthermore, we will show that, in the discrete case, k-sequential CAR is, like ordinary CAR, everything, so that no data evidence can distinguish CAR from k-sequential CAR. We will argue that, in many contexts, the assumption that the coarsening mechanism is k-sequential CAR rather than CAR has greater substantive plausibility, in which case the data should be analyzed under a k-sequential CAR model. We now formalize these ideas.

Definition. We say that the random sets $\mathcal{X}_1, \ldots, \mathcal{X}_k, \mathcal{X}$ with each $\mathcal{X}_m$ and $\mathcal{X} \in \mathcal{E} \setminus \{\emptyset\}$ form a k-sequential coarsening of a random variable X if for $m = 0, \ldots, k+1$, $\mathcal{X}_m \subseteq \mathcal{X}_{m+1}$ with probability 1 where $\mathcal{X}_0 \equiv \{X\}$ and $\mathcal{X}_{k+1} \equiv \mathcal{X}$. Thus a 0-sequential coarsening is just the ordinary coarsening studied above.

Definition. A k-sequential coarsening is k-sequentially CAR if, for $m = 1, \ldots, k$, the conditional distribution of $\mathcal{X}_m$ given $\mathcal{X}_{m-1}$ does not depend on the particular realization of $\mathcal{X}_{m-1}$ except through the fact that it is compatible with $\mathcal{X}_m$. In the discrete case, this means $\Pr(\mathcal{X}_m = A \mid \mathcal{X}_{m-1} = B)$ is the same for all B in the support of $\mathcal{X}_{m-1}$ such that $B \subseteq A$.

Throughout we suppose we only observe the final random set $\mathcal{X} = \mathcal{X}_{k+1}$ of a k-sequential coarsening. One might hope that $\mathcal{X}$ is a CAR coarsening of X (in the previous sense) if $\mathcal{X}_1, \ldots, \mathcal{X}_{k+1}$ is k-sequentially CAR. However, this is not the case as the following example indicates.

Example. Suppose $X = (Y, W)$ where Y and W take values in $\{0, 1\}$. Let $E_{\cdot w} = \{(0, w), (1, w)\}$ denote the event that $W = w$ but Y is not observed. Let $E_{y\cdot} = \{(y, 0), (y, 1)\}$ denote an event in which W is not observed. Finally, let the entire sample space $E = \{(0,0), (0,1), (1,0), (1,1)\}$ be the event that neither Y nor W is observed. Now suppose that $\Pr(\mathcal{X}_1 = \{(y, w)\} \mid X = (y, w)) = c_1(w)$, $\Pr(\mathcal{X}_1 = E_{\cdot w} \mid X = (y, w)) = 1 - c(w)$ so that the probability that Y is not observed in $\mathcal{X}_1$ depends on the observed value of W. Further, suppose $\Pr(\mathcal{X}_2 = \{(y, w)\} \mid \mathcal{X}_1 = \{(y, w)\}) = c_2(y)$, $\Pr(\mathcal{X}_2 = E_{y\cdot} \mid \mathcal{X}_1 = \{(y, w)\}) = 1 - c_2(y)$ so the probability w is not observed in $\mathcal{X}_2$ depends on the observed value of y in $\mathcal{X}_1$. Suppose $\Pr(\mathcal{X}_2 = E_{\cdot w} \mid \mathcal{X}_1 = E_{\cdot w}) = c_3$ and $\Pr(\mathcal{X}_2 = E \mid \mathcal{X}_1 = E_{\cdot w}) = 1 - c_3$ so the probability that W is not observed in $\mathcal{X}_2$ when Y is not observed in $\mathcal{X}_1$ is constant. It is easy to check that $\mathcal{X}_1, \mathcal{X}_2$ is 1-sequentially CAR, more specifically, it is 1-sequentially MAR. However, $\mathcal{X} = \mathcal{X}_2$ is not a CAR coarsening of X since $\Pr(\mathcal{X}_2 = E_{y\cdot} \mid X = (y, w)) = c_1(w)(1 - c_2(y))$ which depends on w.

However, if X is discrete, k-sequential CAR is still everything as formally stated in the following theorem and its corollary. The theorem assumes that the possible values of the intermediate coarsening $\mathcal{X}_m$ are specified in advance and moreover that each value arises in one particular way from its predecessor $\mathcal{X}_{m-1}$. This was in fact the case in the example just discussed, as the reader can easily check. These a priori restrictions on the intermediate coarsenings ensure that a k-sequential CAR model reproducing a particular distribution of observed coarsened data $\mathcal{X}$ not only exists, but is also essentially unique.

Theorem. Let E be a finite set and let $\mathcal{X}$ be a random nonempty set with distribution f_A, $A \in \mathcal{E}_{k+1} = \mathcal{E} \setminus \{0\}$. Let $\mathcal{E}_0 = \{\{x\} : x \in E\}$ be the subset of $\mathcal{E}$ consisting of all the singletons. Let $\mathcal{E}_1, \ldots, \mathcal{E}_k$ be given subsets of $\mathcal{E}$ such that, for each $m = 0, \ldots, k$ and for each $A \in \mathcal{E}_{m+1}$, A is a disjoint union of specified elements B of $\mathcal{E}_m$. Write $B \in_m A$ if $B \in \mathcal{E}_m$ is one of the specified sets partitioning a given $A \in \mathcal{E}_{m+1}$. Then there exist 'CAR probabilities' π_{mA} for $A \in \mathcal{E}_{m+1}$ for $m = 0, \ldots, k$, and distributions p_{mB} over $B \in \mathcal{E}_m$ for $m = 0, \ldots, k+1$, with $p_{(k+1)A} = f_A$, satisfying (i) for each $B \in \mathcal{E}_m$, $\sum \pi_{mA} = 1$ with summation over $\{A \in \mathcal{E}_{m+1} : B \in_m A\}$; (ii) $\sum p_{mB} = 1$ with summation over all $B \in \mathcal{E}_m$; and (iii) $p_{(m+1)A} = p_{mA}\pi_{mA}$ for each $m \leq k$ and each $A \in \mathcal{E}_{m+1}$ and $p_{mA} = \sum p_{mB}$ with summation over $\{B \in \mathcal{E}_m : B \in_m A\}$. Furthermore, for each $A \in \mathcal{E}_{k+1}$ with $f_A > 0$, π_{kA} and p_{kA} are unique. Similarly, for each $A \in \mathcal{E}_{m+1}$ with $p_{(m+1)A} > 0$, π_{mA} and p_{mA} are unique. Thus if $f_A > 0$ for all singletons $A = \{x\}$, then $p_x = p_{0\{x\}}$, $x \in E$, is a uniquely defined distribution on E.

Proof. This follows by repeated application of the main theorem of Section 2 of our companion paper, starting from $m = k + 1$ downwards.

Corollary. If we draw $\mathcal{X}_0 = \{X\}$ from $p_{0\{x\}}$, $x \in E$, and then recursively draw $\mathcal{X}_{m+1}$ from $\mathcal{E}_{m+1}$ under the conditional law π_{mA} given $\mathcal{X}_m = B$ (where B is the value of $\mathcal{X}_m$ drawn in the previous step and A such that $B \in_m A$), then $\mathcal{X}_1, \ldots, \mathcal{X}_{k+1}$ will be a k-sequential CAR coarsening of X and $\mathcal{X} \equiv \mathcal{X}_{k+1}$ will have marginal distribution f_A.

Thus, for discrete E, one cannot use the data to distinguish whether the underlying coarsening process is CAR or is k-sequentially CAR based on specified sets of intermediate coarsened values $\mathcal{E}_1, \ldots, \mathcal{E}_k$ and relations between them. However, a CAR process compatible with the observed data $\mathcal{X}$ will imply a different marginal distribution for the underlying variable X than will a k-sequential CAR process compatible with the data. Thus the choice of modeling the observed data $\mathcal{X}$ as CAR versus k-sequentially CAR will depend solely on one's prior subject matter beliefs. As the following example due to Robins (1996) shows, the assumption that the coarsening process is k-sequentially CAR can be much more substantively plausible than that it is CAR.

Example. Suppose in a prison, a voluntary HIV test is offered to prisoners. Thirty percent agree to the test, of which fifteen percent are positive. It is believed that the decision to be tested (i) is not independent of HIV status due to the differing views on testing held by HIV-positive versus negative inmates but (ii) is independent of HIV conditional on whether a prisoner has sought previous treatment for an STD in the prison hospital. The goal is to estimate the marginal probability of HIV infection among the inmates. Thus, in order to identify this probability, it is decided to randomly sample hospital records and record previous STD treatment with a prisoner's probability of being included in the hospital record sample depending on whether the prisoner agreed to the HIV test and, if so, his test result. Letting Y and W be the dichotomous indicator variables for HIV infection status and previous STD treatment, the coarsening mechanism is exactly that discussed in the previous example, implying the data is 1-sequential CAR but not CAR.

We note that a k-sequential CAR process and an m-sequential CAR process would also imply different marginal distributions for the underlying variable X and, by our theorem, cannot be distinguished from one another by any data evidence. Indeed, two k-sequential CAR processes based on different sequences of selected sets will imply different marginal distributions for the underlying X and also cannot be distinguished from one another based on the data. Thus, multiple sequential CAR models can be used to conduct a sensitivity analysis to help determine the sensitivity of the distribution of the underlying variable X to assumptions about the coarsening mechanism.

2. MAR and randomized monotone missingness

We now restrict attention from coarsened to missing data, and investigate the possibility that all MAR mechanisms can be constructed by random monotone coarsening at the variables level. Recall that in section 3 of our companion paper we showed that, for CAR, this programme fails. However our counter-example there cannot be used in the new situation.

The sample space E for the underlying data X is now a Cartesian product of k finite sets E_i, $i \in \mathcal{K} = \{1, \ldots, k\}$. Write the coordinates of X as $X = (X_1, \ldots, X_k)$. The only sets A which can be observed, i.e., with $\pi_A > 0$, are Cartesian products either of the whole coordinate E_i or of a singleton $\{x_i\} \in E_i$, $i \in \mathcal{K}$. Knowing $X \in A$ is the same as knowing exactly the values of some of the X_i but knowing nothing about the others. We say that the first X_i are *observed*, the others are *missing*. In this situation, coarsening at random is called *missing at random* (MAR), Rubin (1976) or Little and Rubin (1987), and one can define *monotone missingness*, *randomized monotone missingness*, and *Markov randomized monotone missingness* analogously to the CAR case. In monotone schemes, rather than choosing at each stage (if we do not stop) a partition of the current set, we now choose a further variable to observe.

The counter-example which we gave in the CAR case to the conjecture that all CAR mechanisms can be represented by randomized monotone coarsening schemes does not give a counter-example to the analogous conjecture for MAR. It took us some effort to find a new counter-example for MAR, though at the end, as the reader will discover, the counter-example is quite simple. We will first show that any set of MAR probabilities which admit a randomized monotone representation, also admit a Markov randomized monotone representation. So it is sufficient to investigate the possible 'Markov monotone missingness' or MMM property of a given MAR system. This can be done by solving a certain *linear programming problem.* So one can find out, in a finite number of steps, if a specific MAR system admits an MMM representation. Finding a representation comes down to solving a set of linear equalities in more unknowns than there are equations (so typically many solutions exist) subject to a collection of linear inequalities (which rule out some or even all solutions).

Recall we have $X = (X_1, \ldots, X_k)$ where X_i takes values x_i in a finite set E_i. Let $\mathcal{R}$ denote the set of labels of actually observed variables: so $\mathcal{R}$ is a random subset of $\mathcal{K}$ (the letter $\mathcal{R}$ stands for 'recorded', the other variables are missing). Observation of $\mathcal{X}$ is equivalent to observation of the pair $(\mathcal{R}, X_{\mathcal{R}})$ where $X_{\mathcal{R}}$ denotes the vector (a subvector of X) with components X_i for $i \in \mathcal{R}$. Possibly $\mathcal{R}$ is empty. Under MAR, the probability to observe the variables $\mathcal{R}$, given $X = x$, only depends on the X_i with $i \in \mathcal{R}$, or:

$$\Pr\{\mathcal{R} = R | X = x\} = \Pr\{\mathcal{R} = R | X_R = x_R\}$$

We denote this probability by $\pi_R^{x_R}$. Thus a MAR mechanism is determined by a system of probabilities $\pi_R^{x_R}$ such that for all $x \in E = E_1 \times \ldots \times E_k$,

$$\sum_{R \subseteq \mathcal{K}} \pi_R^{x_R} = 1 \qquad (1)$$

Now we specify a randomized monotone missingness (RMM) mechanism. This consists of a set of probability rules for choosing a sequence of distinct indices $I_1, I_2, \ldots, I_K \in \mathcal{K}$ where $0 \leq K \leq k$ is the random number of variables finally observed and $\mathcal{R} = \{I_1, \ldots, I_K\}$. We shall summarize the rules in a big collection of probabilities $(\widetilde{\psi})$; later we will write (ψ), without the tilde, to specify a (less complicated) Markov mechanism. To keep the formulas short, we write $i_{<n}$ for $i_1, \ldots, i_{n-1}$. The *set* of elements $\{i_1, \ldots, i_{n-1}\}$ is then denoted by $\{i_{<n}\}$. The *vector* or *sequence* $(i_1, \ldots, i_{n-1})$ may be written $(i_{<n})$ for emphasis. We drop the curly or round brackets when the distinction between set and sequence is not crucial. When we write events like '$I_n = i_n$' it is implicitly understood that '$K \geq n$' also holds, and conversely, '$K < n$' implies 'I_n is undefined'.

Suppose at the n'th stage we have already determined $I_{<n} = i_{<n}$ and have observed $X_{<n} = x_{<n}$. The order in which the variables were selected is included in the expression $i_{<n}$. Next, with conditional probability

$$\begin{aligned} \widetilde{\psi}_{i_n | i_{<n}}^{x_{i_{<n}}} &= \Pr\{\text{At stage } n \text{ go on to observe variable } i_n \mid \text{past}\} \\ &= \Pr\{I_n = i_n | I_{<n} = i_{<n}, X_{i_{<n}} = x_{i_{<n}}\}, \end{aligned}$$

we choose and observe another variable X_{i_n}, $i_n \notin \{i_{<n}\}$. With the comple-

mentary probability

$$1 - \sum_{i_n} \widetilde{\psi}^{x_{i<n}}_{i_n|i_{<n}} = \Pr\{I_n \text{ is not defined} \mid \text{past}\}$$

we stop and set $K = n-1$ and thus $\mathcal{R} = \{I_{<n}\}$.

Now define

$$\widetilde{\phi}^{x_{i<n}}_{i_{<n},i_n} = \Pr\{I_{<n} = i_{<n}, I_n = i_n | X_{i_{<n}} = x_{i_{<n}}\}$$

and

$$\begin{aligned} \widetilde{\xi}^{x_{i<n}}_{i_{<n}} &= \widetilde{\phi}^{x_{i<n-1}}_{i_{<n-1},i_{n-1}} - \sum_{i_n} \widetilde{\phi}^{x_{i<n}}_{i_{<n},i_n} \\ &= \Pr\{K = n-1, I_{<n} = i_{<n} | X_{i_{<n}} = x_{i_{<n}}\}. \end{aligned} \tag{2}$$

Then

$$\widetilde{\psi}^{x_{i<n}}_{i_n|i_{<n}} = \widetilde{\phi}^{x_{i<n}}_{i_{<n},i_n} / \widetilde{\phi}^{x_{i<n-1}}_{i_{<n-1},i_{n-1}} \tag{3}$$

showing that the RMM system can be described in terms of the system of probabilities $(\widetilde{\phi})$ instead of the system $(\widetilde{\psi})$. An advantage of this is that the MAR probabilities (π) can be expressed *linearly* in $(\widetilde{\phi})$: for any given $R \subseteq \mathcal{K}$ of size $n-1$,

$$\pi^{x_R}_R = \sum_{i_{<n}:\{i_{<n}\}=R} \widetilde{\xi}^{x_{i<n}}_{i_{<n}} \tag{4}$$

where the $(\widetilde{\xi})$ are given by (2). The $(\widetilde{\phi})$ are also convenient because they can be chosen freely up to a collection of *linear* restrictions:

$$\left.\begin{aligned} \widetilde{\phi}^{x_{i<n}}_{i_{<n},i_n} &\geq 0 \\ \sum_{i_n} \widetilde{\phi}^{x_{i<n}}_{i_{<n},i_n} &\leq \widetilde{\phi}^{x_{i<n-1}}_{i_{<n-1},i_{n-1}} \end{aligned}\right\} \tag{5}$$

where for $n = 1$ the right hand side of the last inequality is replaced by the number 1.

It is easy to check that any system $(\widetilde{\phi})$ satisfying (5) leads via (2) and (4) to a system (π) satisfying (1) and $\pi^{x_R}_R \geq 0$ for all R and x. Equation (3) shows how to calculate the actual decision rule probabilities $(\widetilde{\psi})$ from the $(\widetilde{\phi})$. Conversely, $(\widetilde{\phi})$ can be computed recursively from $(\widetilde{\psi})$ by rewriting (3) as

$$\widetilde{\phi}^{x_{i<n}}_{i_{<n},i_n} = \widetilde{\phi}^{x_{i<n-1}}_{i_{<n-1},i_{n-1}} \widetilde{\psi}^{x_{i<n}}_{i_n|i_{<n}}$$

where for $n = 1$ this is interpreted as $\widetilde{\phi}^{\cdot}_{\cdot,i_1} = \widetilde{\psi}^{\cdot}_{i_1|\cdot}$.

Our question: 'does any MAR mechanism admit representation by a RMM scheme?' boils down therefore to the question: given a system of nonnegative numbers (π) satisfying (1), do the linear equations (2) and

(4) admit a solution $(\widetilde{\phi})$ satisfying the linear constraint (5)? We first show that if a solution exists, then one can also find a Markov monotone missing (MMM) solution by which we mean that the decision probabilities $\widetilde{\psi}_{i_n|i_{<n}}^{x_{i_{<n}}}$ (to observe at stage n variable i_n, given the previous sequence of choices and observed values) are invariant under permutations of $i_{<n}$.

3. Markov monotone missingness

Under what we call Markov monotone missingness (MMM), the decision probabilities $\widetilde{\psi}$ are invariant under permutations. We denote then, for a subset R of $n-1$ elements of $\mathcal{K}$,

$$\psi_{i_n|R}^{x_R} = \Pr\{I_n = i_n \mid \{I_{<n}\} = R, X_R = x_R\},$$

$$\phi_{R,i_n}^{x_R} = \Pr\{I_n = i_n, \{I_{<n}\} = R \mid X_R = x_R\}.$$

So $\phi_{R,i_n}^{x_R}$ is the *sum*, over all permutations $i_{<n}$ of the elements of R, of $\widetilde{\phi}_{i_{<n},i_n}^{x_R}$; while $\psi_{i_n|R}^{x_R}$ is *equal* to each of the formerly introduced $\widetilde{\psi}_{i_n|i_{<n}}^{x_R}$ for all permutations of R.

Now suppose a given MAR system (π) has a, not necessarily Markov, randomized monotone missingness (RMM) representation characterized by decision probabilities $(\widetilde{\psi})$ or equivalently by a system $(\widetilde{\phi})$. *Define* for each set R of $n-1$ elements the quantities

$$\phi_{R,i_n}^{x_R} = \sum_{i_{<n}:\{i_{<n}\}=R} \widetilde{\phi}_{i_{<n},i_n}^{x_{i_{<n}}}. \tag{6}$$

Combining (2) and (4) gives

$$\pi_R^{x_R} + \sum_{j\notin R} \phi_{R,j}^{x_R} = \sum_{j\in R} \phi_{R\setminus\{j\},j}^{x_{R\setminus\{j\}}}. \tag{7}$$

Moreover, *define*

$$\psi_{i_n|R}^{x_R} = \phi_{R,i_n}^{x_R} \Big/ \sum_{j\in R} \phi_{R\setminus\{j\},j}^{x_{R\setminus\{j\}}}. \tag{8}$$

The new (ψ) form a system of MMM decision rules. The new (ϕ) can be recovered from (ψ) by recursive use of (8), and since (7) holds we have hereby constructed a Markov representation for the original RMM-generated MAR probabilities (π). Here we use the fact that for a Markov system, described equivalently by a system (ϕ) or a system (ψ) because of the equivalence (8), the MAR probabilities can be calculated from (7); this fact follows from (2) and (4) again.

In conclusion, if a given MAR system can be represented by a randomized monotone missingness scheme, then it can also be represented by a Markov monotone missingness scheme.

4. Is every MAR system MMM?

In view of the results so far, the question to be answered is now: given any system of numbers $(\pi_R^{x_R})$ such that

$$\pi_R^{x_R} \geq 0 \qquad \forall R, x_R \tag{9}$$

$$\sum_{R \subseteq \mathcal{K}} \pi_R^{x_R} = 1 \qquad \forall x \tag{10}$$

do there exist $(\phi_{R,j}^{x_R})$ for $j \notin R \subseteq \mathcal{K}$ such that

$$\phi_{R,j}^{x_R} \geq 0 \qquad \forall R,\ x_R,\ j \notin R \tag{11}$$

$$\pi_R^{x_R} + \sum_{j \notin R} \phi_{R,j}^{x_R} = \sum_{j \in R} \phi_{R\setminus\{j\},j}^{x_{R\setminus\{j\}}} \qquad \forall R,\ x_R \tag{12}$$

If so we can define

$$\psi_{j|R}^{x_R} = \phi_{R,j}^{x_R} / \sum_{j \in R} \phi_{R\setminus\{j\},j}^{x_{R\setminus\{j\}}} \tag{13}$$

and the system $(\psi_{j|R}^{x_R})$ is a set of Markov monotone missingness decision probabilities which generates $(\pi_R^{x_R})$. Note that when R is the empty set $\emptyset$, (12) is rewritten as

$$\pi_\emptyset^{\cdot} + \sum_{j \in \mathcal{K}} \phi_{\emptyset,j}^{\cdot} = 1.$$

Because of this, on adding (12) over all $R \subseteq \mathcal{K}$ for given x, we obtain (10). Note, when $R = \emptyset$, equation (13) is rewritten as $\psi_{j|\emptyset}^{\cdot} = \phi_{\emptyset,j}^{\cdot}$.

Collect all $\pi_R^{x_R}$ and $\phi_{R,j}^{x_R}$ into column vectors $\boldsymbol{\pi}$, $\boldsymbol{\phi}$. Equation (12) can be rewritten in matrix form as $A\boldsymbol{\phi} = \mathbf{b}$ where $\mathbf{b}$ coincides with $\boldsymbol{\pi}$ except in the '$\emptyset$' component, where $\pi_\emptyset^{\cdot}$ is replaced by $1 - \pi_\emptyset^{\cdot}$. A is a matrix of 0's, 1's and -1's. The question is now: for given A and $\mathbf{b}$, does there exist $\boldsymbol{\phi} \geq \mathbf{0}$ (componentwise inequality) such that $A\boldsymbol{\phi} = \mathbf{b}$? This problem can be solved in a specific instance by transforming it into a linear program, solvable in a finite number of steps by the simplex algorithm. Note that we also have $\mathbf{b} \geq \mathbf{0}$ and let $\mathbf{1}$ denote a vector of 1's of appropriate length. Consider the problem:

$$\text{maximize } -\mathbf{1}^\top \mathbf{z}$$

over $\mathbf{z} \geq \mathbf{0}$ subject to $A\mathbf{x} + \mathbf{z} = \mathbf{b}$, for some $\mathbf{x} \geq \mathbf{0}$; equivalently, the standard LP problem

$$\text{maximize } (-\mathbf{1}^\top \vdots \mathbf{z}^\top) \begin{pmatrix} \mathbf{z} \\ \mathbf{x} \end{pmatrix}$$

$$\text{over} \begin{pmatrix} \mathbf{z} \\ \mathbf{x} \end{pmatrix} \geq \begin{pmatrix} 0 \\ 0 \end{pmatrix} \text{ subject to } (I \vdots A) \begin{pmatrix} \mathbf{z} \\ \mathbf{x} \end{pmatrix} = \mathbf{b}.$$

Since $\mathbf{b} \geq \mathbf{0}$,

$$\begin{pmatrix} \mathbf{z} \\ \mathbf{x} \end{pmatrix} = \begin{pmatrix} \mathbf{b} \\ \mathbf{0} \end{pmatrix}$$

is a feasible solution, so the simplex algorithm can be initialized. Since $-\mathbf{1}^\top \mathbf{z} \leq 0$ for all $\mathbf{z} \geq \mathbf{0}$, the solution is finite. The simplex algorithm finds it in a finite number of steps. If at the solution the object function $-\mathbf{1}^\top \mathbf{z} = 0$, so also $\mathbf{z} = \mathbf{0}$, then we have found a solution $\boldsymbol{\phi} = \mathbf{x} \geq \mathbf{0}$ to the equations $A\boldsymbol{\phi} = \mathbf{b}$. If conversely there exists a solution $\boldsymbol{\phi} \geq \mathbf{0}$ to $A\boldsymbol{\phi} = \mathbf{b}$ then the maximization problem is solved by $\mathbf{x} = \boldsymbol{\phi}$, $\mathbf{z} = \mathbf{0}$, with maximum 0. So: a maximizer of the linear program yields a maximum value 0 if and only if $A\boldsymbol{\phi} = \mathbf{b}$ admits a nonnegative solution.

MAR systems are represented by vectors $\boldsymbol{\pi}$ satisfying $\boldsymbol{\pi} \geq \mathbf{0}$, $\sum_R \pi_R^{x_R} = 1$ for all x. This means that (for given $E = E_1 \times \ldots \times E_k$) the set of all MAR systems is a *convex, compact polytope.* The set of all MMM systems is the subset of such $\boldsymbol{\pi}$ which can be written as

$$A\boldsymbol{\phi} = \mathbf{b}, \quad b_R^{x_R} = \pi_R^{x_R} \ \ \forall R \neq \emptyset, \quad b_\emptyset^{\cdot} = 1 - \pi_\emptyset^{\cdot}.$$

for some $\boldsymbol{\phi} \geq \mathbf{0}$. This is also a convex, compact polytope, contained in the former one.

The system $A\boldsymbol{\phi} = \mathbf{b}$, $\mathbf{b}$ given, $\boldsymbol{\phi}$ unknown, always has solutions. We have a consistent set of equations with more unknowns than equations. This means that it is not unreasonable, on the basis of just counting parameters, to hope that all MAR systems are MMM. However, the inequality constraint $\boldsymbol{\phi} \geq \mathbf{0}$ may rule out many or even all solutions of $A\boldsymbol{\phi} = \mathbf{b}$.

Our geometric picture gives a procedure for determining whether or not, for given $E = E_1 \times \ldots \times E_k$, all MAR systems are MMM: inspect the *extremal points*—the *vertices*—of the set of MAR probabilities. If each vertex has an MMM representation, then the whole set does. Conversely, if any set of MAR probabilities does not admit an MMM representation, then at least one vertex also yields such a counterexample.

The next question is: how can we enumerate the extremal points $\boldsymbol{\pi}$ in the convex, compact polytope of $\boldsymbol{\pi}$ satisfying (9) and (10)? One way to do this is to select a sufficient number of the inequalities (9), turn them into equalities, and try to solve the resulting augmentation of the equalities (10). The procedure is 'finite' but already in the smallest non-trivial case—$k = 3$ binary variables—computationally unfeasible, since millions of cases have to be examined. We need to take account of symmetries and other special features to reduce the number of cases to be investigated.

Let us first study a special subcollection of extremal points: *$\boldsymbol{\pi}$ that are vectors of* 0*'s and* 1*'s only.* If we can find a counter-example here, we are done already. These extremal points, as CAR models, are deterministic (or Coarsened Completely at Random): the coarsening is generated by a fixed partition of the sample space.

In the case of 3 binary variables we enumerated all deterministic extremal points in MAR space, up to symmetries. Taking any particular $\pi_R^{x_R}$ equal to

1 forces all other $\pi_R^{x_R}$ in the equations (10) in which the given one appears to equal zero. A $\pi_R^{x_R}$ with R 'small' appear in many equations—for instance, $\pi_\emptyset^{\cdot}$ appears in all—so the enumeration is quite easy, for instance, by ordering according to the minimum number of elements in R with $\pi_R^{x_R} = 1$. We found in this way just 17 different types (i.e., up to permutation of variables and/or values of variables).

Of these 17, all but one were monotone. The exception had exactly

$$\pi_{23}^{11} = \pi_{13}^{10} = \pi_{12}^{00} = \pi_{123}^{101} = \pi_{123}^{010} = 1.$$

In words, if variables 1 and 2 both take the value 0, variable 3 is not observed; if variables 1 and 3 take the values 1 and 0 respectively, variable 2 is not observed; and if variables 2 and 3 both take the value 1, variable 1 is not observed. Otherwise all three variables are observed. Note that each variable has a chance to be observed, but also each variable has a chance *not* to be observed. That makes it obvious that this MAR system does not have a monotone missingness representation since there is no way any monotone procedure (selecting variables to be observed one by one, without knowing their values in advance) can start.

The existence of this example produces uneasiness as to the interpretation of MAR. Consider a computer simulation model, in which first X is generated, and then $\mathcal{R}$ (hence also $X_\mathcal{R}$) according to given MAR probabilities (π). Let us suppose that after X has been realised, the computer has to generate the random indices $\mathcal{R}$. The computer program is allowed to look at values of components of X one by one, and on the basis of these values decide which values to report. We have shown that MAR probabilities exist which, in order to be simulated, require the computer to look at values of components of X which are *omitted* from the collection of observed variables $\mathcal{R}$. Still, given $\mathcal{R} = R$ and $X_R = x_R$, $X_{\mathcal{K}\setminus R}$ has the same distribution as when one only conditions on $X_R = x_R$ so the information about $X_{\mathcal{K}\setminus R}$ which was used in the procedure is not in the least revealed 'statistically'.

We believe that the only plausible natural MAR mechanisms have a randomised monotone representation. If this belief is correct then, in practical applications, an analyst should perform a test of the hypothesis that the missingness process has such a representation. Robins and Gill (1996) provide an appropriate test statistic. If the test rejects, likelihood analysis under the MAR assumption should be avoided altogether or understood to be at best an approximation to a more appropriate analysis under a non MAR model. We note that the set of RMM probabilities is a closed subset of the set of MAR probabilities, so there exist tests that are consistent under the alternative.

Research remains to be done to support these arguments. Development of special methods for use in RMM models and for testing the RMM assumption only makes sense if the non-RMM MAR models are not highly special rarities as the number of variables and the number of their values becomes larger. As the number of variables increases, are the non-RMM

MAR models more and more of an exception or more and more the rule? On the other side, how convincing is our claim that the notion of RMM captures exactly the class of honest MAR mechanisms (the MAR mechanisms which one can expect to meet with in practice)?

Bibliography

R.D. Gill, M.J. van der Laan and J.M. Robins (1997), Coarsening at random: characterizations, conjectures, counter-examples, *Proc. First Seattle Symposium on Biostatistics: Survival Analysis*, ed. D.Y. Lin, Springer-Verlag.

R.J.A. Little and D.B. Rubin (1987), *Statistical Analysis with Missing Data*, Wiley, New York.

J.M. Robins (1996), Non-response models for the analysis of non-monotone non-ignorable missing data. *Statististics in Medicine*, Special Issue, to appear.

J.M. Robins and R.D. Gill (1996), Non-response models for the analysis of non-monotone ignorable missing data, *Statistics in Medicine.*

D.B. Rubin (1976), Inference and missing data, *Biometrika* **63**, 581–592.

Authors:

R.D.Gill, Mathematical Institute, University Utrecht, PO Box 80010, 3508 TA Utrecht, Netherlands; email: gill@math.ruu.nl

J.M. Robins, Depts of Epidemiology and Biostatistics, Harvard School of Public Health, 677 Huntington Road, Boston Ma. 02115, USA; email: robins@hsph.harvard.edu

Addresses for Contact Authors

(in the same order as the papers)

David R. Cox
Department of Statistics and
Nuffield College
Oxford, OX1 1NF, U.K.
email: drcox@vax.ox.ac.uk

Ross L. Prentice
Division of Public Health Sciences
Fred Hutchinson Cancer Research Center
1124 Columbia Street, MP-1002
Seattle, WA 98104, U.S.A.
email: lroberts@cclink.fhcrc.org

L. J. Wei
Department of Biostatistics
Harvard School of Public Health
677 Huntington Avenue
Boston, MA 02115, U.S.A.
email: wei@sdac.harvard.edu

Jerald F. Lawless
Department of Statistics and Actuarial Science
University of Waterloo
Waterloo, Ontario N2L 3G1, Canada
email: jlawless@setosa.uwaterloo.ca

Terry M. Therneau
Department of Health Sciences Research
Mayo Clinic
200 First Street S.W.
Rochester, MN 55905, U.S.A.
email: therneau@mayo.edu

David Oakes
Department of Biostatistics
University of Rochester Medical Center
601 Elmwood Avenue, Box 630
Rochester, NY 14642,U.S.A.
email: oakes@metro.bst.rochester.edu

David Clayton
MRC Biostatistics Unit
Institute of Public Health
University Forvie Site
Robinson Way
Cambridge CB2 2SR, U.K.
email: david.clayton@mrc-bsu.cam.ac.uk

Jon A. Wellner
Department of Statistics, Box 354322
University of Washington
Seattle, WA 98195, U.S.A.
email: jaw@stat.washington.edu

Nicholas P. Jewell
Division of Biostatistics and
Department of Statistics
University of California
Berkeley, CA 94720, U.S.A.
email: jewell@stat.berkeley.edu

Zhiliang Ying
Department of Statistics
Hill Center, Busch Campus
Rutgers University
Piscataway, NJ 08855, U.S.A.
email: zying@stat.rutgers.edu

John J. Crowley
Division of Public Health Sciences
Fred Hutchinson Cancer Research Center
1124 Columbia Street, MP-557
Seattle, WA 98104, U.S.A.
email: johnc@orca.fhcrc.org

Susan S. Ellenberg
Center for Biologics Evaluation and Research
Divisions of Biostatistics and Epidemiology
Food and Drug Administration
1401 Rockville Pike, HFM-210
Rockville, MD 20852, U.S.A.
email: ellenberg@a1.cber.fda.gov

Peter A. Lachenbruch
Center for Biologics Evaluation and Research
Divisions of Biostatistics and Epidemiology
Food and Drug Administration
1401 Rockville Pike, HFM 220
Rockville, MD 20852, U.S.A.
email: lachenbruch@a1.cber.fda.gov

Richard D. Gill
Mathematical Institute
University Utrecht
P. O. Box 80 010
3508 TA Utrecht, Netherlands
email: gill@math.ruu.nl

List of Referees
(in alphabetical order)

Norman Breslow, *University of Washington*
David Clayton, *Medical Research Council*
John Crowley, *Fred Hutchinson Cancer Research Center*
Scott Emerson, *University of Washington*
Lloyd Fisher, *University of Washington*
David Glidden, *Harvard University*
Patricia Grambsch, *University of Minnesota*
Robert Gray, *Dana Farber Cancer Institute*
Frank Harrell, Jr., *University of Virginia*
Li Hsu, *Fred Hutchinson Cancer Research Center*
Jian Huang, *University of Iowa*
Michael Hughes, *Harvard University*
Mark van der Laan, *University of California*
John Lachin, *George Washington University*
Xiao-Li Meng, *University of Chicago*
Ian McKeague, *Florida State University*
Søren Feodor Nielsen, *University of Copenhagen*
David Oakes, *University of Rochester*
Margaret Pepe, *Fred Hutchinson Cancer Research Center*
Ross Prentice, *Fred Hutchinson Cancer Research Center*
Dan Rabinowitz, *Columbia University*
Peter Sasieni, *Imperial Cancer Research Fund*
Steven Self, *Fred Hutchinson Cancer Research Center*
Charles Spiekerman, *University of Washington*
Cyntha Struthers, *University of Waterloo*
Terry Therneau, *Mayo Clinic*
L. J. Wei, *Harvard University*
Zhiliang Ying, *Rutgers University*

Lecture Notes in Statistics

For information about Volumes 1 to 48 please contact Springer-Verlag

Vol. 49: S.L. Lauritzen, Extremal Families and Systems of Sufficient Statistics. xv, 268 pages, 1988.

Vol. 50: O.E. Barndorff-Nielsen, Parametric Statistical Models and Likelihood. vii, 276 pages, 1988.

Vol. 51: J. Hüsler, R.-D. Reiss (Editors). Extreme Value Theory, Proceedings, 1987. x, 279 pages, 1989.

Vol. 52: P.K. Goel, T. Ramalingam, The Matching Methodology: Some Statistical Properties. viii, 152 pages, 1989.

Vol. 53: B.C. Arnold, N. Balakrishnan, Relations, Bounds and Approximations for Order Statistics. ix, 173 pages, 1989.

Vol. 54: K.R. Shah, B.K. Sinha, Theory of Optimal Designs. viii, 171 pages, 1989.

Vol. 55: L. McDonald, B. Manly, J. Lockwood, J. Logan (Editors), Estimation and Analysis of Insect Populations. Proceedings, 1988. xiv, 492 pages, 1989.

Vol. 56: J.K. Lindsey, The Analysis of Categorical Data Using GLIM. v, 168 pages, 1989.

Vol. 57: A. Decarli, B.J. Francis, R. Gilchrist, G.U.H. Seeber (Editors), Statistical Modelling. Proceedings, 1989. ix, 343 pages, 1989.

Vol. 58: O.E. Barndorff-Nielsen, P. Bl¾sild, P.S. Eriksen, Decomposition and Invariance of Measures, and Statistical Transformation Models. v, 147 pages, 1989.

Vol. 59: S. Gupta, R. Mukerjee, A Calculus for Factorial Arrangements. vi, 126 pages, 1989.

Vol. 60: L. Gyorfi, W. Härdle, P. Sarda, Ph. Vieu, Nonparametric Curve Estimation from Time Series. viii, 153 pages, 1989.

Vol. 61: J. Breckling, The Analysis of Directional Time Series: Applications to Wind Speed and Direction. viii, 238 pages, 1989.

Vol. 62: J.C. Akkerboom, Testing Problems with Linear or Angular Inequality Constraints. xii, 291 pages, 1990.

Vol. 63: J. Pfanzagl, Estimation in Semiparametric Models: Some Recent Developments. iii, 112 pages, 1990.

Vol. 64: S. Gabler, Minimax Solutions in Sampling from Finite Populations. v, 132 pages, 1990.

Vol. 65: A. Janssen, D.M. Mason, Non-Standard Rank Tests. vi, 252 pages, 1990.

Vol 66: T. Wright, Exact Confidence Bounds when Sampling from Small Finite Universes. xvi, 431 pages, 1991.

Vol. 67: M.A. Tanner, Tools for Statistical Inference: Observed Data and Data Augmentation Methods. vi, 110 pages, 1991.

Vol. 68: M. Taniguchi, Higher Order Asymptotic Theory for Time Series Analysis. viii, 160 pages, 1991.

Vol. 69: N.J.D. Nagelkerke, Maximum Likelihood Estimation of Functional Relationships. V, 110 pages, 1992.

Vol. 70: K. Iida, Studies on the Optimal Search Plan. viii, 130 pages, 1992.

Vol. 71: E.M.R.A. Engel, A Road to Randomness in Physical Systems. ix, 155 pages, 1992.

Vol. 72: J.K. Lindsey, The Analysis of Stochastic Processes using GLIM. vi, 294 pages, 1992.

Vol. 73: B.C. Arnold, E. Castillo, J.-M. Sarabia, Conditionally Specified Distributions. xiii, 151 pages, 1992.

Vol. 74: P. Barone, A. Frigessi, M. Piccioni, Stochastic Models, Statistical Methods, and Algorithms in Image Analysis. vi, 258 pages, 1992.

Vol. 75: P.K. Goel, N.S. Iyengar (Eds.), Bayesian Analysis in Statistics and Econometrics. xi, 410 pages, 1992.

Vol. 76: L. Bondesson, Generalized Gamma Convolutions and Related Classes of Distributions and Densities. viii, 173 pages, 1992.

Vol. 77: E. Mammen, When Does Bootstrap Work? Asymptotic Results and Simulations. vi, 196 pages, 1992.

Vol. 78: L. Fahrmeir, B. Francis, R. Gilchrist, G. Tutz (Eds.), Advances in GLIM and Statistical Modelling: Proceedings of the GLIM92 Conference and the 7th International Workshop on Statistical Modelling, Munich, 13-17 July 1992. ix, 225 pages, 1992.

Vol. 79: N. Schmitz, Optimal Sequentially Planned Decision Procedures. xii, 209 pages, 1992.

Vol. 80: M. Fligner, J. Verducci (Eds.), Probability Models and Statistical Analyses for Ranking Data. xxii, 306 pages, 1992.

Vol. 81: P. Spirtes, C. Glymour, R. Scheines, Causation, Prediction, and Search. xxiii, 526 pages, 1993.

Vol. 82: A. Korostelev and A. Tsybakov, Minimax Theory of Image Reconstruction. xii, 268 pages, 1993.

Vol. 83: C. Gatsonis, J. Hodges, R. Kass, N. Singpurwalla (Editors), Case Studies in Bayesian Statistics. xii, 437 pages, 1993.

Vol. 84: S. Yamada, Pivotal Measures in Statistical Experiments and Sufficiency. vii, 129 pages, 1994.

Vol. 85: P. Doukhan, Mixing: Properties and Examples. xi, 142 pages, 1994.

Vol. 86: W. Vach, Logistic Regression with Missing Values in the Covariates. xi, 139 pages, 1994.

Vol. 87: J. Müller, Lectures on Random Voronoi Tessellations.vii, 134 pages, 1994.

Vol. 88: J. E. Kolassa, Series Approximation Methods in Statistics.viii, 150 pages, 1994.

Vol. 89: P. Cheeseman, R.W. Oldford (Editors), Selecting Models From Data: AI and Statistics IV. xii, 487 pages, 1994.

Vol. 90: A. Csenki, Dependability for Systems with a Partitioned State Space: Markov and Semi-Markov Theory and Computational Implementation. x, 241 pages, 1994.

Vol. 91: J.D. Malley, Statistical Applications of Jordan Algebras. viii, 101 pages, 1994.

Vol. 92: M. Eerola, Probabilistic Causality in Longitudinal Studies. vii, 133 pages, 1994.

Vol. 93: Bernard Van Cutsem (Editor), Classification and Dissimilarity Analysis. xiv, 238 pages, 1994.

Vol. 94: Jane F. Gentleman and G.A. Whitmore (Editors), Case Studies in Data Analysis. viii, 262 pages, 1994.

Vol. 95: Shelemyahu Zacks, Stochastic Visibility in Random Fields. x, 175 pages, 1994.

Vol. 96: Ibrahim Rahimov, Random Sums and Branching Stochastic Processes. viii, 195 pages, 1995.

Vol. 97: R. Szekli, Stochastic Ordering and Dependence in Applied Probability. viii, 194 pages, 1995.

Vol. 98: Philippe Barbe and Patrice Bertail, The Weighted Bootstrap. viii, 230 pages, 1995.

Vol. 99: C.C. Heyde (Editor), Branching Processes: Proceedings of the First World Congress. viii, 185 pages, 1995.

Vol. 100: Wlodzimierz Bryc, The Normal Distribution: Characterizations with Applications. viii, 139 pages, 1995.

Vol. 101: H.H. Andersen, M.Højbjerre, D. Sørensen, P.S.Eriksen, Linear and Graphical Models: for the Multivariate Complex Normal Distribution. x, 184 pages, 1995.

Vol. 102: A.M. Mathai, Serge B. Provost, Takesi Hayakawa, Bilinear Forms and Zonal Polynomials. x, 378 pages, 1995.

Vol. 103: Anestis Antoniadis and Georges Oppenheim (Editors), Wavelets and Statistics. vi, 411 pages, 1995.

Vol. 104: Gilg U.H. Seeber, Brian J. Francis, Reinhold Hatzinger, Gabriele Steckel-Berger (Editors), Statistical Modelling: 10th International Workshop, Innsbruck, July 10-14th 1995. x, 327 pages, 1995.

Vol. 105: Constantine Gatsonis, James S. Hodges, Robert E. Kass, Nozer D. Singpurwalla(Editors), Case Studies in Bayesian Statistics, Volume II. x, 354 pages, 1995.

Vol. 106: Harald Niederreiter, Peter Jau-Shyong Shiue (Editors), Monte Carlo and Quasi-Monte Carlo Methods in Scientific Computing. xiv, 372 pages, 1995.

Vol. 107: Masafumi Akahira, Kei Takeuchi, Non-Regular Statistical Estimation. vii, 183 pages, 1995.

Vol. 108: Wesley L. Schaible (Editor), Indirect Estimators in U.S. Federal Programs. viii, 195 pages, 1995.

Vol. 109: Helmut Rieder (Editor), Robust Statistics, Data Analysis, and Computer Intensive Methods. xiv, 427 pages, 1996.

Vol. 110: D. Bosq, Nonparametric Statistics for Stochastic Processes. xii, 169 pages, 1996.

Vol. 111: Leon Willenborg, Ton de Waal, Statistical Disclosure Control in Practice. xiv, 152 pages, 1996.

Vol. 112: Doug Fischer, Hans-J. Lenz (Editors), Learning from Data. xii, 450 pages, 1996.

Vol. 113: Rainer Schwabe, Optimum Designs for Multi-Factor Models. viii, 124 pages, 1996.

Vol. 114: C.C. Heyde, Yu. V. Prohorov, R. Pyke, and S. T. Rachev (Editors), Athens Conference on Applied Probability and Time Series Analysis Volume I: Applied Probability In Honor of J.M. Gani. viii, 424 pages, 1996.

Vol. 115: P.M. Robinson, M. Rosenblatt (Editors), Athens Conference on Applied Probability and Time Series Analysis Volume II: Time Series Analysis In Memory of E.J. Hannan. viii, 448 pages, 1996.

Vol. 116: Genshiro Kitagawa and Will Gersch, Smoothness Priors Analysis of Time Series. x, 261 pages, 1996.

Vol. 117: Paul Glasserman, Karl Sigman, David D. Yao (Editors), Stochastic Networks. xii, 298, 1996.

Vol. 118: Radford M. Neal, Bayesian Learning for Neural Networks. xv, 183, 1996.

Vol. 119: Masanao Aoki, Arthur M. Havenner, Applications of Computer Aided Time Series Modeling. ix, 329 pages, 1997.

Vol. 120: Maia Berkane, Latent Variable Modeling and Applications to Causality. vi, 288 pages, 1997.

Vol. 121: Constantine Gatsonis, James S. Hodges, Robert E. Kass, Robert McCulloch, Peter Rossi, Nozer D. Singpurwalla (Editors), Case Studies in Bayesian Statistics, Volume III. xvi, 487 pages, 1997.

Vol. 122: Timothy G. Gregoire, David R. Brillinger, Peter J. Diggle, Estelle Russek-Cohen, William G. Warren, Russell D. Wolfinger (Editors), Modeling Longitudinal and Spatially Correlated Data. x, 402 pages, 1997.

Vol. 123: D. Y. Lin and T. R. Fleming (Editors), Proceedings of the First Seattle Symposium in Biostatistics: Survival Analysis. xiii, 308 pages, 1997.

GPSR Compliance

The European Union's (EU) General Product Safety Regulation (GPSR) is a set of rules that requires consumer products to be safe and our obligations to ensure this.

If you have any concerns about our products, you can contact us on ProductSafety@springernature.com

In case Publisher is established outside the EU, the EU authorized representative is:

Springer Nature Customer Service Center GmbH
Europaplatz 3
69115 Heidelberg, Germany

Batch number: 10704711

Printed by Printforce, the Netherlands